5G Second Phase Explained

5G Second Phase Explained

The 3GPP Release 16 Enhancements

Jyrki TJ. Penttinen
GSMA North America

This edition first published 2021

Registered Offices
John Wiley & Sons, Inc., 111 River Street, Hoboken, NJ 07030, USA
John Wiley & Sons Ltd, The Atrium, Southern Gate, Chichester, West Sussex, PO19 8SQ, UK

Editorial Office
The Atrium, Southern Gate, Chichester, West Sussex, PO19 8SQ, UK

For details of our global editorial offices, customer services, and more information about Wiley products visit us at www.wiley.com.

Wiley also publishes its books in a variety of electronic formats and by print-on-demand. Some content that appears in standard print versions of this book may not be available in other formats.

Library of Congress Cataloging-in-Publication Data
Names: Penttinen, Jyrki, 1967- author. | John Wiley & Sons, Inc., publisher.
Title: 5G second phase explained : the 3GPP release 16 enhancements / Jyrki Teppo Juho Penttinen, GSMA North America.
Description: Hoboken, NJ : John Wiley & Sons, Inc., 2021. | Includes bibliographical references and index.
Identifiers: LCCN 2020047582 (print) | LCCN 2020047583 (ebook) | ISBN 9781119645504 (hardback) | ISBN 9781119645559 (pdf) | ISBN 9781119645535 (epub) | ISBN 9781119645566 (ebook)
Subjects: LCSH: 5G mobile communication systems. | Long-Term Evolution (Telecommunications).
Classification: LCC TK5103.25 .P46 2021 (print) | LCC TK5103.25 (ebook) | DDC 621.3845/6—dc23
LC record available at https://lccn.loc.gov/2020047582
LC ebook record available at https://lccn.loc.gov/2020047583

Cover image: © Nicomenijes/Getty Images
Cover design by Wiley

Set in 9.5/12.5pt STIX Two Text by Integra Software Services Pvt. Ltd, Pondicherry, India
Printed and bound by CPI Group (UK) Ltd, Croydon, CR0 4YY

C9781119645504_260321

Contents

About the Author

Dr Jyrki T.J. Penttinen, the author of *5G Second Phase Explained*, started his activities in the mobile communications industry in 1987 by evaluating 1G and 2G radio networks. After he obtained his MSc (EE) grade from the Helsinki University of Technology (HUT) in 1994, he worked for Telecom Finland (Sonera and TeliaSonera Finland) and Xfera Spain (Yoigo) on 2G and 3G radio and core network architectures and performance aspects. In 2002 he established Finesstel Ltd, carrying out consultancy and technical training projects in Europe and the Americas during 2002–2003. Afterwards, he worked for Nokia and Nokia Siemens Networks in Mexico, Spain, and the United States from 2004 to 2013. During this time with mobile network operators and equipment manufacturers, Dr Penttinen was involved in operational and research activities related to system and architectural design, standardization, training, and technical management. His focus was on the radio interface of GSM, GPRS/EDGE, UMTS/HSPA, and DVB-H. From 2014 to 2018, in his position as program manager with G+D Mobile Security Americas, USA, his focus areas included mobile and IoT security and innovation with a special emphasis on 5G.

Since 2018, he has worked for GSMA North America as Senior Technology Manager assisting operator members with the adoption, design, development, and deployment of GSMA specifications and programs.

Dr Penttinen obtained his LicSc (Tech) and DSc (Tech) degrees from HUT (currently known as Aalto University, School of Science and Technology) in 1999 and 2011, respectively. In addition to his main work, he has given lectures and authored technical articles and books such as *5G Explained* (2019), *Wireless Communications Security* (2017), *The LTE-Advanced Deployment Handbook* (2016), *The Telecommunications Handbook* (2015), *The LTE/SAE Deployment Handbook* (2011), and *The DVB-H Handbook* (2009). More information on his publications and articles can be found at his LinkedIn profile, www.linkedin.com/in/jypen, and at his author's page at Amazon, www.amazon.com/author/jype.

Preface

5G has been a reality since 2019, after some early deployments of isolated 5G networks that were already partially compliant with the 3GPP technical standards. Since the publication of the very first complete set of Release 15-based 5G standards, the number of 5G radio networks has been increasing steadily. According to the forecast of the GSMA, 5G will account for as many as 1.2 billion connections by 2025. Along with the Release 16 standards that the 3GPP released in 2020, we can start enjoying gradually the full 5G experience with fast deployment of advanced features.

My previous *5G Explained* book from Wiley, published in 2019, described the key functionalities as per 3GPP Release 15, a.k.a. the *first phase* of 5G with the focus on security and deployment aspects. Release 15 forms a foundation for 5G and facilitates fast deployment via many intermediate architectural options, while Release 16 makes 5G fully equipped with a variety of enhanced features and functions.

As can be interpreted from the accelerated development schedules of the 5G era, mobile communication technologies evolve faster than ever. Along with such important additions of Release 16, this new book thus complements the foundations laid down by the previous *5G Explained* book, including key descriptions of the features defined in the *second phase* of 5G. The focus of this book is on new key use cases, enhanced security, and deployment aspects. These two books serve as a complementing set of references and form an up-to-date resource for demystifying 5G architecture and functions.

These *5G Explained* books thus summarize the latest knowledge regarding the key features and functionality of the first and second phases of 5G, and provides readers with a common-sense summary of specifications and other information sources. I believe this modular approach is beneficial for network deployment, device designing, and education of personnel and students interested in telecommunication domains.

As has been the case with my previous books published by Wiley, I would highly appreciate all your feedback. For any questions and feedback, please do not hesitate to contact me directly via my LinkedIn profile at www.linkedin.com/in/jypen, and please feel free to comment on my related 5G blog at www.5g-simplified.com, which I use to summarize related updates of selected topics of these *5G Explained* books.

Jyrki Penttinen
Atlanta, GA, US

Acknowledgments

This book is a result of countless hours I have spent exploring 3GPP specifications and other relevant information sources to better understand the up-to-date architecture, functioning, and principles of the 5G system. Because the new representative of mobile generations has advanced at such a fast pace, the task has been highly fascinating yet challenging, especially the balancing of time. I thus want to express my warmest thanks for all the support and patience I have received from my wife Celia as well as my close family, Katriina, Pertti, Stephanie, Carolyne, and Miguel. I am also most thankful for the support of my colleagues and peers as well as all those who have provided me with feedback to my publications.

I also want to express my warmest gratitude to the Wiley team for their professional but gentle approach, which has ensured the successful delivery of this book.

Jyrki Penttinen
Atlanta, GA, US

Abbreviations

1G	1st generation of mobile communication systems
2G	2nd generation of mobile communication systems
3G	3rd generation of mobile communication systems
3GPP	3rd Generation Partnership Project
4G	4th generation of mobile communication systems
5G	5th generation of mobile communication systems
5GC	5G Core
5GS	5G System
5WWC	Wireless and Wireline Convergence for 5G system architecture
A/D	Analogue to Digital
AAA	Authentication, Authorization, and Accounting
AAA-P	AAA Proxy
AAA-S	AAA Server
AAS	Active Antenna System
ADAS	Advanced Driver Assistance System
AES	Advanced Encryption Standard
AF	Application Function
AGW	Access Gateway (IMS)
AI	Artificial Intelligence
AKA	Authentication and Key Agreement
AL	Application Layer (SMS)
ALG	Application Level Gateway (IMS)
AM	Acknowledged Mode
AMF	Access and Mobility Management Function
AMPS	Advanced Mobile Phone Service (1G)
AN	Access Network
ANR	Automatic Neighbor Cell Relation
AoA	Angle of Arrival
AoD	Angle of Departure
API	Access Point Identifier
API	Application Programming Interface
APN	Access Point Name
AR	Augmented Reality

ARP	Allocation and Retention Priority
ARPF	Authentication Credential Repository and Processing Function
AS	Access Stratum
AS	Application Server
ATSSS	Access Traffic Steering, Switch and Splitting
AUSF	Authentication Server Function
BBF	Broadband Forum
BBU	Baseband Unit
BER	Bit Error Rate
BGCF	Breakout Gateway Control Function
BH	Backhaul
BL	Bandwidth reduced Low complexity UE
BRG	Broadband Residential Gateway (5G)
BSS	Business Support System
C2	Command and Control
CA	Carrier Aggregation
CAG	Closed Access Group
CAM	Cooperative Awareness Message
CAPEX	Capital Expenditure
CAPI	Common north-bound APIs (EPC-5GC)
CAPIF	Common API Framework for 3GPP northbound APIs
CAS	Cell Acquisition Subframe
C-Core	Cloud Core
cdma2000	Code Division Multiple Access 2000 (3G)
CDR	Charging Data Record
CHEM	Coverage and Handoff Enhancements for Multimedia
CHF	Charging Function
CI	Certificate Issuer
C-IoT	Cellular IoT
CM sub	Connection Management Sublayer
CM	Connection Management
CMAS	Commercial Mobile Alert System
CO	Cloud Orchestrator
CoMP	Coordinated Multi-Point
CORD	Central Office Re-architected as Data Center
COTS	Commercial Off-the-Shelf
COUNT	Counter (security sequence)
CP	Control Plane
CP	Control Protocol (SMS)
CPA	Certified Public Accountants
CPC	Cyber-Physical Control
CPRI	Common Public Radio Interface
C-RAN	Cloud RAN
CRG	Cable Residential Gateway (5G)
CriC	Critical Communications

CS	Circuit Switched
CSC	Communication Service Customer
cSEPP	Consumer's SEPP
CSFB	Circuit-Switched Fallback
CSI	Channel-State Information
CSI-RS	Channel-State Information Reference Signal
CSP	Communication Service Provider
CU	Centralized Unit
CU-CP	Centralized Unit, Control Plane
CUPS	Control and User Plane Separation
CU-UP	Centralized Unit, User Plane
CU-UP	CU User Plane
C-V2X	Cellular V2X
C-V2X	Cellular Vehicle-to-Everything
CWDM	Coarse Wavelength Division Multiplexing
D/A	Digital to Analogue
D2D	Device-to-Device
DANOS	Disaggregated Network Operating System
DAPS HO	Dual Active Protocol Stack-based Handover
DC	Dual Connectivity
DCSP	Data Centre Service Provider
DFT-s-OFDM	Discrete Fourier Transform spread OFDM
DL	Downlink
DLDC	Downlink Dual Carrier
DLOA	Digital Letter of Approval
DMRS	Demodulation Reference Signal
DN	Data Network
DNN	Data Network Name
DNS	Dynamic Name Server
DRB	Data Radio Bearer
DRX	Discontinuous Reception
DSF	Data Storage Function
DSS	Dynamic Spectrum Sharing
DU	Distributed Unit
DWDM	Dense Wavelength Division Multiplexing
E CID	Enhanced Cell ID
EAP	Extensible Authentication Protocol
EC-GSM-IoT	Extended Coverage GSM IoT
eCPRI	Evolved Common Public Radio Interface
eDual	Enhanced Dual Connectivity
EE	Energy Efficiency
EIR	Equipment Identity Register
eLCS	Enhanced Location Service
eMBB	Evolved Mobile Broadband
eMBMS	Evolved MBMS

eMIMO	Enhanced MIMO
eNB	Evolved NodeB (4G)
EN-DC	E-UTRA–NR Dual Connectivity
en-gNB	5G-RAN node for the EN-DC
ENUM	Electronic Number Mapping System
EPC	Evolved Packet Core (4G)
EPS	Evolved Packet System (4G)
ER	EAP Re-authentication
eSIM	Embedded SIM
E-SMLC	Evolved Serving Mobile Location Centre
eSSP	Embedded Smart Secure Platform
ETN	Edge Transport Node
ETSI	European Telecommunication Standards Institute
eUICC	Embedded UICC
EUM	eUICC Manufacturer
E-UTRA	Evolved UMTS Terrestrial Radio Access (4G)
eV2X	Enhanced Vehicle-to-Everything
FB	Fallback
FCC	Federal Communications Commission (USA)
FDA	Food and Drug Administration (USA)
FDD	Frequency Division Duplex
FEC	Forward Error Coding
FeMBMS	Further Enhanced MBMS
FF	Form Factor (SIM)
FH	Fronthaul
FMC	Fixed-Mobile Convergence (BBF)
FN-BRG	Fixed Network Broadband Residential Gateway (5G)
FN-CRG	Fixed Network Cable Residential Gateway (5G)
FN-RG	Fixed Network Residential Gateway (5G)
FQDN	Fully Qualified Domain Name
FR1	Frequency Range 1
FR2	Frequency Range 2
FRMCS	Mobile Communications System for Railways
FWA	Fixed Wireless Access
GBR	Guaranteed Bit Rate
GMLC	Gateway Mobile Location Centre
GMT	Group Message Delivery
gNB	Next Generation NodeB (5G)
GNSS	Global Navigation Satellite System
GPRS	General Packet Radio Service
GPS	Global Positioning System
GPSI	Generic Public Subscription Identifier
gPTP	Generalized Precision Timing Protocol
GSA	Global Mobile Suppliers Association
GSM	Global System for Mobile Communications (2G)

GSMA	GSM Association
GSM-R	GSM Railway
GST	Generic Network Slice Template
GUAMI	Globally Unique AMF Identifier
GUTMA	Global UTM Association
HAP	High Altitude Platform
HARQ	Hybrid Automatic Repeat Request
HB	High Band
HD	High Definition
HIBS	High Altitude IMT Base Stations
HLS	High Layer Split
HO	Handover
HPLMN	Home Public Land Mobile Network
HR	Home Routed
hSEPP	Home Security Edge Protection Proxy
HSPA	High Speed Packet Access (3G)
HSS	Home Subscription Server
HTTP	Hypertext Transfer Protocol
HW	Hardware
IAB	Integrated Access and Backhaul
IAB-MT	Mobile Terminating Integrated Access and Backhaul
IATN	Inter-Area Transport Node
IBCF	Interconnection Border Control Function
ICI	Inter-Carrier Interference
I-CSCF	Interrogating Call Session Control Function
IEEE	Institute of Electrical and Electronics Engineers
IETF	Internet Engineering Task Force
I-IoT	Industrial IoT
IKE	Internet Key Exchange
IMEI	International Mobile Equipment Identity
IMPI	IP Multimedia Private Identity
IMPU	IP Multimedia Public Identity
IMS	IP Multimedia Subsystem
IMSI	International Mobile Subscriber Identity
IMT	International Mobile Telecommunication
IMT-2000	International Mobile Telecommunications (3G)
IMT-2020	International Mobile Telecommunications (5G)
IMT-Advanced	International Mobile Telecommunications (4G)
IoT	Internet of Things
IP	Internet Protocol
IPUPS	Inter-PLMN UP Security
IPX	Internet Protocol Packet Exchange
IS-95	Interim Standard (2G)
ISD	Inter-Site Distance
ISI	Inter-Symbol Interference

I-SMF	Intermediate SMF
iSSP	Integrated Smart Secure Platform
ITU	International Telecommunications Union
ITU-R	Radio section of the International Telecommunications Union
ITU-T	Telecommunications section of the International Telecommunications Union
I-UPF	Intermediate UPF
JTACS	Japan Total Access Communications System (1G)
KDF	Key Derivation Function
KPI	Key Performance Indicator
LAA	Licensed Assisted Access
LAN	Local Area Network
LB	Low Band
LBO	Local Breakout
LBS	Location-Based Service
LCS	Location Service
LDPC	Low-Density Parity Check
LDS	Local Discovery Service
LDSd	LDS in device
LI	Lawful Interception
LMF	Location Management Function
LOS	Line Of Sight
LPA	Local Profile Assistant
LPAd	LPA in device
LPD	Local Profile Download
LPDd	LPD in device
LPLT	Low Power Low Tower
LPWA	Low-Power Wide Area
LTE	Long Term Evolution (4G)
LTE-A	LTE-Advanced (4G)
LUI	Local User Interface
LUId	LUI in device
M2M	Machine-to-Machine
MAP	Mobile Application Part
MB	Mid-Band
MBMS	Multimedia Broadcast Multicast Service
MC	Mission Critical
MC	Multi-Carrier
MCC	Mobile Country Code
MCData	Mission Critical Data
MCE	Mobile Cloud Engine
MCG	Master Cell Group
MCPPT	Mission-Critical Push-to-Talk
MCS	Modulation and Coding Scheme
MCVideo	Mission Critical Video

MCX	Mission Critical Service
MDT	Minimization of Drive Tests
ME	Mobile Equipment
MEC	Mobile-Edge Computing
MeNB	Master eNB
MeNB	See MN
MGCF	Media Gateway Control Function (IMS)
MGW	Media Gateway (IMS)
MIMO	Multiple In, Multiple Out
mIoT	Massive IoT
MIoT	Mobile IoT (combined NB-IoT and LTE-M)
ML	Machine Learning
MME	Mobility Management Entity
MMF2	Machine-to-Machine Form Factor
mMTC	Massive Machine Type Communications
MMtel	Multimedia Telephony Service
MN	Master Node
MNC	Mobile Network Code
MNO	Mobile Network Operator
MO	Mobile Originated
MOCN	Multi-Operator Core Network
MO-EDT	Mobile Originated Early Data Transmission
MPMT	Medium Power Medium Tower
MPS	Multimedia Priority Service
MR	Multi-Radio
MRB	Media Resource Broker
MRCP	Media Resource Function Processor
MR-DC	Multi-RAT Dual Connectivity
MRF	Media Resource Function
MRFC	Media Resource Function Controller
MS	Mobile Station
MSC	Mobile Switching Center
MSIN	Mobile Subscriber Identification Number
MSISDN	Mobile Station ISDN Number
MSR	Multi-Standard Radio specifications
MT	Mobile Terminal
MT	Mobile Terminated
MTC	Machine Type Communications
MTSI	Multimedia Telephony Service for IMS
MU-MIMO	Multi-User MIMO
N3IWF	Non-3GPP Interworking Function
N5CW	Non-5G-Capable over WLAN
NaaS	Network as a Service
NAI	Network Access Identifier
NAS	Non-access Stratum

NB	NodeB
NBI	Northbound Interface
NB-IoT	Narrow-Band IoT
NCC	Next Hop Chaining Counter
NCR	Neighbor Cell Relations
NE-DC	NR–E-UTRA Dual Connectivity
NEF	Network Exposure Function
NEO	Network Operations
NEP	Network Equipment Provider
NEST	Network Slice Type
NF	Network Function
NFV	Network Functions Virtualization
NFVI	Network Function Virtualization Infrastructure
NG-AP	NG Application Protocol
NGC	Next Generation Core (5G)
ng-eBB	5G Next Generation NodeB (enhanced 4G eNodeB)
NGEN-DC	NG-RAN–E-UTRA-NR Dual Connectivity (also: NE-DC)
NGFI	Next Generation Fronthaul Interface
NGMN	Next Generation Mobile Network
NG-RAN	Next Generation Radio Access Network (5G)
NH	Next Hop
NID	Network Identifier
NIDD	Non-IP Data Delivery
NLOS	Non-line Of Sight
NMO	Network Management and Orchestration
NMT	Nordic Mobile Telephone (1G)
NNI	Network-Network Interface
NOMA	Non-orthogonal Multiple Access
NOP	Network Operator
NPN	Non-public Network
NR	New Radio (5G)
NR-DC	NR–NR Dual Connectivity
NRF	Network Repository Function
NRT	Non-real Time
NR-U	NR on Unlicensed spectrum (5G)
NS	Network Slicing
NSA	Non-standalone
NSaaS	Network Slice as a Service
NSaaSC	NSaaS Customer
NSaaSP	NSaaS Provider
NSC	Network Slice Customer
NSI	Network Slice Instance
NSP	Network Slice Provider
NSSAA	Network Slice Specific Authentication and Authorization
NSSAAF	Network Slice Specific Authentication and Authorization Function

NSSAI	Network Slice Selection Assistance Information
NSSF	Network Slice Selection Function
NTN	Non-terrestrial Network
NTP	National Toxicology Program (USA)
NWDA	Network Data Analytics
NWDAF	Network Data Analytics Function
OAM	Operations Administration and Maintenance
OCP	Open Compute Project
O-CU	O-RAN Central Unit
O-DU	O-RAN Distributed Unit
OFDM	Orthogonal Frequency Division Multiplexing
OLT	Optical Line Terminal
ONAP	Open Network Automation Platform
ONU	Optical Network Unit
OOB	Out of Band leakage
OPEX	Operating Expenditure
O-RAN	Open Radio Access Network
O-RU	O-RAN Radio Unit
OS	Operating System
OSC	Orthogonal Sub-Channel
OSS	Operations Support System
OTDOA	Observed Time Difference of Arrival
P2MP	Point-to-Multipoint
PAPR	Peak-to-Average Power Ratio
PBCH	Physical Broadcast Channel
PCF	Policy Control Function
P-CSCF	Proxy Call Session Control Function
PDCCH	Physical Downlink Control Channel
PDN	Packet Data Network
PDSCH	Physical Downlink Shared Channel
PDU	Packet Data Unit
PEI	Permanent Equipment Identifier
PFD	Packet Flow Description
P-GW	Packet Data Network Gateway
Phy	Physical layer
PLMN	Public Land Mobile Network
PM	Performance Management
PNI-NPN	Public Network Integrated NPN
PoC	Proof of Concept
PON	Passive Optical Network
PRACH	Physical Random Access Channel
PRD	Permanent Reference Document (GSMA)
ProSe	Proximity Service
PRS	Positioning Reference Signal
pSEPP	Producer's SEPP

PSS	Primary Synchronization Signal
PTP	Point-to-Point
PT-RS	Phase-Tracking Reference Signal
PTT	Push-to-Talk
PUCCH	Physical Uplink Control Channel
PUR	Preconfigured Uplink Resource
PUSCH	Physical/Primary Uplink Shared Channel
PWS	Public Warning System
QAM	Quadrature Amplitude Modulation
QCI	QoS Class Identifier
QoE	Quality of Experience
QoS	Quality of Service
QPSK	Quadrature Phase Shift Keying
RA	Random Access
RACH	Random Access Channel
RAN	Radio Access Network
RAT	Radio Access Technology
RCS	Rich Communications Services
RDS	Reliable Data Service
RET	Remote Electrical Tilt
RF	Radio Frequency
RG	Residential Gateway (5G)
RIC	Radio Access Network Intelligent Controller (O-RAN)
RL	Relay Layer (SMS)
RLC	Radio Link Control
RLF	Radio Link Failure
RN	Remote Node
R-NIB	Radio-Network Information Base
RNL	Radio Network Layer
RoI	Return on Investment
ROM	Receive Only Mode
RP	Relay Protocol (SMS)
RRC	Radio Resource Control
RRH	Remote Radio Head
RRM	Radio Resource Management
RRU	Radio Remote Unit
RTP	Real-Time Transport Protocol
RTT	Roundtrip Time
Rx	Receiver
S8HR	S8 Home Routed
SA	Standalone
SA	System Architecture group (3GPP)
SAR	Specific Absorption Rate
SAS	Security Accreditation Scheme
SAS	Service Access Point

SAS-SM	Security Accreditation Scheme for Subscription Management
SAS-UP	Security Accreditation Scheme for UICC Production
SBA	Service-Based Architecture
SBI	Southbound Interface
SCA	Smart Card Association
SCAS	Security Assurance Specification
SCG	Secondary Cell Group
SCM	Security Context Management
SCMF	Security Context Management Function
SCP	Service Communication Proxy
S-CSCF	Serving Call Session Control Function (IMS)
SC-TDMA	Single Carrier Time Division Multiple Access
SCTP	Stream Control Transmission Protocol
SD	Slice Differentiator
SDN	Software Defined Networking
SDO	Standard Development Organization
SDP	Session Description Protocol
SDU	Service Data Unit
SE	Secure Element
SEAF	Security Anchor Function
SEAL	Service Enabler Architecture Layer
SEG	Secure Gateway
SeNB	Secondary eNB
SEPP	Security Edge Protection Proxy
SFN	Single Frequency Network
SgNB	See SN
S-GW	Serving Gateway
SIB	System Information Block
SIDF	Subscription Identifier De-Concealing Function
SIM	Subscriber Identity Module
SINR	Signal-to-Noise and Interference Ratio
SIP	Session Initiation Protocol (IMS)
SLA	Service Level Agreement/Assurance
SLC	SUPL Location Center
SLF	Subscriber Location Function
SLP	SUPL Location Platform
SM	Session Management
SM	Short Message
SMARTER	Services and Markets Technology Enablers
SMC	Security Mode Command
SMC	Short Message Control
SM-DP+	Subscription Manager Data Preparation
SM-DS	Subscription Manager Discovery Server
SMF	Session Management Function
SMR	Short Message Relay

SMS	Short Message Service
SMSF	Short Message Service Function
SN	Secondary Node
SN	Serving Network
SNPN	Stand-Alone Non-Public Network
SNR	Signal-to-Noise Ratio
S-NSSAI	Single NSSAI
SOC	Service Organization Control
SoC	System on Chip
SON	Self-Organizing Network
SPC	SUPL Positioning Center
SPCF	Security Policy Control Function
SRS	Sounding Reference Signal
SRVCC	Single Radio Voice Call Continuity
SSC	Session and Service Continuity
SSP	Smart Secure Platform
SSS	Secondary Synchronization Signal
SST	Slice/Service Type
SUCI	Subscription Concealed Identifier
SUL	Supplementary Uplink
SU-MIMO	Single User MIMO
SUPI	Subscription Permanent Identifier
SUPL	Secure User Plane Location
SW	Software
TA	Tracking Area
TACS	Total Access Communication System (1G)
TAP	Transferred Account Procedure
TAS	Telephony Application Server
TBS	Terrestrial Beacon System
TCAP	Transaction Capabilities Application Part
TDD	Time Division Duplex
TDOA	Time Difference of Arrival
TIF	Transport Intelligent Function
TIP	Telecom Infra Project
TI-SCCP	Transport Independent Signaling Connection Control Part
TL	Transfer Layer (SMS)
TLS	Transport Layer Security
TMA	Tower-Mounted Amplifier
TMA	Telefonía Móvil Automática (1G)
TN	Transport Node
TNAN	Trusted Non-3GPP Access Network
TNAP	Trusted Non-3GPP Access Point
TNGF	Trusted Non-3GPP Gateway Function
TNL	Transport Network Layer
TNS	Time-Sensitive Networking

TP	Transmission Point
TR	Technical Report (3GPP)
TrGW	Transition Gateway (IMS)
TRP	Transmission and Reception Point
TS	Technical Specification (3GPP)
TSN	Time-Sensitive Networking
TSON	Time Shared Optical Network
TT	TSN Translator
TTI	Transmission Time Interval
TWAP	Trusted WLAN Access Point
TWIF	Trusted WLAN Interworking Function
Tx	Transmitter
UAS	Unmanned Aerial System
UAV	Unmanned Aerial Vehicle
UCMF	UE radio Capability Management Function
UDC	Uplink Data Compression
UDM	Unified Data Management
UDR	Unified Data Repository
UDSF	Unstructured Data Storage Function
UE	User Equipment
UI	User Identifier
UICC	Universal Integrated Circuit Card
UL	Uplink
UL-CL	Uplink Classifier
UM	Unacknowledged Mode
UMTS	Universal Mobile Telecommunications System (3G)
UNI	User-Network Interface
UP	User Plane
UPF	User Plane Function
URLLC	Ultra-Reliable Low Latency Communications
USIM	Universal Subscriber Identity Module
UST	Universal SIM Toolkit
UTM	Unmanned Aircraft Systems Traffic Management
UTM	Unmanned Traffic Management
UX	User Experience
V2I	Vehicle-to-Infrastructure
V2V	Vehicle-to-Vehicle
V2X	Vehicle-to-Everywhere
VAMOS	Voice services over Adaptive Multi-user channels on One Slot
vBBU	Virtualized BBU
ViLTE	Video over LTE
VISP	Virtualization Infrastructure Service Provider
VM	Virtual Machine
VNF	Virtual Network Functions
VoLTE	Voice over LTE (4G)

VoNR	Voice over New Radio (5G)
VoWiFi	Voice over Wi-Fi
VPLMN	Visited Public Land Mobile Network
VPN	Virtual Private Network
VR	Virtual Reality
vSEPP	Visited Network Security Edge Protection Proxy
V-SMF	Visited SMF
W-AGF	Wireline Access Gateway Function
WDM	Wavelength Division Multiplexing
WHO	World Health Organization
WiMAX	WirelessMAN-Advanced
WRC	World Radiocommunication Conference
WUS	Wakeup Signal
WWC	Wireless and Wireline Convergence
XR	Extended Reality

1

Introduction

1.1 General

1.1.1 Focus of This Book

The fifth generation of mobile communication became a reality during 2019 as the 3rd Generation Partnership Project (3GPP) released the first set of Release 15 Technical Specifications (TS) and respective equipment, both network elements and mobile devices, to be available for commercial deployments.

Nevertheless, 3GPP Release 15 refers to the very *first phase* of 5G, which provides an initial, "light" version of the renewed system. In terms of 3GPP, the *second phase*, as defined by the Release 16 set of specifications, adds the remaining functionalities, increasing performance and becoming compliant with the strict requirements of International Mobile Telecommunications 2020 (IMT-2020) defined by the ITU-R (the radio section of the International Telecommunications Union). This is an essential step as IMT-2020 sets the reference for the interoperable, full version of the 5G, which all the parties involved with the 5G ecosystem can agree refers to the global and uniform 5G.

While the first phase of 5G is designed to augment the data rates by enhanced Mobile Broadband (eMBB) mode, Release 16 adds needed functionality to support the other base pillars of 5G as defined by the ITU, i.e., massive Machine Type Communications (mMTC) and Ultra Reliable Low Latency Communications (URLLC). The benefit of mMTC is the possibility of tackling a vast number of simultaneously communicating Internet of Things (IoT) devices, which form the very basis for the new connected society concept. URLLC, in turn, provides extremely low latency together with high availability of services for the special needs of critical communications. In addition, Release 16 brings with it more advanced means for highly efficient network management thanks to evolved self-optimizing networks and machine learning platforms.

There is a variety of novelty technologies available for adaptation into system architectures such as Network Functions Virtualization (NFV) and Software Defined Networking (SDN). Virtualization will also change the traditional business models, and open doors for completely new stakeholders such as data center operators and applications supporting Virtual Reality (VR) and Augmented Reality (AR).

5G Second Phase Explained: The 3GPP Release 16 Enhancements, First Edition. Jyrki T.J. Penttinen.

The second phase of 5G is already sufficiently capable of providing a functional and performant platform for highly advanced service types in a dynamic manner by using of a variety of use cases. This happens via Network Slicing (NS), which is available for deployment along with Release 16.

One of the important aspects in this evolution is to guarantee a sufficient level of interoperability between 5G networks for fluent user experiences. 3GPP standards as such are insufficient in this area as we have seen already with previous generations. Thus, there is a need to set guidelines for a feasible, minimum set of features and methods that would work among all operators within the ecosystem. As an example, the GSM Association (GSMA) is in a key position to define such recommendations for, e.g., roaming scenarios for voice and text services as well as for the interworking of packet data connections and subscription management over all the involved networks.

This book presents new key functionalities of Release 16 that complement the first phase of 5G. The book is thus an addition to the contents of the already published *5G Explained* book, providing further descriptions to understand the complete picture of the full version of 5G. Whereas the first book presented the basics, this second book complements it by presenting up-to-date functionalities of Release 16, and some of the indications of the technological topics under development for the forthcoming Release 17 and beyond. This new book adds relevant descriptions in a modular way so that the reader can reference both books.

1.1.2 Generations

A number of countries launched their initial 5G networks by the end of 2019. The year 2019 was in fact of utmost importance for 5G smartphone launches, and the World Radiocommunication Conference 2019 (WRC-19) added and aligned 5G frequency bands for the further optimization of radio.

End-users have been able to use commercial mobile communication networks since the 1980s. The systems at that time were first generation, and offered mainly voice service via analogue channels [1].

1G refers to analogue, automatic mobile networks that handled only voice calls, although data transfer was possible via a data modem adapted to the terminal, or via a handful of devices embedding such functionality into the device itself. The initial systems used vehicle-mounted and portable devices for voice communications. The weight of such devices was typically several kilograms. Some examples of this first phase of 1G were Nordic NMT-450, French Radiocom 2000, Spanish TMA, German Netz-C, the UK's TACS, Japanese JTACS, and American AMPS. As 1G matured, hand-held devices also became popular. The first ones were big and heavy compared to modern devices. An example of this latter phase was the NMT-900 system, which was launched in Nordic countries in 1986–1987.

2G represents digital systems that integrate data services and messaging. Examples of this generation are Global System for Mobile Communications (GSM) and Interim Standard-95 (IS-95). GSM was launched commercially in 1991, and unlike other 2G variants at that time, it was based on a Subscriber Identity Module (SIM) that housed subscription-related data.

SIM has evolved ever since. It is still a useful platform for storing a user's unique key, which is the basis for authentication and authorization of the user, and serves also for radio interface encryption. It is a hardware-based Secure Element (SE). 5G will rely on SIM, too, in one or another form.

2G data speeds were originally as low as 9.6 kb/s, and the service used circuit-switched connectivity. The ETSI/3GPP designed General Packet Radio Service (GPRS) that operators started to deploy in commercial markets as early as 2000, based on the European Telecommunication Standards Institute (ETSI) Release 97. It opened up the era of mobile packet-switched IP data over cellular networks. The data speed has increased along with the further evolution of GSM. Using multislot and multicarrier technologies, speed can nowadays be over 1 Mb/s depending on the service support on the network and device, e.g., by applying dual carrier and multislot techniques such as Downlink Dual Carrier (DLDC). Also, the voice capacity of GSM can be enhanced by offering the same number of voice calls within a reduced spectrum by applying Orthogonal Sub-Channel (OSC) and VAMOS (Voice services over Adaptive Multi-user channels on One Slot (VAMOS) [2].

Due to low spectral efficiency and security, the importance of 2G is decreasing and operators are refarming it for use with other systems. Nevertheless, 2G is still used in many markets for consumer and Machine-to-Machine (M2M) communications such as wireless alarm systems; therefore, only time will tell when 2G will no longer be relevant.

3G was a result of further development of multimedia-capable systems that provided much faster data speeds. 3G is thus a mobile multimedia platform. ITU's IMT-2000 sets the performance requirements for 3G systems. There are various commercial 3G systems such as US-originated cdma2000 and 3GPP-based Universal Mobile Telecommunications Service/High Speed Packet Access (UMTS/HSPA). 3G networks have evolved since their commercial launch at the beginning of 2000, and today they are capable of supporting tens of Mb/s data speeds.

4G continued with the "tradition" of renewed generations. The ITU-R designed a set of IMT-Advanced requirements for 4G systems. There are two commercial systems complying with them: LTE-Advanced (LTE-A) specified as of 3GPP Release 10, and WirelessMAN-Advanced (WiMAX), which is based on the IEEE 802.16 evolution. Oftentimes in the commercial field, the industry considers that Long Term Evolution (LTE) Releases 8 and 9 belong to the 4G era, and there have been operators interpreting even HSPA+ to be part of 4G. Nevertheless, referring strictly to IMT-Advanced, they are merely representatives of 3G technologies. Nowadays, the significance of WiMAX has decreased considerably, leaving LTE-A as the only relevant representative of the 4G era. Today, 4G offers hundreds of Mb/s data speeds.

5G refers to systems beyond IMT-Advanced that comply with the new ITU IMT-2020 requirements. 5G provides much higher data speeds. 3GPP specified the "full version" of 5G, i.e. Release 16, in the second half of 2020. The initial 5G, as defined in 3GPP Release 15, is oftentimes called *phase 1* 5G, whereas Release 16 represents *phase 2* 5G. The latter will comply with the strict requirements of IMT-2020, making it a complete 5G that provides customers with a full set of services and highest performance.

As new generations take over and customers start enjoying their enhanced and much more spectrum-efficient performance, previous generations gradually lose their users and can eventually be decommissioned, as can be seen from the example set by 1G systems in

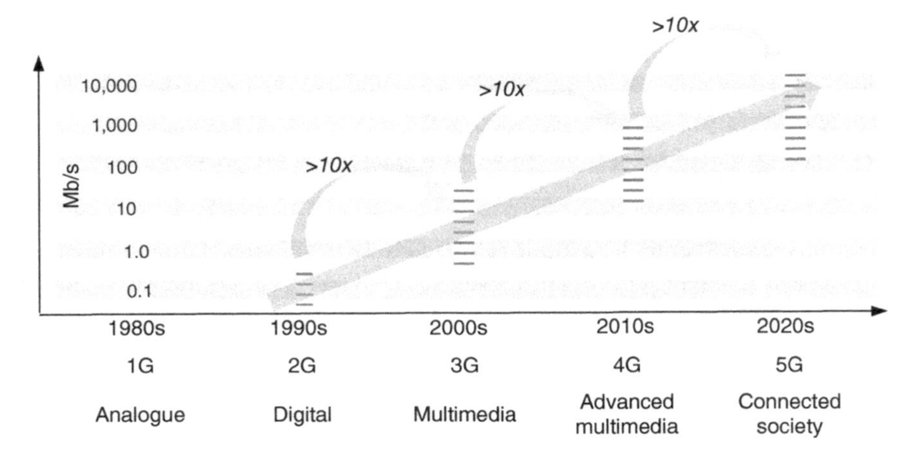

Figure 1.1 Mobile generations vs. downlink data speed evolution.

2000. Although 2G and 3G still have important use bases, including IoT devices, operators may already be considering decommissioning strategies.

Meanwhile, many operators refarm 2G and possibly 3G bands to 4G and 5G to optimize the use of the spectrum. In this transition phase, the already existing base station sites and possibly part of their equipment are reusable, including power supplies and transport lines.

Each generation goes through a series of enhancements during its lifecycle. Since the very initial deployments of 2G, the data services of each generation have continued evolving, providing users with constantly enhancing performance and capacity. As depicted in Figure 1.1, we can see that each new generation has provided at least 10-fold data speed ranges for customers compared to the previous generation.

5G is no exception; so, while LTE-A is capable of delivering some hundreds of Mb/s up to about the 1 Gb/s range, the eMBB mode of 5G can be assumed to handle Downlink (DL) data speeds of around 10–20 Gb/s.

3GPP TS 22.261 presents the service requirements for the next generation of new services and markets for Releases 15, 16, and 17 [3]. The document released in July 2020, V17.3.0 (2020–2007), describes the service and operational requirements for a 5G system, including User Equipment (UE), Next Generation Radio Access Network (NG-RAN), and 5G core network, while the requirements for the Dual Connectivity (DC) between 5G Evolved UMTS Terrestrial Radio Access (5G E-UTRA) and New Radio (NR) in the scenarios for the 5G Evolved UTRA Network (5G E-UTRAN) connected to 4G Evolved Packet Core (EPC) are presented in 3GPP TS 22.278 [4].

1.2 Principles of 5G

5G refers to the fifth generation of mobile communication systems. It represents mobile telecommunication standards beyond 4G LTE, and will comply with the strict IMT-2020 requirements of the ITU-R.

5G provides much faster data speeds with very low latency compared to legacy systems. 5G also supports a higher number of devices communicating simultaneously. 5G is capable of handling much more demanding mobile services than was ever possible before, including tactile Internet and VR applications, which will provide completely new and highly attractive user experiences.

As LTE and its evolution, LTE-A, have been a success story serving a growing base of customers, one might ask why we need yet another generation. The answer follows the same pattern as all the previous generations: as their performance reaches a practical limit, it makes more sense to provide services using more spectral, efficient, and performant new systems instead of trying to enhance legacy platforms.

Capacity has oftentimes been one of the biggest limitations, as the customer base and demand for data consumption continue to grow. Not only is this an issue for consumers, but the increasing popularity of IoT devices has impacted on system design because networks need to support a massive number of intelligent sensors and other devices relying on MTC.

The standardization community foresaw this trend and defined a new generation of specifications. They support up-to-date performance figures that would be challenging to comply with by adding on and developing previous generations further. As a result, 3GPP released a set of the first phase of 5G TS in at the beginning of 2019 based on a completely renewed system architecture.

The new era of the connected society is quickly becoming a reality as operators have started to deploy the initial 5G networks. The GSMA is estimating that 5G will account for 15% of the global mobile industry by 2025, as stated in the Mobile Economy 2019 study [5]. In other words, there will be an estimated 1.4 billion 5G connections in commercial markets within the next five years.

Although the mobile communications industry is eager to offer 5G services already in expedited schedules, it will be some time until we can enjoy the performance of the full version of 5G. Although the current 3GPP Release 15 sets the scene, it only works as an introduction to 5G as ITU envisions it.

The Release 16 specifications give industry the means to deploy second-phase 5G networks that will finally comply with the ITU's IMT-2020 requirements for global 5G. The result not only offers the operators data speeds that outperform previous generations, but the full version of 5G also introduces advanced solutions such as NS, virtualized Network Functions (NFs), support for massive numbers of IoT devices communicating simultaneously, ultra-reliable and low-latency communications, edge computing, and an optimized service-based architecture model. Releases 15 and 16 renew the security architecture, too, to cope with modern cyber-attacks.

Therefore, the early stage of 5G has offered merely a taste of eMBB service via 3GPP's intermediate options that can provide practical Non-Standalone (NSA) scenarios combining 4G and 5G elements, and ease the rollout of the new networks.

Figure 1.2 summarizes some of the key 5G functionalities. There are also other solutions increasing 5G performance such as enhanced radio interface modulation and intelligent multi-array antenna solutions. As soon as operators have deployed new 5G Radio Frequency (RF) bands especially on the high-band spectrum (above 7 GHz), the networks can offer more generous capacity to support wideband radio transmission, which in turn provides users with increased data speeds.

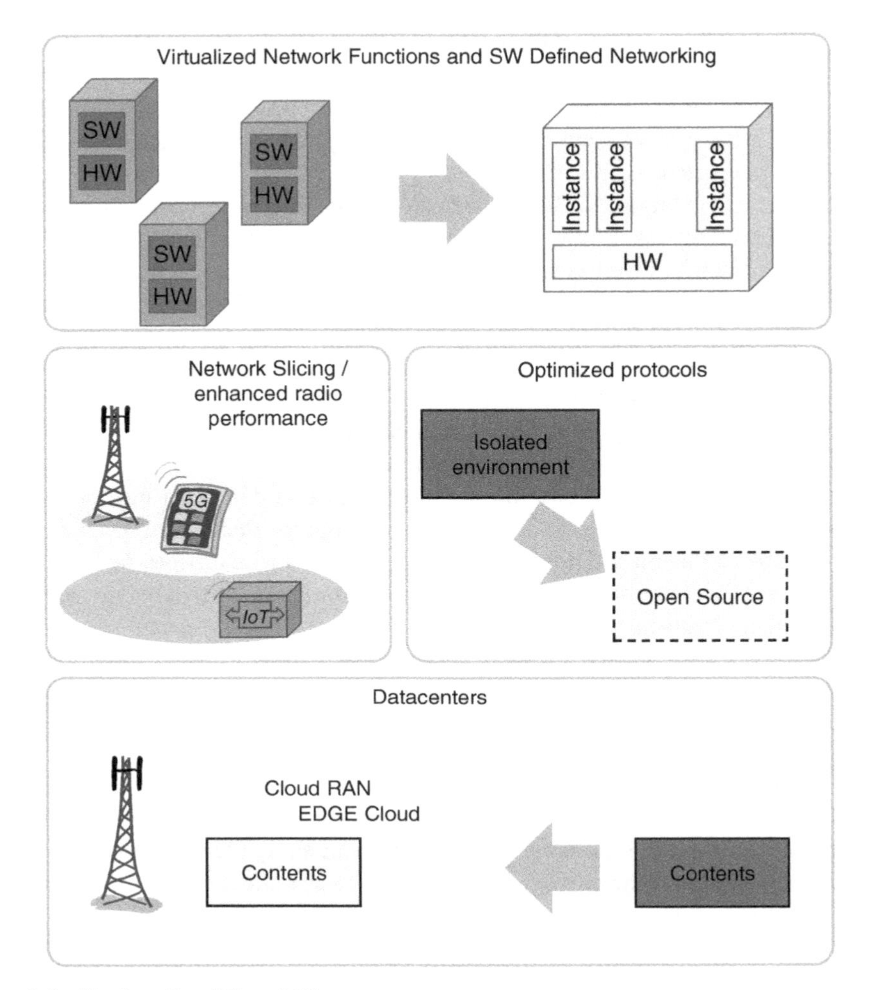

Figure 1.2 Key functionalities of 5G.

The WRC of the ITU in 2019 played an important role in providing concrete frequency band allocation plans at a global level [6]. New frequency bands, such as the ones on 26, 40, and 66 GHz, will pave the way for the increased capacity needed as subscribers start using advanced 5G applications.

5G will include room for NS, which refers to a set of optimized "networks within a network." Depending on the specific needs of the verticals, which are the practical representatives of varying communication profiles such as drones, law enforcement, automotive, self-driving vehicles, critical infrastructure, and smart cities, each one may obtain enough of a suitable network slice to fulfill its specific communications needs.

As an example, a network slice can provide its users with a very high data speed, while another slice can offer low data speeds but high reliability, etc. The previous systems were not able to distinguish between the offered services in this way.

5G is also based increasingly on the open source concept that generates business models and provides opportunities for many new stakeholders to join the developer community.

In practice, 5G will evolve gradually. It will provide enhanced performance as network deployments continue and coverage extends. Meanwhile, previous generations can serve mobile customers for years to come as 5G experiences inevitable outages, especially in the beginning.

5G will offer significant benefits in dense urban areas where small cells and high bands provide the highest data speeds. Also, the closer to users that edge computing is located, the lower the latency. This particularly benefits applications that are delay sensitive, such as autonomous cars.

The downside of high-band cells is their limited coverage for mobile use cases due to increased attenuation on higher frequencies. In fact, 5G small cell may be in range for only a few hundred feet assuming no obstacles are found in the communication line. The construction of such a dense radio network also requires connectivity to the high-capacity core infrastructure. Thus, this deployment strategy is only feasible in limited areas such as a city center where antennas could be installed widely on structures such as light poles. In suburban and rural areas, 5G can rely on lower-frequency bands, which serve much larger areas. While the achieved performance is somewhat lower, it will be much better than any previous generation in the same area.

5G will augment commercial models and business opportunities for the growing number of new stakeholders. One example of new 5G-based businesses is data centers that serve as a platform to process new 5G NF sets. Not all operators may be interested in investing in this type of infrastructure, at least in the beginning, which provides new business opportunities for cloud service providers.

Despite the advanced and complex technology, there are expectations for decently priced consumer devices. According to the latest indications of the industry, we may see 5G smart devices in the US$300 category rather soon. There will be a market for many types of devices, though, from simple 5G-connected sub-US$10 sensors up to complex, high-end VR headsets that could be capable of processing and transferring 360°, 3D audio/video contents.

Further reading on 5G principles can be found at 3GPP TR 22.891 (new services and markets technology enablers), and ITU's IMT-2020 (and beyond) [7, 8].

1.2.1 Open Source

5G architecture relies increasingly on the open source principles. Open source is something people are able to share, modify, and use based on an available design for all. Open source allows the possibility to share the software, investigate the code and its functions, and modify, copy, and distribute it under the software licensing terms. This principle helps expedite 5G development.

The Open Source Initiative (OSI) license is an example of the open source code environment. Of the 5G operators, e.g., AT&T decided to implement an open source platform based on the cloud environment of Kubernetes and OpenStack [9].

5G RAN disaggregates 4G evolved NodeB (eNB) and 5G next generation NodeB (gNB) functionalities into Distributed Unit (DU) and Centralized Unit (CU). The latter can be separated into User Plane (UP) and Control Plane (CP) referred to as CU-CP and CU-UP, respectively, via standardized, interoperable interfaces. The aim of this work is to optimize further the radio resource use and load balancing of separate, hardware-agnostic signaling and user data processing in a virtualized environment.

Open RAN Alliance (O-RAN Alliance) has been involved with the work in cooperation with 3GPP [10]. The goals of this effort are to leverage open source implementations and speed up development and deployments.

Other areas, in the form of workgroups of O-RAN, include the following activities:

- Non-real-time RAN intelligent controller and *A1* interface for radio resource management, procedure and policy optimization, as well as Artificial Intelligence (AI) and Machine Learning (ML) models;
- Near-real-time Radio Intelligent Controller (*RIC*) and *E2* interface architecture on decoupled software implementation of the CP;
- Stack reference design and *E1*, *F1*, and *V1* interfaces, multivendor profile specifications for *F1*, *W1*, *E1*, *X2*, and *Xn* interfaces;
- Open Fronthaul Interfaces for promoting multivendor Distributed Unit-Radio Remote Unit (DU-RRU) interoperability;
- Cloudification and orchestration for decoupling RAN software from the underlying hardware platforms;
- White-box hardware to reduce the cost of 5G deployment via reference design of decoupled software and hardware.

Examples of some other stakeholders and systems promoting open source are Open Compute Project (OCP), which provides telecom data center operators with open platform standards, Disaggregated Network Operating System (DANOS), which is an open networking operating system, and P4, which is an open source initiative for interacting with networking forwarding planes [11].

For more details on the O-RAN model, refer to Chapter 4 and the O-RAN Alliance website [10].

1.2.2 Justifications for 5G

5G networks tackle previous challenges of the increased need for capacity, reliability, and coverage, and provide more fluent user experiences than have been possible in any of the previous generations. 5G, as defined by 3GPP Release 16, is capable of satisfying the needs of current and foreseen environments in the short term, including the most advanced VR applications. At the same time, exponentially enhancing and growing IoT markets require new security measures that 5G can provide.

Without doubt, 5G represents a major step in the mobile communication evolution. The new era will take off gradually providing first limited service areas, while the previous generations, predominantly 4G LTE-A as defined by 3GPP Release 10 and beyond, will take care of the current bulk of traffic within their established service areas.

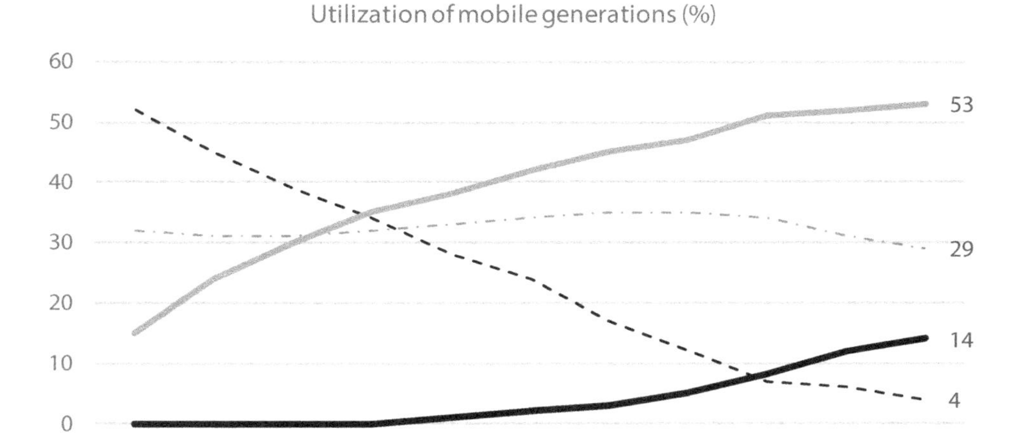

Figure 1.3 Forecast of the share of the utilization of 2G, 3G, 4G, and 5G networks by 2025 as interpreted from the statistics of the GSMA [5].

The older technologies will serve consumers for a long time to come in a parallel fashion, as forecasted based on past data. This refers also to 2G GSM that has been in commercial use since 1991, although 2G and 3G wind-down is inevitable at some stage.

Figure 1.3 shows an estimation of the percentage of users on different generations from 2G to 5G as forecasted by the GSMA. These numbers represent the global view, whereas there will be different types of regions applying diverse deployment strategies. Thus, some countries and operators will have decommissioned 2G and possibly 3G by 2025, while others may rely on them for years to come. For more concrete numbers, and for each region, only time will show how fast 2G, 3G, and 4G will lose consumers' interest as the more capable and spectrally efficient 5G takes over and as the increasing number of modern services and applications start requiring evolved performance to cope with the most fluent user experiences. As an example, a GSMA study released in February 2020 forecast that 5G would account for 15% of the global industry by 2025 [5].

An important step in this evolution is 3GPP Release 16, the ASN.1 implementation specification set available since the second half of 2020 [12]. As soon as it is deployed widely enough, respective devices and networks will represent the full 5G era complying with the ITU's IMT-2020 requirements. Meanwhile, according to 3GPP, the time schedule of 3GPP's final proposal submission of the candidate technology for the IMT-2020 evaluation process has been on track but with some delays [12].

5G will open many new business opportunities for established and completely new stakeholders and participants in the ecosystem. One example of such an environment is related to AR/VR applications. To work fluently, they require very low latency for communications and data processing. Typically, such applications rely on powerful and expensive hardware located at the site. As 5G uses a novelty architecture model of NFV, and it can take advantage of edge computing for offloading data processing and ensuring fast delivery; the same infrastructure can be applied also to offer external services requiring evolved performance, such as AR/VR applications.

By applying AR/VR data rendering to handle optimal sharing of data processing in the edge cloud instead of the user device, it is possible to reduce the handset processing requirements – which, in turn, can provide the possibility of offering advanced services with lower cost multifunctional devices. In fact, this opens new cooperative business models for mobile network operators designing advanced and new added-value services, which were not feasible within the older mobile communications infrastructure.

1.3 Standardization

The ITU-R has defined universal requirements and principles for global 5G under the term IMT-2020.

3GPP TS define the first phase of 5G as per Release 15, and the second phase as per Release 16. Release 17 has been in planning since 2020, and it represents the evolved second phase together with forthcoming Releases. Release 15 is not capable of providing sufficient performance to comply with IMT-2020 requirements, whereas Release 16 does, and future Releases will enhance the performance and feature set further.

There are also other Standard Development Organization (SDO) entities and industry forums contributing to the development of the 5G ecosystem. Table 1.1 lists a snapshot of some of the relevant entities.

1.3.1 Release 16 Key Features

3GPP standardized Release 15 by the end of 2018, and the respective first-phase networks, remarkably NR equipment complying with these standards, entered commercial markets as of early 2019. At the same time, Release 15 also introduced further enhancements to LTE. Some of these include [8]:

- eNB Architecture Evolution for E-UTRAN and NG-RAN. This feature introduces higher-layer functional split architecture of eNB, and brings central unit (LTE-CU) and distributed unit (LTE-DU). The new model provides operators with more efficient integration between the eNB and gNB elements.
- Enhancements for NB-IoT have improved DL transmission efficiency and UE power consumption, Uplink (UL) transmission efficiency and UE power consumption, and enhanced scheduling.
- LTE-NR and NR-NR DC and NR Carrier Aggregation (CA) enhancements. This item supports asynchronous and synchronous NR-NR DC, early measurement reporting, low latency serving cell configuration, activation and setup, and fast recovery.
- Network management tool enhancements include improved Self-Organizing Network (SON), multicarrier operation, and mobility enhancements.

Table 1.1 Some of the key entities standardizing or contributing to 5G evolution.

Entity	Activities in 5G
3GPP	Produces 5G TS as of Release 15
5GAA	5G Automotive Association promotes 5G in an automotive ecosystem
C2C CC	Car-to-Car Communication Consortium, industry forum for Vehicle-to-Vehicle (V2V) technology development
CSA	Cloud Security Alliance promotes security assurance
IETF	Internet Engineering Task Force designs Internet architecture; protocols adapted in 5G
ETSI	Complements and cross-references 3GPP TS, SIM evolution (Smart Secure Platform, SSP); integrated SSP for 5G
Global Standards Collaboration	Enhances global standards. Members: ARIB, ATIS, CCSA, ETSI, IEX, IEEE-SA, ISO, ITU, TIA, TDSI, TTA, and TTC
GlobalPlatform	Universal Integrated Circuit Card (UICC) development
GSMA	Produces industry guidelines for interoperability and roaming, e.g., on remote SIM provisioning; voice and messaging services over 5G and NS interoperability
IEEE	Institute of Electrical and Electronics Engineers produces IEEE 802 series; IoT standards
ISO	International Organization for Standardization works on IT security; smart card standardization; common criteria
ITU	Requirements of 5G in the form of IMT-2020
NGMN	Next Generation Mobile Networks optimizes and guides on advanced network technologies and IoT
NIST	National Institute of Standards and Technology works on cybersecurity framework
OMA	Open Mobile Alliance works on device management; LightweightM2M
O-RAN	Develops cloud-based RAN specifications
SIMalliance	Secure element implementation; eUICC, iUICC
STA	Secure Technology Alliance facilitates the adoption of secure solutions in the United States

The second phase of 5G System (5GS), as defined in Release 16, has improved 5G performance further. The following lists some of the key enhancements of and additions to Release 16:

- Enhancements for Common API Framework (eCAPIF) for 3GPP Northbound APIs;
- Enhancements of the URLLC mode;
- Enhancements to 5G efficiency, including interference mitigation, SON and big data, Multiple In, Multiple Out (MIMO) (eMIMO), location and positioning, power consumption, DC (eDual), device capabilities exchange, and mobility;

- Industrial IoT (I-IoT), including additional 5G NR capabilities such as Time Sensitive Networking (TNS) to serve as a replacement for wired Ethernet in factories;
- Integrated Access and Backhaul (IAB);
- Mobile Communications System for Railways (FRMCS), phase 2;
- New 5G spectrum, including bands above 52.6 GHz;
- Non-Orthogonal Multiple Access (NOMA);
- NR-based access to unlicensed spectrum, including Licensed Assisted Access (LAA) and standalone unlicensed operation;
- Satellite access in 5G;
- Vehicle-to-Everything (V2X) phase 3, including support for platooning, extended sensors, automated driving, and remote driving.

Although consumers will benefit from the faster data speeds of 5G, the verticals are in a special position when using 5G. Industry verticals such as automotive, I-IoT, and entities operating on unlicensed bands are some of the examples of stakeholders to which the 5G offers significant enhancements compared to previous mobile communication generations.

Also, factory automation will have more efficient means to carry out their communications, along with low latency and ultra-reliable communications. Thanks to TSN integration of Release 16, there is a new possibility to replace wired Ethernet altogether from factories. The extension of modes on unlicensed bands can bring fresh businesses and operational modes, based on, e.g., Non-Public Networks (NPNs).

3GPP Release 16 is highly relevant for 5G, and many operators will start deploying the standalone mode, i.e., 5GC, along with it.

Table 1.2 summarizes some key enhancements and additions of Release 16 as interpreted from 3GPP summaries [12] and various 3GPP Technical Reports (TRs) such as 3GPP TS 21.916 [13, 14], and Study on eNB(s) Architecture Evolution for E-UTRAN and NG-RAN [15].

As can be noted from Table 1.2, Release 16 brings many further advancements or completely new items throughout the complete mobile communications network, including enhanced radio systems, additional frequency bands, and a renewed core network with further optimized service-based architecture models. These renovations ensure that 5G as seen by 3GPP is compliant with the strict requirements of the ITU IMT-2020. The latter defines global 5G of which 3GPP produces concrete specifications for the system.

1.3.2 The Phases of 5G

The production of 5G specifications has been a long journey. After initial thoughts of the new generation almost a decade prior to the first implementation, the first of 3GPP's 5G contents was Release 15. 3GPP published the early version of this in December 2018, and continued with two additional Release 15 updates during 2019. 3GPP finalized the work on Release 16 in July 2020, and ASN.1 readiness – referring to the implementation guidelines – in autumn 2020. 3GPP also created the Release 17 content and timeline, which it will finalize during 2021. The goal of Release 17 is to extend the functions and performance for 5G verticals.

Table 1.2 Some of the key features of 3GPP Release 16.

Topic	Summary
Access Traffic Steering, Switch and Splitting	Release 16 supports the Access Traffic Steering, Switch and Splitting feature. It enables traffic steering across multiple access types via a multi-access Packet Data Unit (PDU) session. The access types may include, e.g., 3GPP access, and trusted and untrusted non-3GPP access.
Cellular IoT support and evolution	Evolved I-IoT and URLLC enhancements are important components of Release 16. These add 5G NR capabilities for wired Ethernet replacement in factories and increase the level of compatibility with highly reliable TSN. The support for Cellular IoT (C-IoT) has existed since Release 8 for GSM and LTE systems, but Release 16 presents native 5G IoT support for the first time. Some of the Release 16 features related to the C-IoT include the following [16]: • Core network selection and steering for C-IoT; • Enhanced coverage management; • Frequent small data communication; • Group Message Delivery (GMT) using unicast Non-IP Data Delivery (NIDD); • High-latency communication; • Infrequent small data transmission; • Inter-Radio Access Technology (RAT) mobility support for Narrow-Band IoT (NB-IoT); • Interworking with Evolved Packet System (EPS) for C-IoT; • Monitoring; • Mobile Station ISDN Number (MSISDN)-less Mobile Originated (MO) Short Message Service (SMS); • NB-IoT Quality of Service (QoS) support; • Network parameter configuration API via Network Exposure Function (NEF); • Power-saving functions; • Small data overload control; • Support for expected UE behavior; • Support of Common north-bound APIs (CAPIs) for EPC-5GC interworking; • Support of the Reliable Data Service (RDS).
Enablers for Network Automation Architecture for 5G	5GC includes intelligent automation network analytics. Release 16 brings the extension of the related Network Data Analytics Function (NWDAF). Ref. [14] summarizes the enablers for network automation, including architecture assumptions, use cases and key issues, and respective solutions. The following architectural assumptions are the basis for Release 16 solutions: (1) NWDAF [17] works for centralized data collection and analytics; (2) for instances where certain analytics can be performed by a 5GS NF independently, an NWDAF instance specific to that analytic maybe co-located with the 5GS NF; (3) 5GS NFs and Operations Administration and Maintenance (OAM) decide how to use the data analytics; (4) NWDAF utilizes the existing service-based interfaces to communicate with other 5GC NFs and OAM; (5) a 5GC NF may expose the result of the data analytics to any consumer NF utilizing a service-based interface; (6) the interactions between NFs and NWDAF take place in the same local Public Land Mobile Network (PLMN); (7) solutions cannot be assumed to be aware of NWDAF knowledge about NF application logic; (8) NWDAF and NFs cooperate to contribute to consistent policies, analytics output results, and finally decision-making in the PLMN.

(Continued)

Table 1.1 *(Continued)*

Topic	Summary
Energy efficiency	Energy efficiency of 5G presents concepts, use cases, and requirements for enhancing energy efficiency.
Enhanced support of vertical and Local Area Network (LAN) services	5GC LAN support has resulted in the updated 3GPP TS 22.261, which now includes 5G LAN creation and management, 5G Virtual Private Network (VPN), as well as 5G LAN service authorization, mobility, and service continuity for a 5G LAN-type service.
Enhancement of URLLC	As stated in 3GPP 38.824 [18], the basic support for URLLC was introduced in Release 15 containing Transmission Time Interval (TTI) structures for low latency and methods for improved reliability. Further enhancement of the respective use cases, as well as new use cases with tighter requirements, are important for the NR evolution. The list of improvements includes Release 15-enabled use case improvements, e.g. AR/VR for the entertainment industry; new Release 16 use cases with more demanding requirements, e.g., for factory automation, transport industry, including remote driving, and electrical power distribution.
Features impacting both LTE and NR	Transfer of *Iuant* interface specifications from 25-series to 37-series; introduction of GSM, UTRA, E-UTRA, and NR capability sets to the Multi-Standard Radio specifications (MSR); direct data forwarding between NG-RAN and E-UTRAN nodes for intersystem mobility; eNB architecture evolution for E-UTRAN and NG-RAN; high-power UE (power class 2) for E-UTRA-NR Dual Connectivity (EN-DC) (1 LTE Time Division Duplex (TDD) band + 1 NR TDD band).
LAN-type services	Release 16 supports LAN-type services. They refer to a 5G Virtual Network (VN) group of UE that may communicate privately.
Location and positioning services	5G Release 15 includes a function to retrieve location and velocity of a UE that registers with 5GC. Release 16 brings with it new features like roaming, exposure, and privacy management. Phase 2 5G includes enhancements to 5GC Location Services (5G eLCS), and their related NFs, services, and procedures. 5G eLCS include regulatory and commercial location services, and use cases such as regulatory requirements for emergency services. 3GPP TS 22.261 and TR 22.872 present these updates.
Mission Critical, Public Warning, Railways, and Maritime	Release 16 includes protocol enhancements for Mission Critical Services, Public Warning System, and Future Railway Mobile Communication System, built upon the architecture from Release 15. The enhancements benefit a broad set of verticals such as public safety. As Ref. [19] states, 3GPP has been working on Mission Critical (MC) communications and services as a major driver of the MC industry (emergency services, railway, and maritime). 3GPP started out with Mission-Critical Push To Talk (MCPTT) in Release 13, and the path towards an evolved phase of 5G includes the MCData and MCVideo services, railways and maritime communications (Multimedia Broadcast Multicast Service (MBMS) APIs). At the same time, a platform for MC communications and services evolves along with an increasing number of requirements from the critical communications industry.
Mobility	NR mobility enhancements via reduced interruption time during Handover (HO) and Secondary Cell Group (SCG) change by HO and SCG change with simultaneous connectivity with source cell and target cell; "make-before-break" model, Random Access Channel (RACH)-less HO, and solutions to improve HO reliability and robustness by conditional HO and fast HO failure recovery.

Table 1.1 *(Continued)*

Topic	Summary
NS	NS is a key 5G feature. It enables an operator to support specific use cases with a dedicated set of network resources, which the operator can match with their expected service level instead of offering a uniform performance for all users. Release 16 enhances NS interworking support from EPC to 5GC in the case when UE moves from EPC to 5GC. Slicing provides network slice-based authentication and authorization. Release 16 NS addresses two major limitations of Release 15 in 5GC: enhancement of interworking between EPC and 5GC when UE moves from EPC to 5GC, and support for Network Slice-Specific Authentication and Authorization (NSSAA).
NPNs	Release 16 supports NPNs to discover, identify, select, and control access for the NPN. Release 16 provides service continuity and interoperability between NPN and PLMN for their services. Release 16 covers Standalone Non-Public Networks (SNPN) and Public Network Integrated NPN (PNI-NPN).
NR-related Release 16 features	Two-step RACH for NR achieves the following objectives: (1) A simplified random access procedure reduces the number of interactions between the UE and the network during connection setup and connection resume. It enables a lower CP latency for IDLE and INACTIVE UE. In connected mode, a small amount of data can be sent over the two-step RACH, which results in a lower latency of the UL UP data for connected mode UE. (2) Channel structure of transmitting Physical Random Access Channel (PRACH) and Physical/Primary Uplink Shared Channel (PUSCH) in one step instead of an intermediate network message. Both of these enhancements are applicable to licensed spectrum and NR Unlicensed (NR-U). In Release 15, the EN-DC band combinations include at least one E-UTRA operating band and an NR operating band. In Release 16, the configuration of EN-DC operation needs to be expanded for more simultaneous UL and DL configurations, including band combinations of LTE, Frequency Range 1 (FR1), and Frequency Range 2 (FR2) bands. New configurations emerge from existing bands, and new bands create a potential for several new EN-DC configurations consisting of different DL/UL band combinations. 3GPP specifications will present EN-DC configurations, including NR CA, in a release-independent manner based on 3GPP TS 38.307.
Optimization of radio capabilities signaling	Release 16 enhances the network's signaling about UE radio capabilities and the management of respective mapping. As an example, there is a new UE Capability Management Function (UCMF).
Other cross-Technical Specification Group (TSG) Release 16 features	5G introduces a mechanism to support Single Radio Voice Call Continuity (SRVCC) from 5GS to UTRAN. The definitions cover the following scenarios: (1) operators with both 5G Voice over IP Multimedia Subsystem (IMS) and LTE enabled, but no Voice over LTE (VoLTE); (2) operators with no LTE (or VoLTE); and (3) operators with both 5G Voice over IMS and VoLTE enabled, but the voice service continuity may not be guaranteed if the VoLTE coverage provided by the operators has outages. The Access Traffic Steering, Switch and Splitting (ATSSS) feature enables a multi-access PDU Connectivity Service, which can exchange PDUs between the UE and a data network by simultaneously using one 3GPP access network and one non-3GPP access network and two independent *N3/N9* tunnels.
Redundancy	Release 16 introduces redundancy enhancements, which enable URLLC services for services requiring ultra-reliability. Release 16 specifies support for end-to-end redundancy of the data paths for 5G applications using URLLC services.

(Continued)

Table 1.1 *(Continued)*

Topic	Summary
Satellite Access in 5G	Integration of Satellite Access in 5G to cope with use cases of TR 22.822. The scenarios cover roaming between terrestrial and satellite networks, broadcast and multicast with a satellite overlay, IoT with a satellite network, and temporary use of a satellite component. The cases also include optimal routing and steering over a satellite, global satellite overlay, indirect connection through a 5G Satellite Access network, 5G Fixed Backhaul for an NR-5GC link, 5G Moving Platform Backhaul, 5G to Premises, and offshore wind farms.
Security	Security Assurance Specification (SCAS) for 5G network equipment to help identify threats against 5GS architecture, and to provide functional security requirements and test cases.
Service Enabler Architecture Layer (SEAL)	Release 16 includes common capabilities for a SEAL. It provides a common set of services, including configuration management, group management, and location management to support vertical applications.
Service-Based Architecture (SBA)	Release 16 brings enhancements to service discovery and service routing, e.g., via a new Service Communication Proxy (SCP) NF.
Streaming and TV	The Coverage and Handoff Enhancements for Multimedia (CHEM) feature enables the network to delay or reduce handoffs of a Multimedia Telephony Service for IMS (MTSI) terminal. This can happen by providing the eNB or gNB with additional information about the robustness to packet losses of the negotiated media configurations.
TSN	Release 16 includes support for I-IoT. A major step in this evolution is the support of 5G for TSN architecture. Release 16 enables a 5G network to serve as a bridge to interconnect a TSN network.
Topology	Release 16 enhances the topology for flexible deployment of Session Management Functions (SMFs) and User Plane Functions (UPFs). This enhancement evolves the distributed aspects of the 5GC architecture to serve better the interactions between SMFs and UPFs. As an example, Release 16 introduces new Intermediate SMF (I-SMF) and Intermediate UPF (I-UPF).
UE radio capability signaling optimization	The list of UE radio capability enhancements includes the following: • EN-DC for combined LTE and NR bands. • Enhancements on MIMO for NR for a better MIMO performance, including Multi User MIMO (MU-MIMO). • Enhancements on multibeam operation especially on FR2. • Enhancements on multi-Tx/Rx Point (TRP)/panel transmission such as improved reliability and robustness. • NR intra-band CA defines the CA configurations and their respective RF requirements for contiguous and non-contiguous cases. • UL and DL Channel State Information Reference Signal (CSI-RS) and Demodulation Reference Signal (DMRS) enhancement for Peak to Average Power Ratio (PAPR) reduction.
User identities, authentication, multidevice	The User Identifier (UI) is independent of existing identifiers relating to a 3GPP subscription or UE. It can be used for NSSAA. The related identity provisioning service creates, manages, and authenticates the identities. The 3GPP does not specify this service, though. Instead, the 3GPP does define the interworking of 3GPP systems and external entities for use cases involving authentication of the UI, and for authorizing the user to access a network slice.

Table 1.1 (*Continued*)

Topic	Summary
V2X enhancements	Release 16 enhances V2X services over 5G to facilitate vehicular communications for transport services. Some examples of the new functions include the quality of service support enabling a V2X application, and the change notification of the QoS. As an example, radio conditions expectedly worsening and impacting V2X application can serve as a criterion to increase the distance between vehicles. Release 16 can enable features such as intersection safety and vehicle platooning for self-driving cars.
Wireless and wireline convergence enhancement	Release 16 supports integrated architecture, including wireline access towards a 5GC by using 3GPP interfaces and protocols. The architecture reference model covers wireline access by including a Wireline Access Gateway Function (W-AGF) and support for a Fixed Network Residential Gateway (FN-RG). 5GS is intended for end-to-end communications involving radio, transport, and core networks as well as interconnectivity with the rest of the ecosystem. The 5G architecture facilitates a high level of convergence leveraging modern access mechanisms, including wireline, wireless, and cellular access, and complies with the vision for fixed-mobile convergence. 3GPP TR 23.716 presents a study on wireless and wireline convergence for 5GS architecture. More information on generic development can be found in Refs. [20, 21].

3GPP has worked focusing on use cases and ways to comply with the ITU IMT-2020 vision and requirements for the 5G era. The concrete first phase for 5G began with 3GPP brainstorming use cases as one of the Release 14 work items, and the first definitions for a "light" version of 5G were included in 3GPP Release 15 as a set of TS. These specifications included options for intermediate deployment relying on both 4G and 5G. The concrete form of these options is defined as NSA NR specifications for 5G, while Release 15 also includes the fully end-to-end 5G option referred to as Standalone (SA). Release 15 thus forms the initial approach for 3GPP's submission towards the IMT-2020 process.

The second phase, i.e. Release 16 of 3GPP, complements the initial set by adding needed functionalities for complying fully with the IMT-2020. The original goal was to complete Release 16 by the end of 2019, but 3GPP postponed it until the second quarter of 2020 as presented in Figure 1.4 [22].

The first stage of Release 16 was under construction based on close to 100 study items for consideration in Release 16, and an additional set of items that are more suitable for Release 17. These items covered a variety of topics such as Multimedia Priority Service (MPS), the evolution path of V2X and IoT services, 5G satellite radio access, enhanced security, evolved codecs, 5G's LAN interworking, communications in industry vertical domains, and further advancements for NS.

The selected key services are presented in more detail in Chapter 6 of this book, in Section 6.3, which describes the most important new and enhanced features in Release 16.

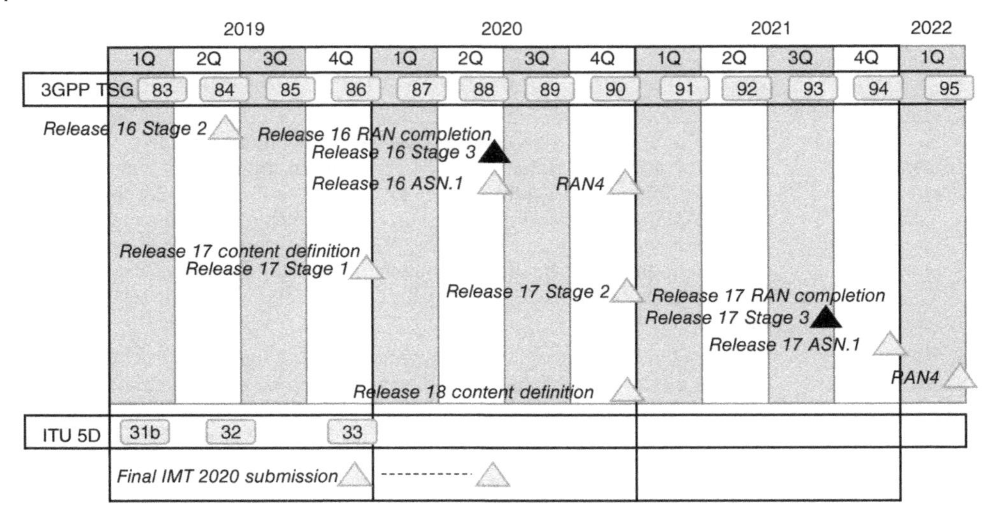

Figure 1.4 The timeline for 3GPP 5G, Release 16 and beyond [23, 24], and the aligned schedule with the ITU IMT-2020 candidate process.

1.3.3 How to Find 5G-Related Specifications

The main entity specifying the concrete 5GS is 3GPP. The first phase of 5G is in the 3GPP Release 15 set of TS. The Release 15 NSA 5G NR mode's ASN.1 freeze was completed in March 2018, whereas the SA 5G NR mode's ASN.1 freeze took place in September 2018 [23].

3GPP specifications are located at the web address: https://www.3gpp.org/DynaReport/xxxxx.htm where xxxxx refers to a desired specification number.

This link leads to 3GPP's specification summary page. By selecting a link on that page, "Click to see all versions of this specification," the complete list of releases and versions of the specification can be displayed. To examine the very latest version, you can download the one with the highest version number, although it is worth noting that it might not represent the most stable specification.

Figure 1.5 presents an example of a 3GPP document. The version numbering of the 3GPP TS and TR follows established principles as detailed in Ref. [25]. They have a specification number consisting of four or five digits, e.g., 05.05 or 25.005. The four-digit format applies to specifications that were developed at ETSI. As 3GPP took over further development of the major part of ETSI specifications in 2000, and keeps adding new ones, the five-digit scheme was adopted instead. The first two digits define the series, which indicates the overall category the specification belongs to; as an example, the abovementioned 05 and 25 define the radio interface. The following two digits are applicable to series ranging from 01 to 13, and the three-digit format is valid for the series from 21 to 55.

The complete and up-to-date list of specifications can be found at the web address: https://www.3gpp.org/ftp/Specs/html-info/status-report.htm

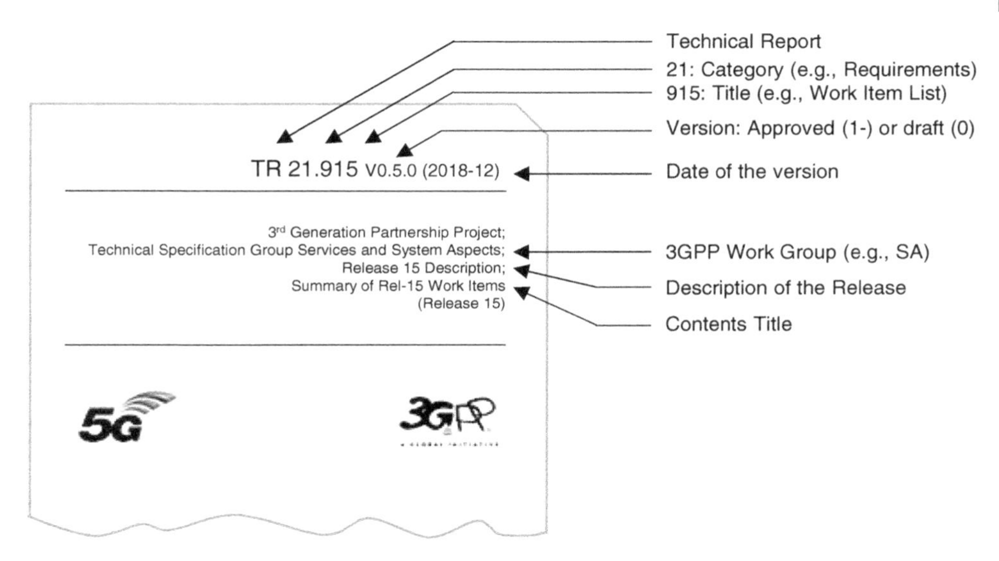

Figure 1.5 Template of 3GPP TS and TR.

A matrix format of 3GPP specifications summarizing the respective working groups can be found at: https://www.3gpp.org/ftp/Specs/html-info/SpecReleaseMatrix.htm

ANNEX 1 summarizes the main thematic categories of the 3GPP specifications and reports relevant to 5G. 5G radio aspects are described in the 38-series, while the core aspects are found from a set of specifications. The functional and architectural descriptions of 5G and other generations are described in the 22-, 23-, and 24-series.

Figure 1.5 depicts an example of the 3GPP 5G TR, and how to interpret the title fields.

3GPP TR 21.915 is in fact the root source of information to understand the new aspects of Release 15. It includes the basis for 5G as well as enhancements for the previous mobile communication generations of 3GPP.

1.3.3.1 Standardization and Regulation of Phase 2

The standardization of 5G is primarily done by 3GPP. There are many other industry forums and SDOs involved in the development, too, such as IETF, GSMA, SIMalliance, and GlobalPlatform.

1.3.3.2 SIM Evolution

The SIM continues to evolve in the 5G era. In fact, the traditional Form Factors (FFs) of the SIM card will not disappear by default even in the case of 5G mobile devices, but are in the portfolio complementing the new models.

As an example, the evolution of subscription management is being considered by ETSI, GSMA, SIMalliance, GlobalPlatform, and Smart Card Association (SCA), all of which complement the base specifications of 3GPP. 3GPP, in turn, references the smart card definitions set by these SDOs.

1.3.4 Release 17

3GPP Release 17 has been under development throughout 2020–2021. 3GPP working groups have discussed the potential delays of Release 17 because of changes in the ways of working from physical to virtual meetings. Most concretely, at TSG#87e held in March 2020, the following Release 17 timeline was agreed:

- Rel-17 Stage 3 freeze September 2021.
- Rel-17 ASN.1 and OpenAPI specification freeze: December 2021.

Nevertheless, as 3GPP finalized Release 16 in July 2020, the forthcoming work items of Release 17 have already been identified. 3GPP TR 21.917 describes Release 17 work items. 3GPP will approve and publish them at the end of the Release 17 production schedule. Meanwhile, draft versions will be available. The work items include the following as indicated in Ref. [24]:

- 5GC enhancements, including edge computing, proximity services, location services, and MPS;
- 5G network automation phase 2;
- Architectural evolution of 5G, including satellite components, access traffic steering, switch and split support, 5G wireless and wireline convergence, UPF enhancement for control, and 5G SBA;
- Architectural evolution of eNB, including further control and UP split;
- Device enhancements, including low-complexity NR devices, power-saving enhancements, and Multi-SIM;
- Enhancements on IoT, including I-IoT, NB-IoT, and LTE-MTC;
- Enhancements to Minimization of Drive Tests (MDT);
- Evolved services, including interactive services, V2X, 5G LAN-type services;
- IAB enhancements;
- Multi-radio Diameter Credit Control Application (DCCA) enhancements (related to online charging);
- NPN enhancements;
- NR QoE, MIMO, sidelink, sidelink relay, positioning and coverage enhancements, and NR over Non-Terrestrial Networks (NTN);
- RAN slicing enhancements and NS, phase 2;
- RF enhancements, including existing waveform on 52.6–71 GHz frequency bands and Dynamic Spectrum Sharing (DSS) enhancements;
- SON enhancements;
- Services enhancements, including 5G MBMS;
- Small data enhancements;
- Study items on IoT over NTN and NR Extended Reality (XR);
- Support of Unmanned Aerial Systems (UAS);
- URLLC enhancements.

Please refer to the details of the Release 17 work plan at the 3GPP web page [26].

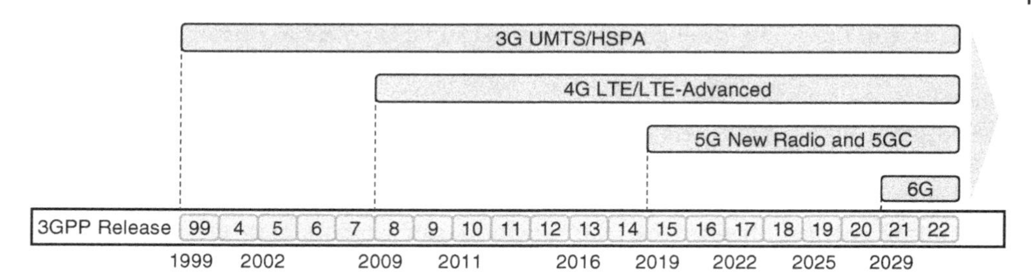

Figure 1.6 3GPP Release roadmap towards 6G.

1.3.5 Later Phases and 6G

As has been the "tradition," there will be new releases steadily after Release 17. Release 18 is the next one, and paves the way for the 6G era that will start concretizing later by the end of the decade.

Figure 1.6 depicts the historic and estimated future release schedules.

As the 5G specifications evolve, there is already work going on to collect requirements for the next generation beyond 5G. There seems to be a tradition that new generations appear in each decade. 6G, or the sixth generation of mobile communications, may well follow this rule of thumb.

The next step beyond the "visible horizon" could be the sixth generation of mobile communications, 6G. Before it will become a standardization, there is plenty of work to do in the 5G era. The first phase of 5G was planned to take off during 2019, and 5G with its forthcoming releases will serve us for years to come. Network deployment of 3GPP Release 16-based second phase 5G began during the latter half of 2020. After that, 5G will evolve through several 3GPP releases, similarly to that already seen with the previous systems; 3GPP is already collecting ideas for Release 17 and beyond.

While 5G starts taking over and will serve us throughout the 2020s, the ITU is already foreseeing the future needs. The timing coincides with the familiar cyclic pattern; a new generation is deployed as soon as the new decade begins. ITU has recently started discussions on how mobile communications would look in 2030. The group dealing with this topic is called the ITU-T Focus Group Technologies for Network 2030, FG NET-2030 [27].

The FG NET-2030 was established by the ITU-T Study Group number 13 in 2018, and its aim was to investigate expected capabilities of networks for the year 2030 and beyond. The study includes future network architecture, requirements, use cases, and capabilities of the networks.

The task is interesting as there are already many scenarios foreseen that might be challenging to cope with evolved 5G, including holographic-type communications, ultra-fast critical communications, and high-precision communication demands of emerging market verticals.

The task of the group is to figure out the most suitable potential network architectures and respective mechanisms for the identified communication scenarios, and document the findings in the form of Network 2030.

The group is not restricted to existing solutions but their evolution is part of the research. Although there is not too much concrete information available on the expected outcome yet, one of the basic principles of the Network 2030 system will be to ensure backwards compatibility so that the networks can support both existing and new applications.

In addition to the abovementioned gap and challenge analysis, the next step of the FG NET-2030 includes the collection of all aspects of Network 2030, including vision, requirements, architecture, novel use cases, and evaluation methodology. As a part of the activities, the group will provide guidelines for a standardization roadmap and establish liaisons with other SDOs.

For those who are interested, participation in the FG NET-2030 is open to all. More information on the initiative can be found at the web page of the group [27].

1.4 Introduction to the Book

This book is aimed at technical personnel of operators and equipment manufacturers, as well as telecom students. Previous knowledge of mobile communications, as well as 3GPP Release 15 (e.g. based on the already published *5G Explained* by Wiley), would help to capture the most detailed messages of the book, but the modular structure of the chapters ensures that the book is useful also for readers not yet familiar with the subject.

Readers in mobile equipment engineering, security, network planning and optimization, as well as application development teams will benefit most from its contents as it details highly novel aspects of Release 16, which helps to deliver essential information in a compact and practical way.

The book is designed primarily for specialists interested in capturing the new key aspects of 5G Release 16 and understanding both the differences between the first phase of 5G as per Release 15 and the key aspects for device planning, network deployment, and app development based on 3GPP Release 16 and accompanying guidelines of the industry.

This *5G Explained* book focuses on Release 16. It forms a complete package with the previously published *5G Explained* book, and the reader can use both in unison to understand the complete picture of 5G.

Chapters 1 and 2 form the introductory module, which will be useful for both technical and non-technical readers with or without preliminary knowledge of existing mobile communications systems. Chapters 3–5 present the technical description and are directed at advanced readers with some knowledge of mobile communications, while Chapters 6–8 represent the planning module and are meant for seasonal subject matter experts.

Figure 1.7 presents the main contents of the book to ease navigation between the modules. The modules and chapters are independent of each other, so readers can study the contents in any preferred order, although it is recommendable to read the book in a chronological order.

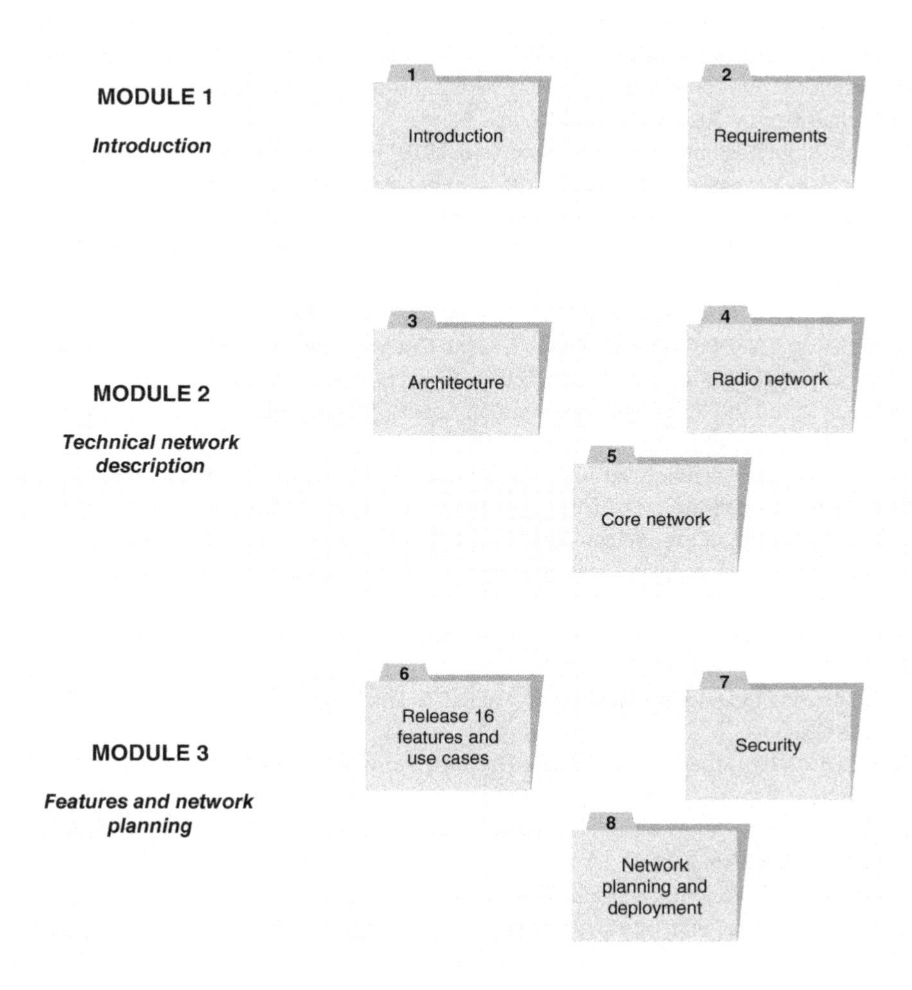

Figure 1.7 Contents of this book.

References

1 Penttinen, J., "Mobile Generations Explained," Interference Technology, 20 November 2015. [Online]. Available: https://interferencetechnology.com/mobile-generations-explained. [Accessed July 2019].
2 Penttinen, J., Calabrese, F.D., Lasek, S. et al. "Performance of Orthogonal Sub Channel with Dynamic Frequency and Channel Allocation," in *Wireless Telecommunications Symposium*, New York, USA, 13–15 April 2011.
3 3GPP, "TS 22.261, Service Requirements for the 5G System," 3GPP.
4 3GPP, "TS 22.278, Service Requirements for the Evolved Packet System (EPS), Release 17," 3GPP, December 2019.

5 GSMA, "New GSMA Study: 5G to Account for 15% of Global Mobile Industry by 2025 as 5G Network Launches Accelerate," GSMA, 25 February 2019. [Online]. Available: https://www.gsma.com/newsroom/press-release/new-gsma-study-5g-to-account-for-15-of-global-mobile-industry-by-2025. [Accessed 30 June 2019].
6 ITU, "World Radiocommunication Conference 2019 (WRC-19), Sharm el-Sheikh, Egypt, 28 October to 22 November 2019," ITU, 2019. [Online]. Available: https://www.itu.int/en/ITU-R/conferences/wrc/2019/Pages/default.aspx. [Accessed 6 July 2019].
7 3GPP, "TR 22.891, Feasibility Study on New Services and Markets Technology Enablers," 3GPP, 2016.
8 ITU, "Session 7: 5G Networks and 3GPP Release 15," Suva, Fiji, 17–19 October2020.
9 Wagner, M., "AT&T Building 5G Network on an Open Source Foundation," LightReading, 12 February 2019. [Online]. Available: https://www.lightreading.com/open-source/openstack/atandt-building-5g-network-on-an-open-source-foundation/d/d-id/749405. [Accessed 6 July 2019].
10 O-RAN Alliance, "Operator Defined Next Generation RAN Architecture and Interfaces," O-RAN Alliance [Online]. Available: https://www.o-ran.org. [Accessed 6 July 2019].
11 5G Americas, "The status of Open Source for 5G," 5G Americas, February 2019. [Online]. Available: https://www.5gamericas.org/the-status-of-open-source-for-5g. [Accessed 23 November 2020].
12 3GPP, "3GPP Release 16." [Online]. Available: http://www.3gpp.org/release-16. [Accessed 13 July 2020].
13 3GPP, "Release 16 Description; Summary of Rel-16 Work Items (Release 16), V0.4.0," 3GPP, March 2020.
14 3GPP, "TR 23.791, Study of Enablers for Network Automatization for 5G (Release 16)," 3GPP, June 2019.
15 3GPP, "TR 37.876 – Study on eNB(s) Architecture Evolution for E-UTRAN and NG-RAN, V0.4.0," 3GPP, March 2020.
16 McNamee, A., "GPP Release 16 and What This Means for 5G – Part 1," Openet, 8 May 2020. [Online]. Available: https://www.openet.com/blog/3gpp16-1. [Accessed 17 August 2020].
17 3GPP, "TS 23.503; Network Data Analytics Function," 3GPP.
18 3GPP, "TR 38.824: Study on Physical Layer Enhancements for NR Ultra-Reliable and Low Latency Case (URLLC), Release 16," 3GPP, 2019.
19 5G Ensure, "3GPP Progress on Mission Critical Services," EU, 21 June 2017. [Online]. Available: https://www.5gensure.eu/news/3gpp-progress-mission-critical-services. [Accessed 28 July 2020].
20 ITU, "5G-Convergence; BBF Initiative and Related Cooperation Activities," ITU, 2017.
21 Broadband Forum, "5G Fixed-Mobile Convergence (MR-427)," Broadband Forum, July 2018.
22 3GPP, "Summary after TSG-RAN#80," 3 July 2018. [Online]. Available: https://www.3gpp.org/ftp/Information/presentations/presentations_2018/RAN80_webinar_summary(brighttalk)extended.pdf.
23 3GPP, "Release 15," 3GPP, 26 April 2019. [Online]. Available: https://www.3gpp.org/release-15. [Accessed 13 July 2020].

24 3GPP, “Release 17,” 3GPP, 3 July 2020. [Online]. Available: https://www.3gpp.org/release-17. [Accessed 17 July 2020].
25 3GPP, “Specification Numbering,” 3GPP, 2020. [Online]. Available: https://www.3gpp.org/specifications/specification-numbering. [Accessed 27 September 2020].
26 3GPP, “Work Plan,” 3GPP, 2020. [Online]. Available: https://www.3gpp.org/specifications/work-plan. [Accessed 20 September 2020].
27 ITU, “Focus Group on Technologies for Network 2030,” ITU, 2018. [Online]. Available: https://www.itu.int/en/ITU-T/focusgroups/net2030/Pages/default.aspx. [Accessed 5 July 2019].

2

Requirements

2.1 Overview

This chapter summarizes technical requirements for 5G systems as per the 3rd Generation Partnership Project (3GPP) Release 16 and the International Telecommunications Union (ITU) International Mobile Telecommunications 2020 (IMT-2020). The latter was originally presented in Ref. [1], and in its finalized form in Ref. [2]. The concrete requirement set for the ITU candidate evaluation was presented at the end of 2017 as stated in [3], and the candidates for the actual evaluation process are presented in Ref. [4]. Among the other contributors, the 3GPP is one of the key Standard Development Organizations (SDOs). As per the 3GPP specifications, the key documents for 5G performance are TS 22.261, which summarizes the service and operational requirements for a 5G system, including User Equipment (UE), Next Generation Radio Access Network (NG-RAN), and 5G core network (5GC), whereas the requirements for Evolved UMTS Terrestrial Radio Access New Radio (E-UTRA-NR) Dual Connectivity (DC) are found in TS 22.278.

The 3GPP finalized Release 16 on July 2020 for formal ITU review, whereas the preceding Release 15 specifications served as an intermediate phase to expedite 5G deployments by offering basic performance for 5G users during 2019–2020.

This chapter focuses on the requirements of the ITU, and presents some of the most important related specifications.

2.2 Background

As has been already a "tradition," the ITU has taken the role of defining the mobile communications generations. This has been the situation for 3G and 4G, after the success of the first analogue generation and the second digital generation. The ITU defines 3G as a set of radio access and core technologies forming systems capable of complying with the requirements of IMT-2000 (3G) and IMT-Advanced (4G). Based on the number of end-users, Universal Mobile Telecommunications System (UMTS) and its evolution up to advanced High Speed Packet Access (HSPA) is the most popular 3G system, while Long Term Evolution (LTE)-Advanced, as of 3GPP Release 10, is the most utilized 4G technology.

5G Second Phase Explained: The 3GPP Release 16 Enhancements, First Edition. Jyrki T.J. Penttinen.

As for the industry, the Next Generation Mobile Network (NGMN) Alliance represents the interests of mobile operators, device vendors, manufacturers, and research institutes. The NGMN is an open forum for participants to facilitate the evaluation of candidate technologies suitable for the evolved versions of wireless networks. One of the main aims of the forum is to pave the way for the commercial launch of new mobile broadband networks. Some practical methods for this are the production of a commonly agreed technology roadmap as well as user trials [5].

So, the ITU is still acting as the highest authority to define global and interoperable 5G requirements for mobile communications systems in such a way that the requirements are agreed by all the stakeholders. This is to avoid different interpretations by the industry when deploying and marketing the networks.

With the ITU setting the scenery, the practical standardization work results in the ITU-5G-compliant technical specifications created by the mobile communications industry. One of the most active standardization bodies for mobile communications technologies is the 3GPP, which has created standards for Global System for Mobile Communications (GSM), UMTS/HSPA, and LTE/LTE-Advanced. At present, the 3GPP is actively creating advanced standards that are aimed to comply with ITU's 5G requirements.

In addition to the 3GPP, there are many other standardization bodies and industry forums that contribute to 5G technologies. Complete end-to-end 5G systems are not under construction in such a large scale as is done by the 3GPP, but, e.g., a large number of recommendations by IEEE are used as a base for 5G (as well as any other existing mobile communication technology).

2.3 Development of the Ecosystem

2.3.1 New Needs

Due to the vast business opportunities relying on a new, more performant platform for the delivery of increased capacity with faster and reduced lag, it is obvious that the 5G industry is willing to advance as fast as technically feasible. The elemental step in this process is the setting up of new, standard-based networks and equipment at a global level.

All the essential stakeholders have demonstrated their interest in 5G, including chipset and module makers, network and device manufacturers, service providers, and Mobile Network Operators (MNOs). There are also plenty of existing and new verticals expressing their desire for future 5G networks to help execute their functions more efficiently. As envisioned by the ITU in their connected society concept, 5G will be an elemental and important communications and data transfer platform for the whole of society, including smart cities, critical infrastructure, telematics, and a huge number of Internet of Things (IoT) devices, to mention only a few. This, in turn, will enhance the ecosystem, with its infrastructure, services, and applications.

As the 3GPP specifications evolve and the respective devices and solutions become available, there will be new applications developed as well. This is the familiar pattern we have already witnessed throughout all the previous generations, since the first one back in the 1980s.

For consumers, the clearest benefit of 5G at the beginning of the 5G era was the increased data speed providing faster download speeds. Initial data rates initially were close to 1 Gb/s in ideal, non-congested radio coverage, whereas theoretical single-user download speeds will rise to 20 Gb/s as dictated by the ITU IMT-2020 for the respective evolved broadband use cases.

The initial networks of 3GPP Release 15-compliant installations are based on the 4G core infrastructure. This phase uses the intermediate Non-Standalone (NSA) options combining 4G and 5G communications.

5G also defines the Standalone (SA) architecture, which is the ultimate goal of many operators. This includes sole 5G radio and core segments, and eventually it will be able to offer the maximum performance of 5G.

Low latency is beneficial in Augmented Reality (AR) and Virtual Reality (VR) applications. AR means that the device superimposes a computer-generated virtual image on the user's view of the real environment. This composite view is an interesting technology area that can be used in a variety of environments by mixing computer-generated and real views, such as a car being driven while accompanying information is superimposed on a screen. Combining the 5G system's low latency with the high data speed and high reliability, such applications may become popular, including vehicle communications that use advanced entertainment and control systems.

Beyond the vastly increasing data speeds and low latency for special environments, there are many verticals – user types – that may benefit from 5G. The combination of evolved Mobile Broadband (eMBB), Ultra-Reliable Low Latency Communications (URLLC), and Massive Machine-Type Communications (mMTC) will be highly relevant in such environments as smart cities and those using advanced cloud edge services.

2.3.2 Enhanced 5G Functionality

The 5G system updates completely the traditional network architectural model. Whereas the traditional SA network elements rely on their own software per device, the 5G network is based on common hardware and selected software instances running on top of it and serving different 5G network functions. This optimizes the performance of the network and makes the adoption of new services, such as edge computing, possible much faster than in the "traditional," non-virtualized networks.

As a consequence, the 5G core and radio network functions can be processed in the cloud environment of the data centers, so we will see an increasing number of cloud providers offering their capacity to MNOs. As soon as these advanced functionalities are available for consumers, it will be possible to take advantage of the low latency and high data speeds much more efficiently than has been the case with 4G networks.

2.4 Introduction to Requirements

While the 3GPP produces the technical specifications of 5G, the ITU Recommendation ITU-R M.2083 sets the reference presenting the high-level requirements for 5G. It also defines IMT-2020 overall aspects and enhanced capabilities under the title "IMT

Vision; Framework and Overall Objectives of the Future Development of IMT for 2020 and Beyond" [6].

IMT-2020 supports a set of deployment scenarios for different environments and service capabilities. The ITU foresaw 5G as an enabler for a seamlessly connected society as of 2020. The idea of 5G is to connect people via a set of "things": data, applications, transport systems, and cities in a smart-networked communications environment.

The idea of 5G was presented as early as 2012 when the ITU-R initiated a program to develop IMT for the year of 2020 and beyond. The ITU-R Working Party (WP) 5D has been paving the way for IMT-2020, including investigation of the key elements of 5G in cooperation with the mobile broadband industry and other stakeholders interested in 5G.

ITU-R's 5G vision for the mobile broadband-connected society was agreed in 2015. This vision was considered as instrumental and it served as a solid foundation for the World Radiocommunication Conference 2019 (WRC-19), which decided the international allocation of the additional 5G frequency spectrum.

ITU-R Rec. M.2083 binds multiple documents that form the basis for 5G systems (Table 2.1). This publication includes IMT Vision (framework and overall objectives of the future development of IMT for 2020 and beyond), future technology trends of terrestrial IMT systems during 2015–2020, technical feasibility of IMT in bands above 6 GHz, framework and overall objectives of the future development of IMT-2000, and systems beyond IMT-2000, among others.

Table 2.1 Key requirements of the ITU IMT-2020, based on M.2083 [7].

Attribute	Value	Examples (scenario dependent)
Dense areas	1 million devices/km^2	eMBB, sensor networks, ad-hoc broadband, massive IoT
Real-time latency	Down to 1 ms (Release 15) and 0.5 ms (Release 16)	Tactile Internet, industry use cases, critical communications; according to TR 38.913, New Radio (NR) should support latencies down to 0.5 ms Uplink/Downlink (UL/DL) for URLLC
User mobility	Up to 500 km/h	Very fast-moving vehicles such as bullet trains
Throughput	10–20 Gb/s/user	Mobile broadband; in eMBB, target maximum for peak data rate is 20 Gb/s in DL and 10 Gb/s in UL
Ultra-high reliability	Up to 99.999% (Release 15) and 99.9999% (Release 16)	Remote surgery, remote control of objects such as drones, lifeline communications
Ultra-low cost	Sub-US$10 devices	Low-cost IoT, data offloading, network function virtualization, network slicing
Energy efficiency	Battery life up to 10 years	Considerably longer battery life than in previous generations

2.5 World Radiocommunication Conference

The ITU organizes WRCs to discuss and decide global allocations for Radio Frequency (RF) bands, including those of 5G. The work aligns the efforts of international frequency administrations.

WRC events are organized every 3 to 4 years. The latest WRC was organized on 28 October to 22 November 2019, and it decided further 5G frequency allocations [8]. WRC reviews and revises global radio regulations. The international treaty governs use of the RF spectrum as well as geostationary satellite and non-geostationary satellite orbits.

During the preparation of each WRC, the ITU Council takes into account recommendations of the previous WRCs, and coordinates the meeting for member states. WRC then revises the radio regulations and associated frequency assignment and allotment plans. The Conference Preparatory Meeting (CPM) will prepare a consolidated report to be used in support of the work of the conferences [9]. The World and Regional Radiocommunication Conferences consider contributions from administrations, the Radiocommunication Study Groups, and other sources related to regulatory, technical, operational, and procedural matters.

WRC is a formal forum for setting general rules for 5G and other radiocommunication frequency allocations, and is relevant for supporting 5G deployment strategies. The allocation plans are also the most important base for forthcoming 5G frequency auctions.

The ITU regions are divided into three main zones: Asia, Europe, and Americas regions. The overall allocation plan differs from region to region. For 5G and previous generations, there are both global and regional bands to be considered by regional authorities. As a consequence, to be a global variant, 5G UE needs to support a common set of internationally compatible frequencies from all three regions.

In the era of 5G, the task is not straightforward as the support of multiple bands increases the complexity, cost, and physical dimensions of the device. The optimization of an adequate set of terminal frequencies is thus one of the important optimization tasks of device manufacturers, and the outcome of the WRC has its impact on future planning.

2.6 Building Blocks of 5G: eMBB/URLLC/mMTC

5GC and NR form the 5G System (5GS). It enhances capacity and performance, and complies with the demanding requirements for eMBB, mMTC, and URLLC.

eMBB, mMTC, and URLLC are the elemental building blocks, or dimensions (Table 2.2), of 5G as indicated in the ITU IMT-2020, which sets the reference for the new era [10]. Figure 2.1 summarizes their respective capabilities.

Figure 2.2 depicts these dimensions. It also presents additional remarks of practical, expected 5G service usage environments as interpreted from the Technical Report (TR) of the 3GPP, TR 22.891 (Release 14). It details 74 feasibility studies on new 5G services and technology enablers of the market.

5G is able to support a variety of *use cases* independently from each other. Each of these use cases requires different performance values in terms of area traffic capacity, peak data speed, user's experienced data speed, spectrum efficiency, mobility, latency, connection

Table 2.2 Description of 5G dimensions.

Dimension	Description
URLLC	For the highest reliability and lowest latency applications
mMTC	For the densest sensor and other IoT environments
eMBB	For the highest data rate applications
Management	For the evolved, Machine Learning (ML)-based self-optimized network
Vehicle-to-Everywhere (V2X)	For vehicle communications

density, and energy efficiency. These characteristics can be referred to as attributes, and their values dictate their capabilities. Table 2.2 summarizes the first phase characteristics or dimensions.

The offering of these attributes with an adequate Quality of Service (QoS) is possible thanks to new 5G technologies such as virtualized 5G network functions, software-defined networking, edge computing, and network slicing.

URLLC: This dimension provides the lowest latency values and serves adequately for fulfilling the needs of critical communications. Some potential use cases of this category include drone flight control, localized factory automation, and wide-area smart city functions. Although the base for the URLLC framework was identified already in 3GPP Release 15 NR, Release 16 adds and enhances further the URLLC features.

mMTC: This dimension is designed to support a considerable number of simultaneously communicating IoT devices and applications. LTE Releases 13 and 14 define already Low-Power Wide Area (LPWA) Cellular IoT (C-IoT) modes, which are Narrow-Band IoT (NB-IoT) and LTE-M. Please note that as per the notation of the GSM Association (GSMA),

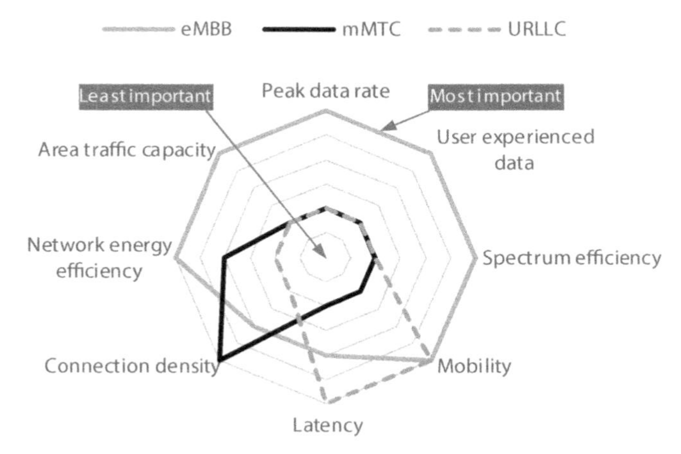

Figure 2.1 5G dimensions and their capabilities as per the ITU IMT-2020. As an example, the connection density is the most important capability for mMTC, whereas the lowest possible latency and mobility are of utmost importance for URLLC.

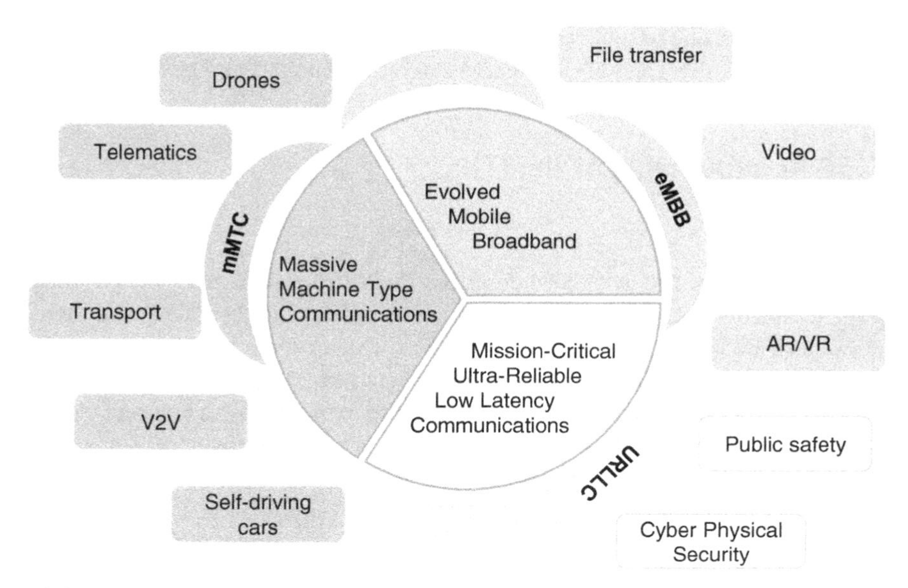

Figure 2.2 The building blocks of 5G. eMBB, URLLC, and mMTC are the main dimensions and usage types of 5G as defined by the ITU IMT-2020.

the generic term for NB-IoT and LTE-M is Mobile IoT (MIoT). These already existing modes comply largely also with the 5G mMTC requirements, whereas Release 16 will enhance their performance to ensure full compliancy with IMT-2020 expectations [11].

eMBB: This dimension provides very high data speeds clearly exceeding 4G speeds, and is included in the 5G system as of Release 15. The respective first-phase deployments focus on this category until URLLC and mMTC will be fully available later in the Release 16 timeframe. Even Release 16 deployments enhance further the 5G performance; this intermediate phase already paves the way for much more demanding mobile broadband use cases such as AR/VR applications.

As the network functions become more complex and dynamic, **network management and orchestration** will rely increasingly on the evolved Self-Optimizing Network (SON) concept, Artificial Intelligence (AI), and ML. The 3GPP has designed and enhanced SON already for previous generations, and the same principles apply in the 5G era, although network slicing with its high dynamics requires even more advanced versions of SON.

V2X is a good example of an environment where some of the most critical 5G performance requirements are needed, especially if self-driving vehicles are involved. It also is an example of an automotive vertical that would benefit from the maximum performance that is a combination of all the dimensions, i.e., URLLC, mMTC, and eMBB.

In addition to 5G-based car communications, the 3GPP is extending Vehicle-to-Vehicle (V2V) to cover special environments such as maritime communications. As GSM and its GSM-R (Railway) communications mode may gradually turn out to be obsolete along with the probable decline of many 2G networks, 5G would, in fact, be quite a logical base for future railway communications and control platforms.

More information on the topic can be found at 3GPP TR 22.891 (new services and markets technology enablers) and ITU-R Rec. M.2083 (IMT for 2020 and beyond).

2.7 5G Requirements of the ITU

2.7.1 Process

ITU Recommendation ITU-R M.2083 describes IMT-2020 overall aspects. It will facilitate the development of enhanced capabilities compared to those found in ITU Recommendation ITU-R M.1645. IMT-2020 has a variety of aspects, which extend greatly the requirements compared to previous mobile communication generations [3, 12].

The ecosystem and respective performance relate to users, manufacturers, application developers, network operators, as well as service and content providers. This means that there are many deployment scenarios supported by IMT-2020 with a multitude of environments, service capabilities, and technological solutions [1].

The ITU represents the highest authority in the field of defining mobile system generations. The overall vision of 5G, according to the ITU, is presented in Ref. [13]. The ITU foresees 5G to function as an enabler for a seamlessly connected society in the 2020 timeframe and beyond. The high-level idea of 5G is to bring together people via a set of "things": data, applications, transport systems, and cities in a smart-networked communications environment. The ITU and the respective, interested partners believe that the relationship between IMT and 5G are elements that make it possible to deploy the vision in practice by relying on mobile broadband communications.

Within the ITU, the WP 5D of the ITU-R has been working on the ITU's expected timeframe paving the way for IMT-2020, including investigation of the key elements of 5G in cooperation with the mobile broadband industry and other stakeholders interested in 5G.

One of the elemental parts of this development has been the 5G vision of the ITU-R for a mobile broadband-connected society, which was agreed in 2015. This vision was considered instrumental and served as a solid foundation for the WRC-19. This was a major event for decisions on 5G frequency bands, including the additional spectrum that will be required in different regions for the massively increasing mobile communications traffic.

From various teams considering 5G evolution, one of the most significant ones is the ITU WP 5D. It has the role of investigating study areas and deliverables towards IMT for 2020 and beyond via a multitude of activities, such as workshops and seminars for information sharing within the industry and standardization entities. Some of the more specific work item categories include:

- Vision and technology trends. These aspects also include market, traffic, and spectrum requirements for the forthcoming 5G era;
- Frequency band channeling arrangements and spectrum sharing and compatibility. This item consists of investigations and reporting;
- IMT specifications. This item also contains related technical works;
- Support for IMT applications and deployments.

The ITU IMT-2020 requirements form the foundation of internationally recognized 5G systems. The respective ITU Radiocommunications Bureau Circular Letters work by announcing an invitation for standardization bodies to submit their formal 5G technical proposals for the ITU WP 5D evaluation of their compliance with IMT-2020 requirements. This process follows the principles applied in the previous IMT-Advanced for selecting 4G systems. Prior to candidate evaluation, the WP 5D finalized the performance requirements by the end of 2017 and formed the evaluation criteria and methodology for the assessment of IMT-2020 radio interface.

After the candidate submission, the WP 5D evaluated the proposals during 2018–2020. The work was based on independent, external evaluation groups, and the process was completed during 2020 along with a draft ITU-R Recommendation that contained detailed specifications for NR interfaces.

2.7.2 Documents

The key sources of information for ITU 5G development can be found in ITU-R Rec. M.2083 [7], which collects the documents for forming 5G:

- M.2083-0 (09/2015). IMT Vision – "Framework and overall objectives of the future development of IMT for 2020 and beyond." This document contains the recommendations for the future development of IMT for 2020 and beyond, and defines the framework and overall objectives of future development considering the roles that IMT could play to better serve the needs of networked societies. It also includes a variety of detailed capabilities related to foreseen usage scenarios, as well as objectives of the future development of IMT-2020 and existing IMT-Advanced. It is based on Recommendation ITU-R M.1645;
- Report ITU-R M.2320 – Future technology trends of terrestrial IMT systems addresses the terrestrial IMT technology aspects and enablers during 2015–2020. It also includes aspects of terrestrial IMT systems related to WRC-15 studies;
- Report ITU-R M.2376 – Technical feasibility of IMT in bands above 6 GHz summarizes the information obtained from the investigations related to the technical feasibility of IMT in the bands above 6 GHz as described in ITU-R Rec. 23.76 [14];
- Recommendation ITU-R M.1645 – Framework and overall objectives of the future development of IMT-2000 and systems beyond IMT-2000;
- Recommendation ITU-R M.2012 – Detailed specifications of the terrestrial radio interfaces of International Mobile Telecommunications Advanced (IMT-Advanced);
- Report ITU-R M.2320 – Future technology trends of terrestrial IMT systems;
- Report ITU-R M.2370 – IMT Traffic estimates for the years 2020 to 2030;
- Report ITU-R M.2376 – Technical feasibility of IMT in bands above 6 GHz;
- Report ITU-R M.2134 – Requirements related to technical performance for IMT-Advanced radio interface(s).

The ITU has published the minimum 5G requirements in the document IMT-2020 Technical Performance Requirements, available at [1]. The most important minimum set of technical performance requirements defined by the ITU provide a means for consistent definition, specification, and evaluation of the candidate IMT-2020 Radio Interface Technologies (RITs), or a Set of Radio Interface Technologies (SRITs).

There is a parallel, ongoing development for the ITU-R recommendations and reports related to 5G, including the detailed specifications of IMT-2020. As a highest level global authority of such requirements for a new mobile communications generation, the ITU aims, with the production of these requirements, to ensure that IMT-2020 technologies can comply with the objectives of the IMT-2020. Furthermore, the requirements set the goal for the technical performance that the proposed set of RITs must achieve for being called ITU-R's IMT-2020 compliant 5G technologies.

The ITU will evaluate the IMT-2020-compliant 5G candidate technologies based on the following documents and for the development of IMT-2020:

- Report ITU-R IMT 2020 M-Recommendation for Evaluation;
- Report ITU-R IMT-2020 M-Recommendation for Submission.

Recommendation ITU-R M.2083 contains eight *key capabilities* for IMT-2020, and functions as a basis for the technical performance requirements of 5G. It should be noted that the key capabilities have varying relevance and applicability as a function of different use cases within IMT-2020.

In summary, the ITU's minimum radio interface requirements include the 5G performance requirements as presented in the next section. More specific test evaluation is described in the IMT-2020 evaluation report of the ITU-R [15].

2.7.3 Peak Data Rate

The peak data rate (b/s) refers to the maximum possible data rate assuming ideal, error-free radio conditions that are assigned to a single mobile station with all the available radio resources, excluding the resources for physical layer synchronization, reference signals, pilots, guard bands, and guard times. With the term W representing bandwidth and E_{sp} referring to a peak spectral efficiency, the user's peak data rate $Rp = WE_{\mathrm{sp}}$. The total peak spectral efficiency is obtained by summing the value per each applicable component frequency bandwidth. This requirement is meant to evaluate the eMBB use case, for which the minimum DL peak data rate is 20 Gb/s, whereas the value for UL is 10 Gb/s.

2.7.4 Peak Spectral Efficiency

The peak spectral efficiency (b/s/Hz) normalizes the peak data rate of a single mobile station under the same ideal conditions over the utilized channel bandwidth. The peak spectral efficiency for the DL is set to 30 b/s/Hz, whereas the value for the UL is 15 b/s/Hz.

2.7.5 User Experienced Data Rate

The user experienced data rate is obtained from the 5%-point of the Cumulative Distribution Function (CDF) of the overall user throughput, i.e., correctly received Service Data Units (SDUs) of the whole data set in layer 3 during the active data transfer. If the data transfer takes place over multiple frequency bands, each component bandwidth is summed up over the relevant bands, and the user experienced data rate $R_{\mathrm{user}} = WE_{\mathrm{s\text{-}user}}$. This equation refers to the channel bandwidth multiplied by the fifth percentile user spectral efficiency. The ITU requirement for the user experienced data rate in the DL is 100 Mb/s, whereas it is 50 Mb/s for the UL.

2.7.6 Fifth Percentile User Spectral Efficiency

Fifth percentile user spectral efficiency refers to the 5%-point of the CDF of the normalized user data throughput. The normalized user throughput (b/s/Hz) is the ratio of correctly received SDUs in layer 3 during a selected time divided by the channel bandwidth. This requirement is applicable to the eMBB use case, and the requirement values for DL and UL, respectively, are the following:

- **Indoor hotspot**: 0.3 b/s/Hz (DL) and 0.21 b/s/Hz (UL);
- **Dense urban**: 0.225 b/s/Hz (DL) and 0.15 b/s/Hz (UL), applicable to the Macro TRxP layer of a Dense Urban eMBB test environment;
- **Rural**: 0.12 b/s/Hz (DL) and 0045 b/s/Hz (UL), excluding the Low Mobility Large Cell (LMLC) scenario.

2.7.7 Average Spectral Efficiency

Average spectral efficiency can also be called spectrum efficiency as has been stated in the ITU Recommendation ITU-R M.2083. It refers to the aggregated throughput taking into account the data streams of all the users. More specifically, the spectrum efficiency is calculated via the correctly received SDU bits on layer 3 during a measurement time window compared to the channel bandwidth of a specific frequency band divided further by the number of TRxPs, resulting in the value that is expressed in b/s/Hz/TRxP. The ITU requirement values for DL and UL for the eMBB use case, respectively, are the following:

- **Indoor hotspot**: 9 b/s/Hz/TRxP (DL) and 6.75 b/s/Hz/TRxP (UL);
- **Dense Urban for Macro TRxP layer**: 7.8 b/s/Hz/TRxP (DL) and 5.4 b/s/Hz/TRxP (UL);
- **Rural (including LMLC)**: 3.3 (DL) and 1.6 (UL).

2.7.8 Area Traffic Capacity

Area traffic capacity refers to the total traffic throughput within a certain geographic area, and is expressed in Mb/s/m^2. More specifically, the throughput refers to the correctly received bits in layer 3 SDUs during a selected time window. If the bandwidth is aggregated over more than one frequency band, the area traffic capacity is a sum of individual bands.

2.7.9 Latency

User plane latency refers to the time it takes for the source sending a packet in radio protocol layer 2/3 and the destination receiving it on the respective layer. The latency is expressed in milliseconds. The requirement for user plane latency is 4 ms for the eMBB use case, while it is 1 ms for URLLC. The assumption here is an unloaded condition in both DL and

UL without users other than the observed one, while the packet size is small (zero payload and only Internet Protocol (IP) header).

Control plane latency, in turn, refers to the transition time it takes to change from the idle stage to the active stage in the URLLC use case. The requirement for control plane latency is a maximum of 20 ms, and preferably 10 ms or less.

Release 16 introduces an additional latency requirement of 0.5 ms.

2.7.10 Connection Density

Connection density refers to the total number of 5G devices that can still comply with the target QoS level within a geographical area that is set to 1 km^2, with a limited frequency bandwidth and the number of TRxPs, the variables being the message size, time, and probability for successful reception of the messages. This requirement applies to the mMTC use case, and the minimum requirement for connection density is set to 1 000 000 devices per km^2.

2.7.11 Energy Efficiency

The high-level definition of 5G energy efficiency indicates the capability of the RIT and SRIT to minimize the RAN energy consumption for the provided area traffic capacity. Furthermore, device energy efficiency is specifically the capability of the RIT and SRIT to optimize the consumed device modem power down to a minimum that still suffices for adequate quality of the connection. For the energy efficiency of the network as well as the device, the support of efficient data transmission is needed for the loaded case, and the energy consumption should be the lowest possible for the cases when data transmission is not present. For the latter, the sleep ratio indicates the efficiency of the power consumption. Energy efficiency is relevant for the eMBB use case, and the RIT and SRIT need to have the capability to support a high sleep ratio and long sleep duration.

2.7.12 Reliability

The reliability of 5G in general refers to the ability of the system to deliver the desired amount of packet data on layer 2/3 within the expected time window with high success probability, which is dictated by channel quality. This requirement is applicable to the URLLC use cases.

More specifically, the reliability requirement in 5G has been set to comply with the successful reception of a 32-bit PDU on layer 2 within a 1 ms period with a 1×10^{-5} success probability referring to 99.999%. This requirement is applicable to the edge of the cell in urban macro-URLLC, assuming 20 bytes of application data and relevant protocol overhead.

Release 16 also adds a requirement for the 99.9999%.

2.7.13 Mobility

The 5G mobility requirement refers to the maximum mobile station speed in such a way that the minimum QoS requirement is still fulfilled. There is a total of four mobility classes defined in 5G: (1) Stationary with 0 km/h speed; (2) Pedestrian with 0–10 km/h; (3)

Vehicular with 10–120 km/h; and (4) High Speed Vehicular with speeds of 120–500 km/h. The applicable test environments for the mobility requirement are Indoor Hotspot eMBB (Stationary, Pedestrian), Dense Urban eMBB (Stationary, Pedestrian, and Vehicular 0–30 km/h), and Rural eMBB (Pedestrian, Vehicular, High-Speed Vehicular).

2.7.14 Mobility Interruption Time

Mobility interruption time refers to the duration of the interruption in the reception between the UE and BS, including RAN procedure execution, Radio Resource Control (RRC) signaling, or any other messaging. This requirement is valid for eMBB and URLLC use cases and is set to 0 ms.

2.7.15 Bandwidth

Bandwidth in 5G refers to the maximum aggregated system bandwidth and can consist of one or more RF carriers. The minimum supported bandwidth requirement is set to 100 MHz, and the RIT/SRIT shall support bandwidths up to 1 GHz for high-frequency bands such as 6 GHz. Furthermore, the RT/SRIT shall support scalable bandwidth.

2.8 The Technical Specifications of the 3GPP

The initial 5G is defined according to 3GPP Release 15 specifications. After the first phase, 5G, as described in Release 16, will comply with the demanding ITU IMT-2020 requirements.

While the 3GPP produces concrete specifications for 5G, there are other SDOs and industry forums contributing to the ecosystem of 5G. One of such entities is the GSMA, which, among other tasks, works with member organizations on the interoperability of networks and develops services on top of the 3GPP definitions. Some examples are Voice over LTE (VoLTE) and Rich Communications Services (RCS), which will be valid also in the 5G era in NSA deployments, and in a renewed form of Voice over New Radio (VoNR) in SA networks.

Tables 2.3 and 2.4 summarize some 5G documents of the 3GPP and GSMA. These documents include further references for more detailed exploration of 5G.

2.8.1 Releases

There are two paths within Release 15 of the 3GPP: one defining the ITU-R-compliant 5G, while the other route continues developing LTE under the further stage of LTE-Advanced (which can be referred to as LTE-A Pro). The overall 5G as defined via Release 15 of the 3GPP is described in Ref. [19].

The development of the 5G radio interface NR has two phases. The first phase is an intermediate step relying on the 4G infrastructure, referred to as the NSA scenario, while the final, native, and fully 5G-based core and radio system is referred to as SA. The respective features can be found in the 3GPP Release summary [20].

Table 2.3 Key 5G documents and specifications of the 3GPP.

TS/TR	Title	Area
TR 21.916	Summary of Release 16 Work Items	Generic
TR 22.891	Feasibility study on new services and market technology enablers	Services
TS 23.501	System architecture of the 5G system	Core
TS 23.502	Procedures for the 5G system	Functionality
TR 25.903	Deployment aspects	Network planning
TS 33.501	Security architecture	Security
TS 38.101	NR; UE radio transmission and reception	Radio
TS 38.104	NR; BS radio transmission and reception	Radio
TS 38.201	NR; general description	Radio

On the path towards 3GPP Release 15, which is the first set of technical specifications defining 5G, the 3GPP has studied plenty of items to comply with the high-level 5G requirements of the ITU-R. The items are related to the enhancement of LTE, as well as completely new topics related to 5G. Some of these study items are:

- Enablers for network automation for 5G. This item relates to automatic slicing network analysis, based on Network Data Analytics (NWDA).
- Enhanced LTE bandwidth flexibility. This item aims to further optimize the spectral efficiency within the bandwidths of 1.4–20 MHz.
- Enhanced VoLTE performance. This item refers to maintaining voice call quality sufficiently high to delay Single Radio Voice Call Continuity (SRVCC), thus optimizing the signaling and network resource utilization.
- NR support on non-terrestrial networks. This item refers to the channel models and system parameters respective to the non-terrestrial networks, to support satellite systems as one part of the 5G ecosystem.
- NR-based access to unlicensed spectrum. This item is related to the advanced phase of Release 15 by investigating further the feasibility of License Assisted Access (LAA) both below and above the 6 GHz frequency band, such as 5, 37, and 60 GHz.

Table 2.4 Some of the 5G-related Permanent Reference Documents (PRDs) and guidelines of the GSMA.

PRD	Title	Area
NG.116	Generic 5G Network Slice Template [16]	Network slicing
NG.113	5G Roaming Guidelines [17]	Interworking
Guideline	5G Implementation Guidelines [18]	Implementation

- System and functional aspects of energy efficiency in 5G networks. This item refers to the overall topic of Energy Efficiency (EE). The subtopics include EE Key Performance Indicators (KPIs) relevant to the 5G system, including the ones identified by the ETSI TC EE, ITU-T SG5, and ETSI NFV ISG, as well as the feasibility of operation and maintenance to support 5G EE and innovation in general for better energy optimization.
- Uplink Data Compression (UDC). This is a feature that would be implemented between Evolved NodeB (eNB) and UE. The benefit of UDC is compression gain especially for web browsing and text uploading, and the benefits also cover online video performance and instant messaging. The benefit is thus capacity enhancement and better latency in the UL direction especially in challenging radio conditions [21].
- VR. This item refers to the innovation of potential use cases and their respective technical requirements on VR.

2.8.2 Security Requirements for 5G

The 3GPP has defined a set of security requirements for 5G in TS 33.501. The generic security requirements of 3GPP TS 33.501 state that the UE must include protection against bidding-down attack and that the network must support subscription authentication and key agreement.

For more details on the security requirements of 5G, please refer to the contents of the security requirements in "5G Explained: Security and Deployment of Advanced Mobile Communications" [22], and 3GPP TS 33.501 [23].

2.9 NGMN

The NGMN is an initiative that also contributes, among many other technology areas, to 5G development.

The 5G vision of the NGMN is the following: "5G is an end-to-end ecosystem to enable a fully mobile and connected society. It empowers value creation towards customers and partners, through existing and emerging use cases, delivered with consistent experience, and enabled by sustainable business models."

For complying with these aspects, Ref. [24] considers the following items in the list of requirements as NGMN Alliance understands the priorities:

- Devices;
- Enhanced services;
- Network deployment, operation, and management;
- New business models;
- System performance;
- User experience.

2.9.1 User Experience

User Experience (UX) refers to the end-user experience while consuming a single or multiple simultaneous service. In the ever-increasing competitor environment, this is one of the key aspects that differentiate the stakeholders and can either fortify or break the customer relationship.

Especially for the 5G era, the enhanced networks need to deliver smooth and consistent UX independently of time or location. The KPIs contributing to the UX include the minimum needed data rate and maximum allowed latency within the service area, among a set of various other parameters that are configurable, including their allowed ranges of variations, by each MNO. These service-dependent KPIs and their respective values are investigated by many stakeholders at present, including the ITU-R, which designs the minimum requirements for 5G systems; therefore, they may be called ITU-compliant 5G (IMT-2020) systems.

The UX data rate is dictated by the vision of providing broadband mobile access everywhere. In terms of use cases of the NGMN, the practical requirement is to provide at least 50 Mb/s peak data rate for a single user over the whole service area of 5G. The NGMN emphasizes that this value must be provided constantly over the planned area, including the cell edge regions.

The UX mobility requirements of the NGMN is a set of use cases. They include the following:

- 3D connectivity for, e.g., aircraft, for providing advanced passenger services during the flight comparable with the UX on the ground.
- A high-speed train with a velocity of over 500 km/s. 5G must support data transmission during train trips, including High Definition (HD) video streaming and video conferencing.
- Moving hot spots. This category refers to the extension of the static hot spot concept to cope with high-mobility demands in the 5G era. As dictated by the NGMN, 5G must complement the stationary mode of planning of capacity by adding the non-stationary, dynamic, and real-time provision of capacity.
- Remote computing. This category includes both stationary (e.g., home office) and mobile (e.g., public transportation) environments.

Other important 5G UX-related requirements, per the NGMN Alliance, are the system performance in general, connection and traffic density, spectrum efficiency, radio coverage area, resource and signaling efficiency, and system performance.

In addition to these requirements for broadband access everywhere and high user mobility, the NGMN has formed an extensive list of other cases and their requirements for 5G. These include the following aspects:

- Broadcast-like services. This category extends the traditional model of information broadcast such as TV networks by adding means for more advanced interaction.
- Extreme real-time communications. These cases represent the environment for the most demanding real-time interactions, and for full compliance may require more than one attribute among extremely high data throughput, high mobility, and critical reliability. A real-world example in this category includes the tactile Internet, which requires tactile controlling and audio-visual feedback. Robotic control falls into this category.
- Lifeline communications, such as disaster relief and emergency prediction. Along with 5G deployments, there will be increasing expectations for mobile systems to support

lifeline communications in all situations and in wide areas. Operators would need to optimize 5G for robust communications in case of natural disasters such as earthquakes, tsunamis, floods, and hurricanes.
- Massive IoT. The vision beyond 2020 indicates a need to support a considerably higher number of all types of devices compared to current markets, such as smart wearables and clothes equipped with a multitude of sensors, sensor networks, and mobile video surveillance applications.
- Ultra-reliable communications. This category includes critical cases such as automated traffic control and driving, i.e., self-driving vehicles, collaborative robots and their respective networking, life-critical health care and people's emergency services such as remote surgery, 3D connectivity for drones with a sufficiently large coverage for ensuring packet transport, and public safety utilizing real-time video from emergency areas.

2.9.2 Device Requirements

The role of ever-evolving and increasingly complex devices is important in the 5G era. The hardware, software, and Operating System (OS) of the devices continue to evolve, and assurance of the correct, fluent, and protected functioning of these is of utmost importance. Not only do the devices as such need to perform correctly, but they also have an increasingly important role in providing a relay element with other 5G devices, especially in the IoT environment, as well as in various consumer use cases.

Table 2.5 summarizes the high-level device requirements as informed by the NMGN Alliance.

Table 2.5 NGMN Alliance's device requirements.

Requirement	Description
MNO controllability	Ability for the network or UE to choose the desired profile depending on the QoS need, element capabilities, and radio conditions; ability for MNO to manage the hardware and software diagnostics, fixes, and updates; ability to retrieve performance data from the UE as a basis for further optimization and customer care.
Multiband and multimode	A sufficiently wide support of RF bands and modes (Time Division Duplex (TDD), Frequency Division Duplex (FDD), and mixed) is needed for efficient roaming scenarios; ability to take advantage of simultaneous support of multiple bands optimizes the performance; aggregation of data from different Radio Access Technologies (RATs) and carriers is required.
Power efficiency	Along with the need to support a vast number of IoT devices that may be in remote areas, devices require enhanced battery efficiency. For consumer devices (aka smart devices), the minimum battery life requirement is three days, while autonomously working IoT devices need to function up to 15 years.
Resource and signal efficiency	5G devices require optimized signaling, which is one of the ways to provide long battery life.

2.9.3 Enhanced Services

There are various special requirements for the enhanced functionality of 5G related to connectivity, location, security, resilience, high availability, and reliability. Connectivity is related to transparency, which is key for delivering a consistent experience in such a heterogeneous environment that 5G will represent.

It is estimated that 5G combines both native 5G networks as well as legacy RATs, especially as the 3GPP LTE system is evolving in a parallel fashion with the 5G system. Furthermore, 5G allows UE to connect simultaneously to more than one RAT at a time, including carrier-aggregated connections. This type of connectivity may also involve other systems such as IEEE 802.11ax, which is a state-of-the-art, high-efficiency Wi-Fi variant.

5G systems optimize automatically respective combinations based on the most adequate achievable UX, which requires a seamless transition between RATs. Furthermore, both inter-RAT and intra-RAT mobility interruption time for all RAT types and technologies needs to be unnoticeable. In addition, intersystem authentication needs to be seamless, including scenarios involving both the internal 3GPP as well as non-3GPP RATs.

Another enhanced service requirement is location, which is key for contextual attributes. More concretely, network-based positioning in 5G should be able to achieve an accuracy of 1–10 m at least 80% of the time, and should be better than 1 m indoors. Driven by high-speed devices, this accuracy needs to be provided in real time. Furthermore, the network-based location of 5G should be able to cooperate with other location technologies. One example of such a cooperation is data delivery between partner sources so that the set of complete information enhances the final accuracy of the location. The NGMN also puts a requirement on the cost of the 5G location-based solution, which should not exceed the currently available partner options based on, e.g., satellite systems or 4G solutions.

The NGMN also has requirements for the security of 5G. It will support a variety of diverse applications and environments, including both human and machine-based communications. This means that there will be an increased amount of sensitive data transferred over 5G networks, and it will need enhanced protection mechanisms beyond the traditional models of protecting communications between nodes and end-to-end chains. This enhancement aims to protect user data, create new business models, and protect users and systems against cybersecurity attacks.

5G will support a wide range of applications and environments, from human-based to machine-based communications, and will deal with a large amount of sensitive data that need to be protected against unauthorized access, use, disruption, modification, inspection, and attacks. Since 5G offers services for critical sectors such as public safety and eHealth, the importance of providing a comprehensive set of features guaranteeing a high level of security is a core requirement for 5G systems.

2.10 Mobile Network Operators

In a parallel fashion with the standardization work of 5G, several MNOs have been testing and developing concepts for 5G. These activities have helped understand the performance of the NR and core network concepts and has eased the evaluation work of the standardization bodies.

Among many others, Verizon has evaluated the 5G concepts and has established a Verizon 5G Technology Forum, V5GTF [25]. It is a cooperative set up with Cisco, Ericsson, Intel, LG, Nokia, Qualcomm, and Samsung with the focus on creating a common and extendable platform for Verizon's 28/39 GHz fixed wireless access trials and deployments. The participating partners have collaborated to create 5G technical definitions for the 5G radio interface of Open Source Initiative (OSI) layers 1, 2, and 3 apart from the 3GPP technical specifications. The specifications also define the interfaces between the UE and network for ensuring interoperability among the supporting network, UE, and chipset manufacturers.

The initial release of 2017 included the V5G.200 series for physical layer 1. The V5G.300 series describes layers 2 and 3 for Medium Access Control, Radio Link Control, Packet Data Convergence Protocol, and Radio Resource Control. The definitions of the V5GTF were an important basis for the early deployment of Verizon's fixed wireless 5G access prior to the release of the first 3GPP 5G specifications.

There were also many other MNOs driving actively for 5G development prior to the commercial Release 15-based deployments, with some of the most active stakeholders being AT&T, Sprint, and DoCoMo.

2.11 Mobile Device Manufacturers

As an example from mobile network equipment manufacturers, Ericsson has identified the following aspects for the requirements [26]:

- High data rates. This aspect refers to considerably increased sustainable data rates in much wider areas compared to previous generations, whereas the focus has been traditionally on the dimensioning of the networks based on peak data rates. 5G needs to support data rates of 10 Gb/s in limited environments such as indoors, whereas several hundreds of Mb/s need to be offered in urban and suburban environments. The idea of 5G is to provide 10 Mb/s data rates almost everywhere.
- Massive system capacity. The system needs to support augmented capacity, and 5G networks need to transfer data with lower cost per bit compared to today's networks. In addition, energy consumption is an important aspect; along with increased data consumption, the energy footprint of the networks also widens, which 5G needs to compensate.

As a comparison, Nokia, in alignment with merged Alcatel Lucent, summarizes expected 5G use cases and requirements in Ref. [27].

5G introduces new solutions, yet it is also an evolution path enhancing the previous mobile generations. One of the new aspects is related to IoT, including Machine-to-Machine (M2M) communications. This results in new businesses and stakeholders relying on the 5G infrastructure.

2.12 Consumer Requirements

As 5G networks and their evolved services become available in wider areas, there will be an increasing number of consumers as well as completely new user segments, or verticals, benefiting from the high capacity and data speed, low latency, and ultra-reliable connectivity of the 5G networks. Such verticals include many profiles such as law enforcement, transport, and health care, as well as vehicle communications and VR applications. Further examples of other segments include the communication and control of drones, remote and secured communication of medical equipment, factory automation, and the enormous number of IoT sensors.

To increase the number of relevant 5G apps, there is a need to increase the interest of the developer community to create services that take advantage of the lower latency and faster data – together with the gradually evolving 5G coverage and capabilities. The related industry initiatives such as the GSMA's Operator Platform could expedite this evolution.

Table 2.6 summarizes some of the foreseen use cases of the 5G networks for consumers.

More information about consumer use cases can be found, e.g., in the Ericsson Consumer & IndustryLab Insight Report [28].

Table 2.6 Some potential uses of 5G for consumers.

Use case	Benefits for consumers
AR/VR	AR and VR applications have been defined as some of the primary use cases of 5G. Not only is the gaming industry expected to offer novelty applications and products based on 5G connectivity and AR/VR, but there may also be plenty of other services such as virtual shopping via mobile (comparing different products from home).
Automotive	5G is designed to serve also as a platform for V2V communications and autonomously driven vehicles, which translates to a better quality of life for consumers.
File download	Much faster data transfer from servers to 5G devices or accompanying devices such as laptops will be possible. 4 K movies are estimated to be typically around 100 Gb in size. The downloading of such content via 5G, in the best case at 20 Gb/s speed, would take around (100 Gb × 8 bits)/(20 Gb/s) = (800 Gb/20 Gb/s) = 40 seconds. At the start of 5G, the practical download speed may be around 2 Gb/s, which means that such contents can be downloaded in 400 seconds, or less than 7 minutes. In comparison, 1080p HD videos are typically approximately 4–8 Gb in size, and would take only a few seconds to download via the theoretical 20 Gb/s speed of 5G, or 30 seconds in the realistic first-phase 5G networks.
Games	5G can tackle the demanding requirements of advanced gaming thanks to the combination of 5G pillars, eMBB, and URLLC.
Multimedia	5G is a logical base for mobile multimedia, which can be applied to countless environments, including entertainment on the go. At the same time, traditional consumer devices may take many other forms such as foldable screens, and are expected to include even holographic projection and 360° cameras in the future.
Smart home	Apart from home Wi-Fi routers, appliances can also benefit from 5G connectivity. One example of the evolved smart home could be the use of 5G Fixed Wireless Access (FWA) for 5G-connected TV, intelligent sensors, and AI-equipped robots. The integrated security of the 5G networks might be highly beneficial to offer an extra layer of protection against intruders.
Streaming	Services such as YouTube transfer only part of the data and keep buffering while the consumer is streaming the contents. 5G will further minimize breakdowns due to failure when downloading. 5G can also provide a more optimized cloud edge for storing the desired contents closer to the user.

2.13 Vertical Requirements

Along with the 5G networks, there will be a variety of verticals, i.e., user types such as enterprises, law enforcement, health care, and many more, that benefit from the enhanced performance of the new generation. In fact, the 5G system can offer personalized service levels for those verticals via specifically adjusted network slices.

Use case refers to a certain situation in which a product or service could potentially be used. eMBB provides use cases for which fast data transfer is beneficial, such as file downloading, audio/video streaming, and multimedia messaging. These use cases are not necessarily critical for the delay while the information is sent or received in a timely manner. eMBB use cases are the ones that users and verticals alike can start experimenting with in the first phase of 5G.

The second phase, meaning Release 16-compliant 5G networks, will provide us with many new non-eMBB use cases. The new phase provides considerably more additional capacity per site, and thus new use cases are based on mMTC and URLLC. These use cases can rely on the cooperation between 5G and wired environments such as time-sensitive network, which is a highly reliable counterpart for mobile 5G [29].

An example of such use cases is telemedicine, which could provide end-to-end connectivity in wide areas for the most critical remote surgery. Another example could be the remote manufacturing control of critical production in a safe and reliable manner over a large area.

Today, these use cases are still merely ideas while operators upgrade their 5G infrastructure and interconnect the 5G radio and core to the new wired infrastructure. Once this work is completed and there are concrete applications and services enabled and available, the verticals can start taking advantage of the mMTC and URLLC modes.

At the same time, these new solutions can serve as a platform for completely new businesses that we might not be able to even visualize quite yet until the enablers are in place. 5G is able to support a variety of use cases independently from each other. Each of these use cases requires different performance values in terms of area traffic capacity, peak data speed, user's experienced data speed, spectrum efficiency, mobility, latency, connection density, and energy efficiency.

The offering of these capabilities with an adequate QoS per each vertical, or user type, is possible thanks to new 5G technologies such as virtualized 5G network functions, software-defined networking, edge computing, and network slicing. The latter means that a 5G network can be segmented in such a way that within the very same radio coverage, there are many different types of 5G networks to be selected from by different verticals.

There are many existing and new verticals that benefit from such personalized service. The following sections summarize the opportunities for some of them.

2.13.1 SME Business

5G may offer many new benefits for small and medium enterprises. Some of the most logical use cases are related to the evolved payment processes that can rely on 5G mobility in a secure manner, fortified by additional layers of security. Furthermore, the 5G system is capable of working as a platform for virtual tactile shopping, drone delivery, hologram calling, and 5G facial recognition payment, to mention a few examples of the analysis by Ericsson [28].

2.13.2 Transport and Traffic

5G can help optimize traffic flows by offering fast and reliable communications and control links between a vast number of sensors, surveillance cameras, and other devices connected to smart city traffic management systems.

Combined with AI and ML, adaptive prediction models, analytics based on environmental variables such as weather forecasts, predicted mass events impacting traffic flows, and seasonal traffic histograms, 5G can provide a highly feasible component to this ecosystem.

There are many research programs set up on the topic. As an example, the smart highway in Atlanta, Georgia, works as an outdoor laboratory to test future solutions, including V2X communications. These intelligent road solutions are still rather far away in the future but 5G can form part of such concepts [30, 31].

2.13.3 Health Care

5G can provide an important communications link for people in rural areas and can save valuable time, as an alternative to a physical doctor's appointment. This is a form of telehealth and remote home monitoring for doctors to assess a patient's state in the form of a video call. In such an environment, latency is not critical but the quality of the connection, including possible imaging contents, is of utmost importance to ensure correct diagnostics.

AT&T has identified important aspects of 5G working as a platform in such an environment. The benefits of the 5G system include the ability to quickly transmit large imaging files, and provide telemedicine, enhanced AR, VR, and spatial computing, as well as reliable, real-time remote monitoring, and the integration of AI. AT&T has concluded that by enabling advanced technologies through 5G networks, healthcare systems are able to improve the quality of care and patient experience, while reducing costs. 5G also works for personalized and preventive care [32].

2.13.4 Critical Infrastructure

Critical infrastructure refers to all the widely utilized systems people rely on as for their well-being. If some of these components fail, it may reduce the quality of life and even jeopardize it. Some examples of critical infrastructure are power grids, energy and water distribution, and telecom networks. It is thus of utmost importance that these components are highly protected against natural disasters, cyber-attacks, and other problems.

The characteristics of 5G, with its high reliability and low latency, make it suitable for controlling critical services and infrastructures, and help public safety, governments, and utility companies to maintain normal functions.

Ericsson has investigated the benefits of 5G on critical infrastructure and presents examples in their use case white paper [33]. As an example, energy and water utilities may connect to a large number of networked 5G devices while performing real-time, autonomous decisions.

The 3GPP has also investigated use cases related to critical infrastructure in the study of scenarios and requirements for 5G access technologies [34]. This publication summarizes the special aspects for public safety and emergency communications as well as public warning and emergency alert systems. Some examples of these are direct Device-to-Device (D2D) communications such as ProSe [35] as well as Mission-Critical Push-to-Talk (MCPPT), Mission Critical Video transfer (MCVideo), and Mission Critical Data transfer (MCData) within those special areas [36].

The mission critical mode of 5G and LTE networks provides public safety operators with ways to use communications on-network using the 5G or LTE network infrastructure, or off-network, or in combined mode. This ensures the needed priority and available capacity for very special environments.

In addition, 5G is planned to provide mechanisms to enable emergency calls, including positioning and location for emergency calls, as well as multimedia priority services. Furthermore, 5G includes public warning services for notifications to users complying with regional regulatory requirements

2.13.5 Aviation and Drones

5G can work as a feasible platform for completely new verticals such as drones (Figure 2.3), which are developing quickly and can perform increasingly complex and autonomous tasks such as packet delivery. They benefit from large, fast, and low-latency communications and control links.

While traditional cellular radio network planning tends to focus on coverage at street level, adaptive beam-forming antennas can serve new environments because drones may require a functional link well above the BS antenna levels. 5G has identified these as an

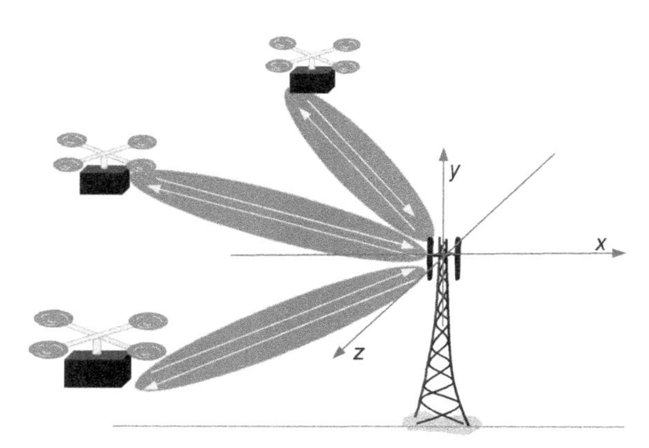

Figure 2.3 The 5G network can provide feasible control and communications channels for drones by offering enhanced, three-dimensional adaptive coverage.

important solution in such environments. As an example, the GSMA and Global UTM Association (GUTMA) have explored ways in which the GUTMA can work to unlock the full value of connectivity in drones by the telecommunications industry [37].

Also, commercial and special flights can benefit from the use cases that 5G networks provide. As an example, as soon as an aircraft has landed, it can upload telematics data from a previous flight into servers via the eMBB network. High speed at this specific moment is essential because oftentimes commercial flights remain on the ground for only a short period before the next flight takes off, and there may be a large amount of collected flight data that need to be backed up to the ground station quickly. Meanwhile, the aircraft may receive files, software packages, and other important content via the URLLC network slice. The integrity of these data as well as the secured contents may be of the utmost importance, which a special 5G network slice can take care of.

2.13.6 Virtual Reality

AR and VR represent a special use case. In fact, AR/VR use cases oftentimes map with a variety of other verticals.

AR means that a real-world view merges with a computer-aided view. There may be several types of human sensors involved such as visual, auditory, and haptic. In the most realistic user experiences, all these may be present.

VR, in turn, builds up a virtual environment. It can be imaginary space or based on a real environment, and human beings can interact within this space. The underlying technologies include 360° panoramic video or a light field forming advanced holographic views.

There are various AR/VR applications such as games, broadcasting, simulated education, virtual health care, and machine assembly training, to mention a few.

AR/VR data processing requires high-performance processors. Processing can be done by SA hardware such as dedicated equipment or a powerful desktop PC, or it can also be offloaded onto cloud. 5G could potentially provide a highly functional platform for cloud by transferring the raw AR/VR data onto it for postprocessing, and for delivery back to the 5G device [38].

2.13.7 Other Verticals

The examples above only scratch the surface of how 5G can benefit verticals. There are many more verticals that may be able to function more efficiently using 5G services, such as retail and financial services (in-built security of 5G, platform for mobile ID), entertainment (low latency, fast data transfer for advanced multimedia), and agriculture (support for a massive number of intelligent rural sensors working autonomously for users), to mention a few.

An example of the efforts to understand the needs of verticals is the EU's 5G-PPP, which is funding several projects, including vertical use cases [39]. Another example is the GSMA, which is coordinating research into network slicing, attributes best benefiting a variety of verticals in the interconnected environment [40].

References

1 ITU, "Minimum Requirements Related to Technical Performance for IMT-2020 Radio Interface(s)," ITU, 23 February 2017. [Online]. Available: https://www.itu.int/md/R15-SG05-C-0040/en. [Accessed 1 March 2017].

2 ITU-R, "Rep. ITU-R M.2410-0, Minimum Requirements Related to Technical Performance for IMT-2020 Radio Interface(s)," ITU, November 2017.

3 ITU, "ITU Agrees on Key 5G Performance Requirements for IMT-2020," 23 February 2017. [Online]. Available: http://www.itu.int/en/mediacentre/Pages/2017-PR04.aspx. [Accessed 1 March 2017].

4 ITU, "Itu-r IMT.2020," ITU-R, 14 July 2020. [Online]. Available: https://www.itu.int/md/R15-IMT.2020-C. [Accessed 28 September 2020].

5 NGMN, "NGMN," NGMN, 4 July 2018. [Online]. Available: https://www.ngmn.org/home.html. [Accessed 4 July 2018].

6 ITU, "M.2083: IMT Vision – 'Framework and Overall Objectives of the Future Development of IMT for 2020 and Beyond'," September 2015. [Online]. Available: https://www.itu.int/rec/R-REC-M.2083. [Accessed 17 August 2020].

7 ITU-R, "ITU-R Recommendation M.2083," ITU-R, 4 July 2018. [Online]. Available: http://www.itu.int/rec/R-REC-M.2083. [Accessed 4 July 2018].

8 ITU, "World Radiocommunication Conference 2019 (WRC-19), Sharm el-Sheikh, Egypt, 28 October to 22 November 2019," ITU, 2019. [Online]. Available: https://www.itu.int/en/ITU-R/conferences/wrc/2019/Pages/default.aspx. [Accessed 6 July 2019].

9 ITU, "ITU," ITU [Online]. Available: https://www.itu.int/en/ITU-R/conferences/wrc/Pages/default.aspx. [Accessed 6 July 2019].

10 ITU-T, "ITU-T Focus Group IMT-2020 Deliverables," 2017. [Online]. Available: https://www.itu.int/dms_pub/itu-t/opb/tut/T-TUT-IMT-2017-2020-PDF-E.pdf.

11 GSMA, "Mobile IoT in the 5G Future – NB-IoT and LTE-M in the Context of 5G," GSMA, 14 May 2018. [Online]. Available: https://www.gsma.com/iot/mobile-iot-5g-future. [Accessed 5 July 2019].

12 ITU, "Press Release: ITU Agrees on Key 5G Performance Requirements for IMT-2020," ITU, 27 February 2017. [Online]. Available: http://www.itu.int/en/mediacentre/Pages/2017-PR04.aspx. [Accessed 4 July 2018].

13 ITU, "ITU Towards 'IMT for 2020 and Beyond'," 2016. [Online]. Available: http://www.itu.int/en/ITU-R/study-groups/rsg5/rwp5d/imt-2020/Pages/default.aspx. [Accessed 7 September 2016].

14 ITU-R, "Itu-R Recommendation M.2376, Technical Feasibility of IMT in Bands Above 6 GHz," ITU, July 2015. [Online]. Available: https://www.itu.int/pub/R-REP-M.2376. [Accessed 4 July 2018].

15 ITU, "Itu-r M.[IMT-2020.EVAL]," ITU, 2017.

16 GSMA, "Generic 5G Network Slice Template V3.0," GSMA, 22 May 2020. [Online]. Available: https://www.gsma.com/newsroom/wp-content/uploads//NG.116-v3.0.pdf. [Accessed 20 September 2020].

17 GSMA, "The 5G Guide," GSMA, April 2019. [Online]. Available: https://www.gsma.com/wp-content/uploads/2019/04/The-5G-Guide_GSMA_2019_04_29_compressed.pdf. [Accessed 20 September 2020].
18 GSMA, "5G Implementation Guidelines," GSMA, July 2019. [Online]. Available: https://www.gsma.com/futurenetworks/wp-content/uploads/2019/03/5G-Implementation-Guideline-v2.0-July-2019.pdf. [Accessed 20 September 2020].
19 3GPP, "Release 15," 3GPP, 4 July 2018. [Online]. Available: http://www.3gpp.org/release-15. [Accessed 4 July 2018].
20 3GPP, "3GPP Features and Study Items," 3GPP, 4 July 2018. [Online]. Available: http://www.3gpp.org/DynaReport/FeatureListFrameSet.htm. [Accessed 4 July 2018].
21 3GPP, "R2-161747, Discussion on Uplink Data Compression," 3GPP, 4 July 2018. [Online]. Available: http://portal.3gpp.org/ngppapp/CreateTdoc.aspx?mode=view&contributionId=687049. [Accessed 4 July 2018].
22 Penttinen, J., "5G Explained: Security and Deployment of Advanced Mobile Communications," Wiley, April 2019. [Online]. Available: https://www.wiley.com/en-gb/5G+Explained:+Security+and+Deployment+of+Advanced+Mobile+Communications-p-9781119275688. [Accessed 20 September 2020].
23 3GPP, "Security Architecture and Procedures for 5G System, Release 15, V.15.1.0," 3GPP, 2018.
24 NGMN, "NGMN 5G White Paper," NGMN Alliance, 2015.
25 Verizon, "Verizon 5G Technical Forum," Verizon, 2018. [Online]. Available: http://www.5gtf.net. [Accessed 28 July 2018].
26 Ericsson, *5G Radio Access*, Ericsson AB, 2015.
27 Nokia, *5G Use Cases and Requirements*, Espoo: Nokia Networks, 2014.
28 Ericsson, "5G Consumer Potential: Busting the Myths Around the Value of 5G for Consumers," May 2019. [Online]. Available: https://www.ericsson.com/498f26/assets/local/reports-papers/consumerlab/reports/2019/5g-consumer-potential-report.pdf.
29 Ericsson, "5G Meets Time Sensitive Networking," 18 December 2018. [Online]. Available: https://www.ericsson.com/en/blog/2018/12/5G-meets-Time-Sensitive-Networking.
30 Quain, J.R., "Digital Trends," 10 August 2018. [Online]. Available: https://www.digitaltrends.com/cool-tech/the-highway-of-the-future-is-being-paved-in-georgia.
31 5GAA, "V2X," 2020. [Online]. Available: https://5gaa.org/5g-technology/c-v2x.
32 AT&T, "5 Ways 5G Will Transform Healthcare," 2020. [Online]. Available: https://www.business.att.com/learn/updates/how-5g-will-transform-the-healthcare-industry.html.
33 Ericsson, "Critical Services and Infrastructure Control," 2020. [Online]. Available: https://www.ericsson.com/en/5g/use-cases/critical-services-and-infrastructure-control.
34 3GPP, "TR 38.913: Study on Scenarios and Requirements for the Next Generation Access Technologies (Release 15)," 3GPP, 2019.
35 3GPP, "TS 23.303: Proximity-Based Services," 3GPP, 2018.
36 3GPP, "3GPP," 20 June 2017. [Online]. Available: https://www.3gpp.org/news-events/1875-mc_services.
37 GSMA, "GUTMA and the GSMA Announce Collaboration to Help Define the Future of Aerial Connectivity," 23 December 2019. [Online]. Available: https://www.gsma.com/iot/news/gutma-and-the-gsma-announce-collaboration-to-help-define-the-future-of-aerial-connectivity.

38 Kelvin Qin, M.Z., "GSMA," April 2018. [Online]. Available: https://www.gsma.com/futurenetworks/wp-content/uploads/2018/07/Network-Slicing-Use-Case-Requirements-fixed.pdf.

39 5GPPP, "5G and Verticals," 2020. [Online]. Available: https://5g-ppp.eu/verticals.

40 GSMA, "5G for Verticals: Reshaping Europe's Mobile Sector Narrative," 6 September 2019. [Online]. Available: https://www.gsmaintelligence.com/research/2019/09/5g-for-verticals-reshaping-europes-mobile-sector-narrative/799.

3

Phase 2 System Architecture and Functionality

3.1 Introduction

3.1.1 General

The Standalone (SA) 5G network uses Service-Based Architecture (SBA). This means that the network presents the feasible network elements as Network Functions (NFs). The NFs offer their services via interfaces of a common framework to any other NFs that are allowed to make use of these provided services.

Network Repository Functions (NRFs) have a special role in this architecture as they allow each NF to discover the services offered by other NFs. This architecture model provides benefits of the modern ways of virtualization and software technologies in deployments.

3rd Generation Partnership Project (3GPP) Release 15, as defined by the System Architecture group SA, describes the 5G system architecture via a set of features and functionality that are essential for deploying operational 5G systems in the field. The essential technical specifications for the complete description are 3GPP TS 23.501, TS 23.502, and TS 23.503 from which stage 2 includes the overall architecture model and principles.

This chapter presents the 5G architecture with some of the most relevant scenarios in roaming and non-roaming environments. The 3GPP 5G system is a major enhancement from previous mobile communication generations. Along with considerably better performance, the complete philosophy of the network architecture receives a major facelift. The most remarkable difference between the old world, New Radio (NR), and core networks is the virtualization of the functionality that provides a much wider set of use cases. Furthermore, the new architecture model provides a fluent modernization of the network and utilization of cloud principles for processing and storing data.

3GPP Release 15-based 5G, referring to the first phase of 5G, introduced NR and 5G Core (5GC) that together formed the 5G System (5GS). 5GS as per Release 15 forms the very foundation of the new era, but it only provides the initial functions and performance, whereas Release 16 introduces the rest of the features and enhances the system sufficiently to comply with the ITU IMT-2020 requirements. This section presents Release 15 and Release 16 5G architecture with new and reutilized elements, connected and cooperative networks, aggregated services, and key functions of evolved networks.

5G Second Phase Explained: The 3GPP Release 16 Enhancements, First Edition. Jyrki T.J. Penttinen.

5G opens new opportunities to the ecosystem by introducing open interfaces as one of the benefits of the virtualization of the network functions. *OpenBTS.org* is an open source software project. According to the statements of this organization, the open source concept is dedicated to revolutionizing mobile networks by substituting legacy telecommunications protocols and traditionally complex, proprietary hardware systems with Internet Protocol (IP) and a flexible software architecture. This architecture is open to innovation by anyone, allowing the development of new applications and services and dramatically simplifying the setting up and operation of a mobile network.

In addition to a variety of benefits of open source, as the source code of network components is now easily available to anyone, it may also be more susceptible to fraud, which needs to be taken care of in the standardization, product development, and deployment of 5G.

Driven by the requirement for ultra-low latency and high bandwidth, both NFs and content must move closer to the subscriber, i.e., as close to the edge of the radio network as feasible. Mobile-Edge Computing (MEC) allows content, services, and applications to be accelerated, increasing responsiveness from the edge. Furthermore, the mobile subscriber's experience can be enriched through efficient network and service operations, based on insight into the radio and network conditions. Those applications are implemented as software-only entities running on top of a virtualization infrastructure.

With the increasing demand of enabling new business models, 5G also needs to reduce the costs of utilization and the offering of services. Network Functions Virtualization (NFV) and Software Defined Networking (SDN) are technologies that help to reduce infrastructure and management expenses significantly. In many cases, however, these mechanisms imply that communication between network entities is migrating from today's proprietary protocols to standard IP-based mechanisms, such as IPsec.

In 5G, as defined by the 3GPP, the SBA is applicable between the Control Plane (CP) NFs of the core network. In 5G, the NFs can store their contexts in Data Storage Functions (DSF), which makes it possible to separate data storage for the User Equipment (UE), Access Network (AN), and Access and Mobility Management Function (AMF). This means that there is no more tight binding of the elements as in previous generations, which makes the functionality more flexible and increases the performance by optimizing the resource utilization. The previous architecture models performed UE-specific transport association, which was not optimal for the UE's serving node change.

5G functionality simplifies changing the AMF instance that serves a UE. The new architectural model also supports enhanced AMF resilience and load balancing. This is a result of AMF functionality, which allows a set of AMFs within the same network slice to handle procedures of any UE. The SBA of 5G refers to the capability of the network to present its elements as NFs in a virtualized environment. It enhances the performance of the network and provides a means for new features.

The new 5G architecture model provides services between NFs using a common hardware that can be based on generic Commercial Off-the-Shelf (COTS) components. In previous generations, the cellular networks have traditionally relied on dedicated SA elements for each network function such as the Long Term Evolution (LTE) Serving Gateway (S-GW), Home Subscription Server (HSS), and Mobility Management Entity (MME).

The initial 5G networks can be deployed by applying a hybrid architecture mode that takes advantage of the 4G core infrastructure, interconnecting to it 5G Next Generation NodeB (gNB) elements. This intermediate mode provides advanced performance compared to the pure 4G network, although the 4G core network does not allow the deployment of many of the 5G-type functions.

The SA 5G architecture model relies on NFV and SDN, and thus execution of the procedures takes place via virtualized NFs. As a result, the SA 5G network provides faster data speeds and more reliable connections, and is easier to manage compared to the older networks. Among a large set of enhanced and new functions, the NRFs have a special role in 5G architecture as they allow the NFs to discover the services of the other NFs in a secure manner.

Furthermore, 5G uses network slicing. It is a concept that consists of a set of features that are able to form virtual cellular networks on the very same areas, yet differentiating the performance characteristics of each. It connects smartphones and other 5G devices. The 5G UE connects to the 5G infrastructure via the *Uu* radio interface. While traditional cellular network architectures define only uniform physical cellular networks, 5G network slicing provides a means for a single Mobile Network Operator (MNO) to form a set of parallel Public Land Mobile Networks (PLMNs) within the same physical area. In other words, slices are comparable with "networks within a network" that can be tailored to the performance figures to satisfy the varying needs of verticals.

The MNO can set up and optimize each slice individually to achieve the best performance for different usage scenarios "on-the-fly." The operator can create and terminate slices dynamically as per the need for required periods. As an example, the operator can set up a specific slice to provide fast data for its users, while another slice may serve a huge number of low-bit-rate sensors within the very same service area. Adjusting adequately the functional and performance parameters, different users benefit from the selection of the most suitable slices. This enhances the user experience, and the MNO has a better chance of optimizing the offered network capacity.

Furthermore, 5G includes an extended Quality of Service (QoS) functionality, which can better differentiate the data flows based on varying priority levels of services. The QoS of 5G uses the already existing ones of the LTE system, and adds new definitions.

Yet another benefit of 5G is the support of a variety of access systems. In addition to the new 5G radio network itself (Next Generation Radio Access Network, NG-RAN), the 5G core network can also serve generic ANs such as Wi-Fi (WLAN) hotspots. As per the Release 15 definitions, the 5G core network is capable of interconnecting with both 3GPP NG-RAN and 3GPP-defined untrusted WLAN networks, whereas Release 16 includes more access options.

3GPP Release 15 defines the initial 5G system architecture and functionalities. Some of the most important Technical Specifications (TS) are 3GPP TS 23.501, TS 23.502, and TS 23.503, which also describe the evolved Mobile Broadband (eMBB) data service, subscriber authentication and authorization, application support, edge computing, IP Multimedia Subsystem (IMS), and interworking with 4G and possible other access systems.

Thus, some of the key differentiators of the 5G networks are network virtualization (making the functions decoupled from the underlying hardware), network slicing (offering special networks over a network), open source (opening doors for new ecosystem stakeholders), and edge (data centers).

3.1.2 Release 16 Development

Release 16 includes around two dozen additional items such as Multimedia Priority Service (MPS), Vehicle-to-Everything (V2X) application-layer services for car-to-car and railway communications, 5G satellite access, LAN support, convergence of wireless and wireline, enhanced terminal positioning, network automation, and evolved radio techniques, security, codecs, and streaming services [1]. All these are beneficial for a connected society using 5G as a platform. The 3GPP working groups are already assessing new features for the forthcoming Release 17, which keeps enhancing 5G. The 3GPP approved these work items in December 2019 [2].

3.1.3 Radio Network

According to 3GPP TS 38.300, gNB is a node providing NR, User Plane (UP), and CP protocol terminations towards the UE, connected via the NG interface to the 5GC.

Physically, gNB elements form a base station, which can refer to equipment shelters housing the transmitters, receivers, power supplies, and other devices required for radio connectivity with mobile devices. The shelter can be a separate building, a room within an already existing construction, or a simple box installed on a wall or pole. Nevertheless, the 5G specifications allow the split of gNB functions, so not all of them are necessarily in the same physical location; instead, part can be housed remotely, e.g., in a cloud that is physically a data center.

3.1.4 Core Network

The 5G core network houses all the needed functionalities for establishing, maintaining, and releasing data and voice calls, and for establishing any other communication links such as messaging between the users and the network's own signaling.

The new 5G core network differs from the previous architectural models significantly. Before, the NFs were processed by dedicated, operator-owned elements for those specific functions, such as subscription authentication or user register. These network devices were SA, individual components with their own hardware and software.

5G is based on virtualized NFs. Each function is thus merely a set of software instances that runs on common, virtualized hardware. This means that instead of having to use a mesh of the traditional and individual network elements, the 5G core networks can be operated in clouds of operators or a third party. Cloud typically refers to the resources offered by data centers. These virtualized networks enhance greatly the performance of 5G compared to previous generations.

Another novelty of 5G is edge computing. This means that the operator can move the desired contents nearer to the user within the core network. If the user wants to see a YouTube video, instead of the user contacting the video server somewhere on the other side of the network, the operator may want to transfer the contents to the network's edge near the user. This reduces the delays when the user keeps downloading and buffering the video on the go.

Another benefit of edge is that the device could offload part of its processing onto it. This would be useful in, say, Virtual Reality (VR) content creation and playback using a terminal that sends the raw data to the edge for processing, and receives the postprocessed data back to be merely displayed.

These devices require much less processing power, enhancing battery life and lowering the cost of the device. If this concept becomes popular, we could start seeing low-cost multimedia devices equipped with 5G connectivity. An example of such a device is a simple visor to display 3D contents of games, movies, and environments mixing Augmented Reality (AR).

Some of the key tasks of the core network are the establishment of voice and video calls, and the delivery of messaging, data, and related signaling. 5G standards ensure that these functions are applicable both in national as well as international environments.

Interoperability is thus an important aspect. 5G networks need to comply with interoperability so that customers can enjoy fluent user experiences in their own home operator's network as well as when roaming other countries. Interoperability is ensured by the globally approved 3GPP standards.

The 3GPP ensures 5G standardization and its evolution. There are also supporting efforts for providing additional guidelines on top of the 5G standards. Some examples are voice calls and messages as defined by the GSMA.

3.1.5 Transport Network

The transport network is a set of fiber optics, other cables, and radio links between the radio and core networks. The transport network is becoming increasingly intelligent along with the demand for service-based quality and geo-redundant service levels. The modern transport network benefits from the virtualization of functions that can optimize the management of the transport.

3.1.6 5G NFs of Release 16

The 5G architecture consists of RAN based on the NR interface, and Next Generation Core (NGC) that has renewed network elements. Access into the 5G network can also take place via the non-3GPP AN. AN refers thus to a general radio base station, including non-3GPP access such as Wi-Fi.

The accessing device is the UE, which is formed by the Mobile Terminal (MT) device and a tamper-resistant Secure Element (SE), which may have a form of "traditional" Universal Integrated Circuit Card (UICC), commonly known as a Subscriber Identity Module (SIM) card, or its evolved variant such as embedded UICC (eUICC). If the UE connects to a Data Network (DN), it can be, e.g., operator services, Internet access, or third party services.

The 5GS architecture consists of a set of NFs of which each has its own defined task. 3GPP TS 23.501 includes a functional description of these NFs that are detailed in Chapters 4 and 5.

Each 5G NF performs tasks as described in 3GPP TS 23.501. Figure 3.1 depicts the mapping of the 4G and 5G key functions, and Table 3.1 compares their roles. In Figure 3.1, EPC refers to the Evolved Packet Core network of 4G, evolved NodeB (eNB) to the 4G base station, and gNB to the 5G base station. The SBA provides gradual 5G network deployments, and MNOs can always take advantage of the latest advancements of the virtualization concept.

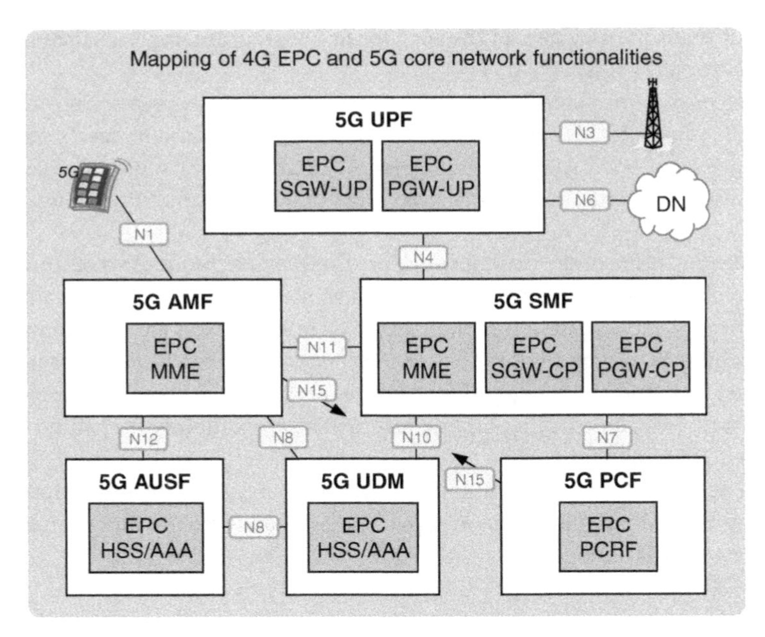

Figure 3.1 Mapping of the key 4G and 5G elements. Releases 15 and 16 bring new components that do not exist in 4G.

Table 3.1 The key components of 5G, and mapping with the 4G LTE system.

5G NF	Description	Mapping with 4G
5G-EIR	Equipment Identity Register	Evolution of LTE EIR
AF	Application Function	LTE Application Server (AS) and GSM Service Control Function (gsmSCF)
AMF	Access and Mobility Management Function	Replaces the LTE Mobility Management Entity (MME)
AUSF	Authentication Server Function	Replaces the LTE MME/AAA (Authentication, Authorization, and Accounting)
CAPIF	*Common Application Programming Interface Framework for 3GPP northbound APIs*	*Not included in 4G; please note that CAPIF is a framework, not a function*
CHF[1]	Charging Function	The LTE Policy and Charging Control (PCC) functions include the Policy and Charging Rules Function (PCRF) for policy control and flow-based charging control, whereas the Policy and Charging Enforcement Function (PCEF) of the S-GW enforces individual IP flow gating and QoS supporting PCRF as stated in 3GPP TS 32.240 [3].
GMLC[1]	*Gateway Mobile Location Centre*	*Location Server (LS); extends the functionality of Release 15 LMF of 5G*
I-SMF[1]	Intermediate Session Management Function	Not included in 4G; extends the functionality of SMF of 5G

Table 3.1 (*Continued*)

5G NF	Description	Mapping with 4G
I-UPF[1]	Intermediate User Plane Function	Not included in 4G; extends the functionality of UPF of 5G
LMF	Location Management Function	LS based on Location Service Client (LCS)
N3IWF	Non-3GPP Interworking Function	3GPP-based enhanced generic AN architecture has tight coupling via rerouting of cellular network signaling through Wi-Fi ANs. Wi-Fi is an example of a non-3GPP WLAN RAN. The 3GPP also provides loosely coupled architecture for Wi-Fi, Interworking Wireless LAN (IWLAN)
NEF	Network Exposure Function	Evolution of the Service Capability Exposure Function (SCEF) and API layer
NRF	NF Repository Function	Part of the evolution of the Domain Name System (DNS)
NSSF	Network Slice Selection Function	New function for 5G-specific network slicing concept (not in 4G)
NSSAAF[1]	Network Slice Specific Authentication and Authorization Function	Not included in 4G; extends the functionality of NSSF of 5G
NWDAF	Network Data Analytics Function	External analytics tools
PCF	Policy Control Function	Evolution of the LTE PCRF
SCP[1]	Service Communication Proxy	Not included in 4G
SEPP	Security Edge Protection Proxy	New element for securely interconnecting 5G networks (not in 4G)
SMF	Session Management Function	Replaces, with the 5G UDR, the LTE S-GW and PDN Gateway (P-GW)
SMSF	Short Message Service Function	Short Message Service Center (SMSC)
TNGF[1]	Trusted Non-3GPP Gateway Function	Not included in 4G; extends the concept of N3IWF
TWIF[1]	Trusted WLAN Interworking Function	Not included in 4G; extends the concept of N3IWF
UCMF[1]	UE radio Capability Management Function	Not included in 4G
UDM	Unified Data Management	Evolution of HSS and UDR
UDR	Unified Data Repository	Evolution of the LTE Structured Data Storage (SDS)
UDSF	Unstructured Data Storage Function	The function comparable with LTE Structured Data Storage Function (SDSF)
UPF	User Plane Function	Replaces, with the SMF, the LTE S-GW and P-GW
W-AGF[1]	Wireline Access Gateway Function	Not included in 4G

Note: [1] NF added in Release 16 [4]. Please note that the cursive components are not considered as NFs.

The tasks of 5G NFs are divided into a UP for data connectivity via UPF, and a CP for signaling via SMF. Please see further details of 5G NFs and interfaces in Chapters 4 and 5.

Along with renewed physical network elements and a variety of mandatory and optional NFs, there are also many new interfaces in 5G. The virtualized architecture model of 5G differs from the legacy systems, and there are also plenty of new interfaces. Please find more details on the radio and core interfaces in Chapters 4 and 5, respectively. The introductory specifications of these can also be found in TS 38.300 (NR), TS 38.401 (RAN), and TS 23.501 (system architecture);

3.2 Release 16 Enhancements

3.2.1 LTE in Release 16

Not only has the 5G system been enhanced along with Release 16, but also the LTE has evolved in a parallel fashion. The key enhancement is related to the architectural evolution of the eNB of the E-UTRAN and NG-RAN, as described in the 3GPP Technical Report (TR) 37.876 [5]. The document details the principle of the eNB functional split architecture at a higher layer. The new architecture is based on the LTE-CU (Central Unit) and LTE-DU (Distributed Unit), which provides better means to integrate the LTE eNB and NR gNB. Some of the evolved Release 16 items for LTE include:

- Coexistence with NR;
- Downlink (DL) transmission efficiency enhancement and UE power consumption enhancement along with early transmission of Mobile Terminated (MT) communications, and early data transmission via support for UE group wake-up signal; Uplink (UL) transmission efficiency enhancement and UE power consumption enhancement along with support of transmission in preconfigured resources in idle and connected mode;
- Early measurement reporting for fast setup of Multi-Radio Dual Connectivity (MR-DC) and Carrier Aggregation (CA);
- Enhanced NB-IoT (Internet of Things);
- Enhanced scheduling for multiple UL and DL transport blocks;
- Enhancement of the LTE-NR, NR-NR DC, and NR CA;
- Fast recovery; if the Master Cell Group (MCG) fails, there is a Secondary Cell Group (SCG) link and the split Signaling Radio Bearers (SRBs) as a means of recovery;
- Improved Multi-Carrier (MC);
- Improved mobility;
- Improved Self-Organizing Network (SON);
- Low-latency cell configuration, setup, and activation;
- Support of asynchronous and synchronous NR-NR DC.

3GPP TS 36.300, Section 24.1, summarizes the scenarios for 5G support for 4G with respective protocol stacks, core network selection and mobility, slicing, access control, and RAN sharing. NG-RAN supports E-UTRA that connects to 5GC. E-UTRA is thus capable of connecting to both EPC and 5GC.

TS 38.300 describes the architecture scenario for E-UTRA connected to 5GC as part of NG-RAN. In that scenario, ng-eNB refers to the radio access of E-UTRA that connects to 5GC. It should be noted though that the term eNB is used equally for both cases in the 3GPP specifications unless there is a specific reason to distinguish between eNB and ng-eNB.

E-UTRA connected to 5GC supports 5G Non-Access Stratum (NAS)message transport as per 3GPP TS 36.300. 5G supports the following 3GPP TS 38.300 functions: 5G security framework except for data integrity protection, access control, flow-based QoS, and network slicing. In addition, the following functions are supported with the exception of the NB-IoT: Service Data Adaptation Protocol (SDAP) (3GPP TS 37.324), NR Packet Data Convergence Protocol (PDCP) (3GPP TS 38.323), and UE in *RRC_Inactive* state. Furthermore, E-UTRA connected to 5GS also supports Cellular IoT (C-IoT) 5GS optimizations for Bandwidth reduced Low complexity (BL) UE, UE in enhanced coverage and NB-IoT UE (3GPP TS 36.300), Mobile Originated Early Data Transmission (MO-EDT) for BL UE, UE in enhanced coverage and NB-IoT UE (3GPP TS 36.300), and transmission using Preconfigured Uplink Resource (PUR) for BL UE, UE in enhanced coverage, and NB-IoT UE (3GPP TS 36.300).

E-UTRA connected to 5GC can also support V2X sidelink communication and NR sidelink communication for UE in the Radio Resource Control (RRC) states of idle, inactive, and connected (TS 38.300).

3.2.2 5G of Release 16

Release 15 brought along a new architectural model based on services. This is possible thanks to network virtualization. Release 16 adds more and enhances the existing functionalities. Some examples of items in Release 16 are:

- 5G fixed wireless backhaul for NR–core link.
- 5G satellite access component requirements for enabling integration of satellite access as per the use cases of TR 22.822, such as roaming between terrestrial and satellite networks, satellite overlay broadcast and multicast, satellite IoT, and optimal routing and steering over a satellite.
- DC (E-UTRA-NR DC, EN-DC) for the new DL and UL EN-DC configurations on LTE and NR bands.
- Enhanced 5G positioning services to better support, e.g., emergency services as per 3GPP TS 22.261 and TR 22.872.
- Enhanced energy efficiency.
- Enhanced multibeam operation on Frequency Range 2 (FR2) operation.
- Enhanced NR mobility via reduced interruption time of handover, e.g., via a handover without Random Access Channel (RACH), conditional handover, and fast handover failure recovery.
- Enhanced security via Security Assurance Specifications (SCAS) for 5G network products helps identify threats to the 5Gsystem architecture; SCAS specifies security requirements and test cases for network equipment implementing 3GPP NFs.
- Evolved NR Multiple In, Multiple Out (MIMO) such as enhanced Multi-User (MU)-MIMO and multi-Tx/Rx Point (TRP) transmission, e.g., for the more robust backhaul.

- LAN support in 5G as per 3GPP TS 22.261, including 5G private virtual network and the creation and management of 5G LAN.
- NR intraband CA for contiguous and non-contiguous use cases ensuring the Radio Frequency (RF) limits.
- Peak to Average Power Ratio (PAPR) enhancements via Channel State Information Reference Signal (CSI-RS) and Demodulation Reference Signal (DMRS).

Furthermore, as stated in 3GPP TS 32.298 and TS 32.291, Release 16 introduces a new Charging Function (CHF) [6, 7]. This offers charging services to relevant NFs. There are various enhancements in 5G DC charging. Charging is achieved by SMF invocation of charging services exposed by the charging function. Functionality is specified in TS 32.255, and contains various configurations and functionalities of SMF, such as:

- Charging capabilities for flexible deployment of AFs such as edge computing.
- Charging continuity for interworking and handover between 5G and EPC.
- Evolution of access-type traffic charging differentiation supporting NG-RAN and untrusted WLAN access.
- Identification of the PLMN in shared RAN for operator settlement.
- Local Breakout (LBO) roaming charging
- 5G support for QoS-based charging, including interoperator use cases in a roaming home-routed scenario.
- Operator application of business case charging differentiation for different network slices.

3.2.3 Fixed-Mobile Convergence

Release 16 enables 5G connection for broadband. One of the entities dealing with the overall evolution of broadband aspects is the Broadband Forum (BBF) [8].

BBF's 5G Fixed-Mobile Convergence (FMC) project provides definitions and recommendations to enable 5G for broadband networks. The Forum has identified the project as one of the critical areas for the success of 5G and to ensure respective market potential. The BBF is researching respective requirements and publishes information such as the 5G FMC white paper [8]. Figure 3.2 depicts the high-level principle of fixed-wireless convergence interpreted from it.

3.2.4 Control and User Plane Separation of EPC Nodes

The 3GPP-defined 5GS has a variety of new concepts. Some of these are CP and UP split, network slicing, and SBA. All these being essential building blocks for complying with the strict requirements of ITU-R IMT-2020, the network slicing is a key enabler supporting multiple different use cases and instantiations of the same functionality.

The 5G MBB use case is an evolution of 4G broadband connectivity. The difference is that this is enabled via the SBA in the 5G era. To take advantage of technical options and market needs, an interim solution provided by the 3GPP is the support of diverging architectures for 5G services.

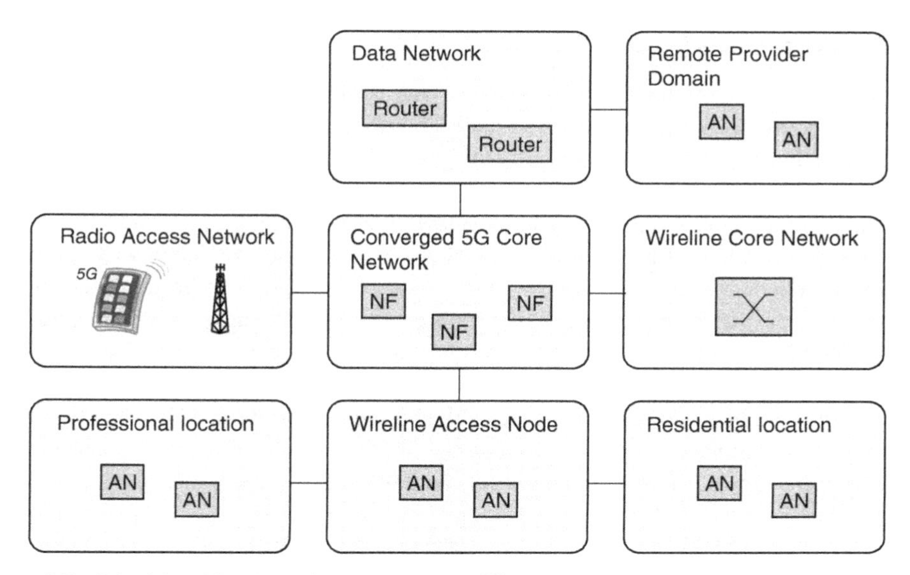

Figure 3.2 Principle of fixed wireless convergence [8].

As indicated in [9], 5GC standardization would ideally define a functional architecture where implemented technologies may evolve and can be replaced when time is adequate to do so. Support of the multivendor environment is one of the important principles for achieving this goal and enhancing independently the UP and CP functionalities. The latter provides the possibility for flexible deployment, allowing variable network configurations via network slices.

In this transition phase, the original 4G EPC will thus change from a signaling point of view. 4G transition towards Control and User Plane Separation (CUPS) splits the S-GW into S-GW-CP and S-GW-UP, for CP and UP, respectively. Equally, the P-GW will be broken into P-GW-CP and P-GW-UP, for CP and UP [10].

The CUPS concept refers to CU and UP separation of EPC nodes and provides the architecture enhancements for the separation of functionality in the EPC's S-GW, P-GW, and Traffic Detection Function (TDF). CUPS is defined as of 3GPP Release 14 and paves the way for gradual 5G adaptation via 4G EPC as the concept enables flexible network deployment and operation via distributed or centralized deployment and independent scaling between CP and UP functions [10].

The benefit of CUPS is reduced latency on application service. This can be done by relying on UP nodes closer to the RAN without impact on the number of CP nodes. The CUPS concept also supports increasing data traffic as the service utilization increases because the UP nodes can be added into the MNO infrastructure without impacting the number of S-GW-C, P-GW-C, and TDF-C elements of the mobile network.

Other benefits include the possibility of adding and scaling the EPC node CP and UP resources independently, which means that the evolution of the CP and UP functions can be done independently. CUPS also enables SDN for optimized UP data delivery.

3.2.5 Java and APIs in 5G

5G is based on the virtualization of NFs. It also uses the principles of open source and SDN. Practical means for information transfer between interfaces is thus needed, as well as hardware-agnostic ways to execute code. Java and APIs offer some of these means.

5G is an end-to-end telecommunication system. It integrates and converges various network types, including wireless and fixed systems. Furthermore, 5G radio, core, and transport alike rely strongly on cloud infrastructure. The former distributed networks and web application services are not able to comply with the strict 5G networking requirements, but the API and Representational State Transfer (REST) concepts can facilitate the new ecosystem to do so.

The role of the API is very important in 5G. Among other indications, the ITU Focus Group IMT-2020 is emphasizing their relevancy in the 5G ecosystem as they facilitate the adaptation of applications and services to deal with the programmable NFs, and help applications to communicate with each other in the highly virtualized mobile communication infrastructure. One of the key statements of the IMT-2020 is the following:

> Operators of IMT-2020 network should expose network capabilities to 3rd party ISPs (Internet Service Providers)/ICPs (Internet Content Providers) via open APIs to allow agile service creation, flexible and efficient use of the capabilities. ... For IMT-2020, there is an increased need on service customization by the service providers whereby some of them will offer their customers the possibility to customize their own services through service-related APIs in order to support the creation, provisioning and management of services.

APIs are, in fact, essential to interconnect systems and share data. APIs can help to reduce the cost of the 5G ecosystem because API-interconnected systems can use common software functions and thus optimize software production.

Another important component in the 5G ecosystem is REST. This is an architectural style designed for distributed hypermedia systems. REST has become important along with the popularity of the geographic web. REST works well for sharing information. RESTful Web services integrate to the web as a transport medium. They are less strict for bandwidth, processing power, and memory compared to earlier models, and they are capable of communicating through firewalls and proxy web servers [11].

REST is thus an adequate component for 5G in the effort to interconnect societies. It eases programming and collaboration between stakeholders and promotes multivendor and multi-operator ecosystems.

In a typical 5G network infrastructure, when offering services to enterprises and end-users, an API Gateway exposes REST APIs to third party applications and partners. The API Gateway serves as an entry point, which routes requests and is able to do protocol conversion. It is beneficial for cooperating parties such as developers who want to deploy their own user interfaces or transfer information via APIs.

Also, Java has an essential role in many 5G functions and procedures. It is a class-based, object-oriented, largely implementation-independent programming language for generic purposes. The aim of Java is to provide a means to design software once, and run the complied Java code on different platforms without the need for further adaptations.

JavaScript Object Notation (JSON) is a standardized, language-independent file format based on human-readable text. It is able to deliver data objects that contain attributes with their values and array data types. As the popularity of XML format is declining, JSON is nowadays a popular data format for asynchronous communications between browser and server. It also is an adequate base for many 5G-related procedures within the 5G system and with external entities.

The mobile communication industry has also used a special version of the Java environment for SIM card production. Referring to the UICC and its Universal Subscriber Identity Module (USIM) application, their Card Operating System (COS) is typically a card vendor's proprietary solution. Because of such different environments, each of the USIM card applications would need to be adjusted for the different operating systems. This would be a waste of time and resources.

To overcome this issue, Java Card Run Time Environment (JCRE) provides an abstraction layer between each card vendor's own Operating System (OS) variant and the apps running on them as Java applets. As the apps are OS agnostic in JCRE, they need to be developed only once and are compatible with any OS that supports the abstraction. It is expected that the same principle will continue in the 5G era.

Java Telephony API (JTAPI) is relevant for the 5G era, too. It supports telephony call control, and is an extensible API designed to scale and can be used in a range of domains from first party call control in a consumer device to third party call control in large distributed call centers [12].

More information on APIs and Java in the 5G era can be found in 3GPP TS 43.019 (SIM API for Java Card), 3GPP TS 31.130 (SIM/USIM API for Java Card), and ETSI TS 143 019 – V5.6.0 (SIM API for Java Card). See also the ITU-T IMT-2020 deliverables document [13].

3.2.6 Identifiers

The 5G system uses renewed identifiers to provide a unique means to recognize subscriptions and network elements in the new network ecosystem. Also, new methods are applied for identifiers such as concealing and Public Key Infrastructure (PKI).

Of the new 5G identifiers, the most important ones are Subscription Permanent Identifier (SUPI), Subscription Concealed Identifier (SUCI), Generic Public Subscription Identifier (GPSI), Globally Unique AMF Identifier (GUAMI), and Permanent Equipment Identifier (PEI). Also, the established identifiers are valid for interoperability, including Mobile Subscriber ISDN number (MSISDN), International Mobile Subscriber Identity (IMSI), International Mobile Equipment Identity (IMEI), and Network Access Identifier (NAI).

For the mapping of identifiers in different scenarios, please refer to 3GPP TS 29.571, which details the 5G data types for subscription, identification, and numbering.

SUPI is a primary identifier in 5G, and forms the foundation for all the key derivation scenarios together with the subscribers' unique K key. The serving network authenticates the SUPI during the authentication and key agreement procedures between the UE and the network. Afterwards, the serving network authorizes the UE to use services through the subscription profile obtained from the home network.

SUCI is the concealed version of the SUPI. As defined in 3GPP TS 33.501, the SUCI is a one-time subscription identifier that contains a concealed subscription identifier such as the Mobile Subscription Identification Number (MSIN). The network forms a new SUCI as per needs.

The SUCI is an optional identifier that is managed by the UICC, that is, the 5G SIM card. The SUCI provides additional security as it hides the permanent user's identification.

The UE generates the SUCI using a protection scheme. It is based on a public key that the user's home network has provisioned beforehand in a secure manner. Based on the indication of the USIM, and dictated by the MNO, forming of the SUCI can be done either by the USIM or the Mobile Equipment (ME). After that, the UE forms a scheme input for the subscription identifier of the SUPI and executes the protection scheme.

The UE does not conceal the home network identifiers, Mobile Country Code (MCC) and Mobile Network Code (MNC). This is because they are needed for home network routing and the protection scheme.

Please note that there is no requirement for protecting the SUPI in the case of an unauthenticated emergency call.

Please refer to 3GPP TS 29.571 for more information on the common data types, and TS 33.501 for the 5G security architecture.

3.2.7 Multicast/Broadcast in 5G

Release 6 was the starting point for the Multimedia Broadcast Multicast Service (MBMS). Ever since, it has been evolving, and it is applicable to 2G, 3G, and 4G. Nevertheless, it has not yet been included in the 5G system architecture in Release 15 or Release 16. The 3GPP is considering to integrate the service into the 5G system in Release 17.

3GPP TR 36.976 V16.0.0 presents the latest summary of LTE-based 5G broadcast as per Release 16. It describes LTE-based 5G terrestrial broadcast comprising a service delivering free-to-air content, a radio network comprising only MBMS-dedicated cells, or a mix of Further enhanced MBMS (FeMBMS) and unicast-mixed cells as transmitters, and Receive Only Mode (ROM) devices and UE supporting FeMBMS as receivers. The 3GPP 36 series includes respective details of radio interface protocols and procedures.

Release 16 TR 36.776 presents a gap analysis comparing Release 14 LTE terrestrial broadcasting capabilities with 5G requirements for dedicated broadcast networks that 3GPP TR 38.913 presents. This analysis found two requirements that are not complied with by the Release 14 LTE Evolved MBMS (eMBMS). These gaps are support for service over a large geographic area, including Single Frequency Network (SFN) with the Inter-Site Distance (ISD) larger than 100 km, such as cells utilizing high-gain rooftop directional antennas, and support for mobility scenarios, including speeds of up to 250 km/h, such as receivers in vehicles using external omnidirectional antennas.

Therefore, the 3GPP defined a new ISD of 125 km in the form of a High Power/High Tower (HPHT) network relying on omnidirectional transmitters. The evaluation included an ISD of 15 km for a Low Power/Low Tower (LPLT) network with sector cells, and an ISD of 50 km in a Medium Power/Medium Tower (MPMT) network with omnidirectional transmitters. The 3GPP also added a requirement to improve Cell Acquisition Subframe (CAS) reception for both large ISD and high-mobility scenarios.

As a result, Section 15.1.1 of 3GPP TS 36.300 presents the network architecture for LTE-based 5G terrestrial broadcast. TS 36.300 and TS 36.440 present the related RAN interfaces for LTE-based 5G terrestrial broadcast. 3GPP TS 23.246 defines the broadcast service announcement and session management procedures, and 3GPP TS 36.300, TS 23.246, and TS 24.116 define the respective ROM devices.

For more information on broadcast and multicast communication enablers for 5G, please refer to the EU report in Ref. [14].

3.3 5G Network Architecture in Release 16

3.3.1 System Architecture

3GPP TS 23.501 describes the 5G system architecture. It contains various architectural models for different use cases such as:

- Non-roaming architecture for UE concurrently accessing two DNs using multiple Packet Data Unit (PDU) sessions with multiple PDU sessions where two SMFs are selected for the two different PDU sessions.
- Non-roaming architecture in the case of concurrent access to two DNs within a single PDU session.
- Non-roaming architecture for NEF.

The interaction between the 5G NFs can be represented in two ways: via the traditional reference point model or service-based model. Service-based representation refers to the NFs within the CP enabling other NFs to access their services. Reference point representation, in turn, depicts the interaction between the NF services in the NFs.

The NFs have thus both functional behavior and interface. Each NF can be deployed as dedicated hardware forming the network element, or as a software instance that relies on hardware. Furthermore, the NF can also be implemented as a virtualized function instantiated on a relevant platform such as in a cloud-based infrastructure.

3.3.2 Non-roaming Reference Architecture

Figure 3.3 depicts an example of the basic architecture of the 5G network in a non-roaming scenario comparing 3GPP TS 23.501 Release 15 and Release 16 [4, 15]. The figure summarizes the main elements and interfaces of 5G. This presentation is based on reference point architecture, and the respective key 5G interfaces are marked between the functional elements. In this case, the Release adds NSSAAF and respective interfaces compared to the Release 15 architecture. Please note that this figure does not contain all the possible 5G NFs, such as the Unstructured Data Storage Function (UDSF), NEF, and NRF, although they are capable of interacting, when appropriate, with the UDSF, UDR, NEF, NRF, and NWDAF.

Data transmission between the UE and the DN takes place via a set of PDUs. The respective pipeline is formed via the UE, AN, or RAN and the UPF, while the rest of the NG, i.e., 5G core network elements, take care of the related signaling.

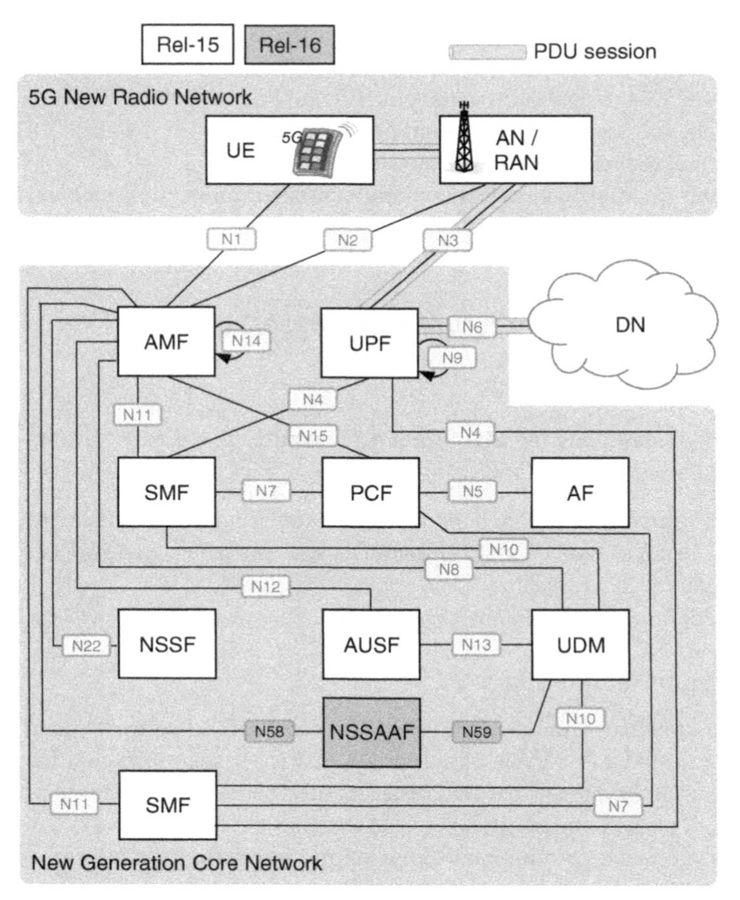

Figure 3.3 Reference point presentation of the 5G system architecture for the non-roaming case [15]. Release 16 adds the highlighted component and interfaces.

The 5G system can also be presented in an SBA model. Service-based interfaces are used within the CP. Figure 3.4 shows the respective reference architecture of the Release 16 5G network as interpreted from 3GPP TS 23.501 comparing Release 15 and Release 16 [4, 15]. In this specific case, the new Release 16 elements are the NSSAAF and SCP. Please note that when the operator deploys SCP, it can be used for indirect communication between NFs and NF services, but the SCP does not expose services itself.

In Figure 3.4, the tasks of the presented elements are the following, whereas a more detailed description of the elements can be found in Chapters 4 and 5, as well as in 3GPP TS 23.501, TS 38.300, and TS 38.401 [15–17]. Release 16 brings new network functions, which are NSSAAF and SCP.

- AF is Application Function, which interacts with the 3GPP core network to provide services such as application influence on traffic routing, access to NEF, and interaction with the policy framework.

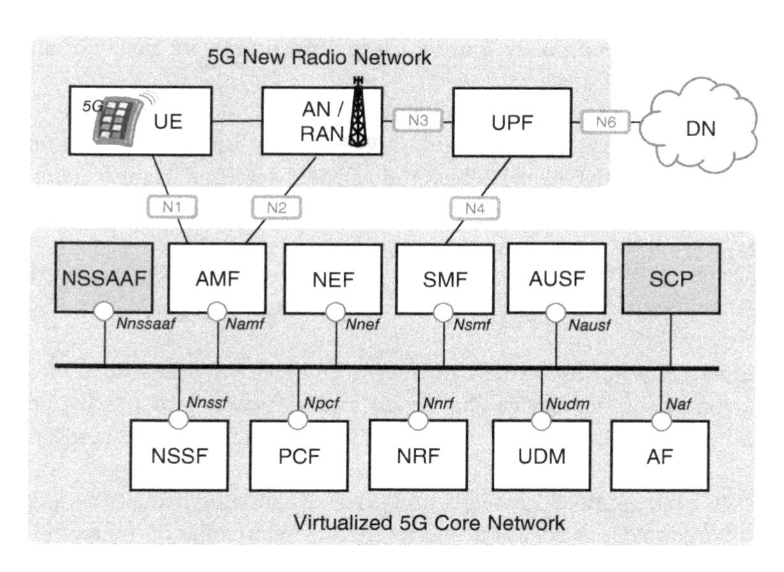

Figure 3.4 Non-roaming 5G system architecture, presented via service-based interfaces. Release 16 adds the highlighted components.

- AMF is Access and Mobility Management Function. It has a multitude of functions, including registration, connection, reachability, and mobility management. It also participates in access authentication and authorization, providing Security Anchor Functionality (SEAF). It interacts with the AUSF and UE, and receives the intermediate key from the UE authentication process. If USIM-based authentication takes place, it retrieves the security material from the AUSF. It also performs Security Context Management (SCM) by receiving a key from the SEAF to derive AN-specific keys. It also stores the UE capability list, performs load balancing, and interacts with the SMF.
- AN/RAN is the 5G NR AN or other AN, including Wi-Fi access. Release 15 defines non-3GPP access, and Release 16 adds further options for other ANs.
- AUSF supports authentication procedures as specified by SA WG3.
- NEF takes care of the exposure of capabilities and events, among many other tasks. It stores and retrieves information as structured data. It also makes secure provision of information from external application to the 3GPP network.
- NRF supports service discovery function and maintains the NF profile of available NF instances and their supported services.
- NSSAAF is a Release 16-specific NF.
- NSSF selects the set of network slice instances serving the UE, determines the allowed Network Slice Selection Assistance Information (NSSAI), configured NSSAI, and the AMF set to be used to serve the UE or a list of candidate AMF(s).

- PCF supports the unified policy framework to govern network behavior and provides policy rules to CP functions.
- SCP is another function introduced in Release 16.
- SMF takes care of session management and roaming functionality, among a multitude of tasks. It establishes, modifies, and releases sessions. It manages the UE's IP addresses and Dynamic Host Configuration Protocol (DHCP) v4/v6 server and client functions.
- UDM generates 3GPP Authentication and Key Agreement (AKA) authentication credentials. It handles user identification by storing and managing SUPI per 5G subscriber. It also manages SMS.
- UE is the 5G UE such as a smart device, IoT device, or other form of 5G-capable device. It comprises the device (hardware and software) and the tamper-resistant secure element such as UICC or its evolved version such as embedded or integrated UICC.
- UPF interacts between the 5G network and external DN, supporting PDU session anchor functionality. Its role is to act as an external PDU session point of interconnect to a DN. UPF service area refers to an area within which the PDU session associated with the UPF can be served by AN and RAN nodes. This happens via the *N3* interface between the AN and the UPF. A UE can establish multiple PDU sessions to the same DN and served by different UPFs terminating *N6*.

For the rest of the NFs that are not presented in this example, please refer to Table 3.1, as well as Chapters 4 and 5.

Figure 3.5 shows the non-roaming architecture for NEF. In this example, the 3GPP interface represents the set of southbound interfaces between the NEF and 5G core NFs.

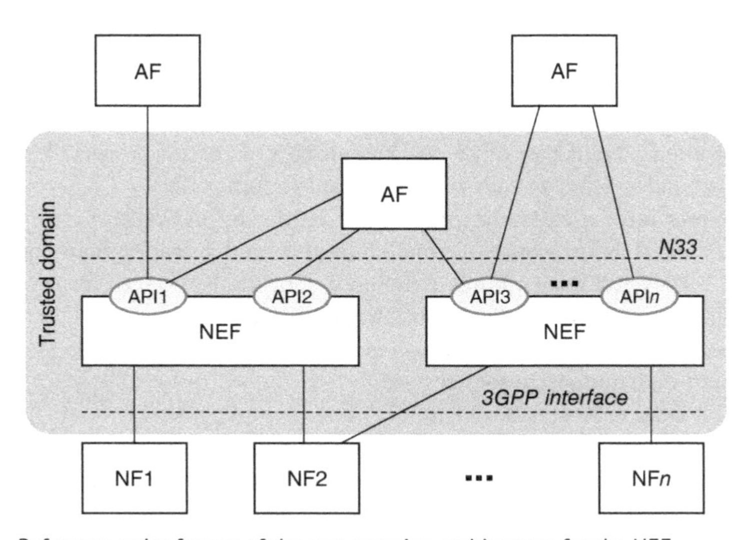

Figure 3.5 Reference point format of the non-roaming architecture for the NEF.

These interfaces include, e.g., *N29* for NEF-SMF and *N30* for NEF-PCF, among other interfaces.

Release 16 extends the northbound interface exposure to third parties via the CAPIF. Please refer to Chapter 5 for more information on the principle of the CAPIF.

3.3.3 Roaming Reference Architecture

Figure 3.6 depicts an example of a 5G roaming case for roaming based on LBO as interpreted from TS 23.501 [4]. In the LBO scenario, Visited Public Land Mobile Network (VPLMN) controls the SMF and all UPF components involved with the PDU session. In this scenario, the UE, which is roaming in the visited network (VPLMN), establishes connection to the DN of the VPLMN, while the Home Public Land Mobile Network (HPLMN) enables connectivity based on the user's subscription information on the UDM, subscriber authentication via the AUSF, and policies via the PCF for this specific UE. The interworking between HPLMN and VPLMN is protected by the home Security Edge Protection Proxy (hSEPP) and the visited Network Security Edge Protection Proxy (vSEPP). The SEPP is a non-transparent proxy and supports message filtering and policing on inter-PLMN CP interfaces, and topology hiding.

In this example, the visited network provides functions for the network slice selection (via NSSF), network access control and mobility management (via AMF), data service management (via SMF), and AF. 5G applies the same principles for the separate UP and CP managed by the UP (via UPF) as in 4G.

In the LBO architecture of 5G, the PCF residing in the VPLMN is able to interact with the AF to generate PCC rules for the services the VPLMN delivers. In that case, the PCF relies on the locally configured policies based on the roaming agreement between the VPLMN and HPLMN operators. Nevertheless, the PCF of the VPLMN has no access to subscriber policy information from the HPLMN.

In this specific example, Release 16 brings a new SCP component. It can communicate indirectly between NF components and NF services within the VPLMN, within the HPLMN, or within both the VPLMN and the HPLMN.

TS 23.501 presents various scenarios for 5G for both home networking and roaming such as the roaming architecture in the case of the home-routed scenario, which involves the new NSSAAF component in the HPLMN. In that case, the UPF components in the home-routed scenario can be used also to support an optional Inter PLMN UP Security (IPUPS) functionality. 3GPP TS 33.501 and Section 5.8.2.14 of TS 23.501 specify the IPUPS.

IPUPS functionality is housed at the border of the operators' networks, and protects the network from invalid inter-PLMN *N9* traffic in home-routed roaming scenarios [15]. This solution allows the UPF to terminate GTP-U *N9* tunnels. The UPFs that support the IPUPS in both VPLMN and HPLMN are controlled by the V-SMF and H-SMF of the respective PDU session. In practice, operators could deploy the IPUPS functionality as a separate NF from the UPF. In that case, the IPUPS serves as a transparent proxy that is capable of reading transparently the *N4* and *N9* interfaces.

Please refer to TS 23.501 for more architectural scenarios of the IPUPS.

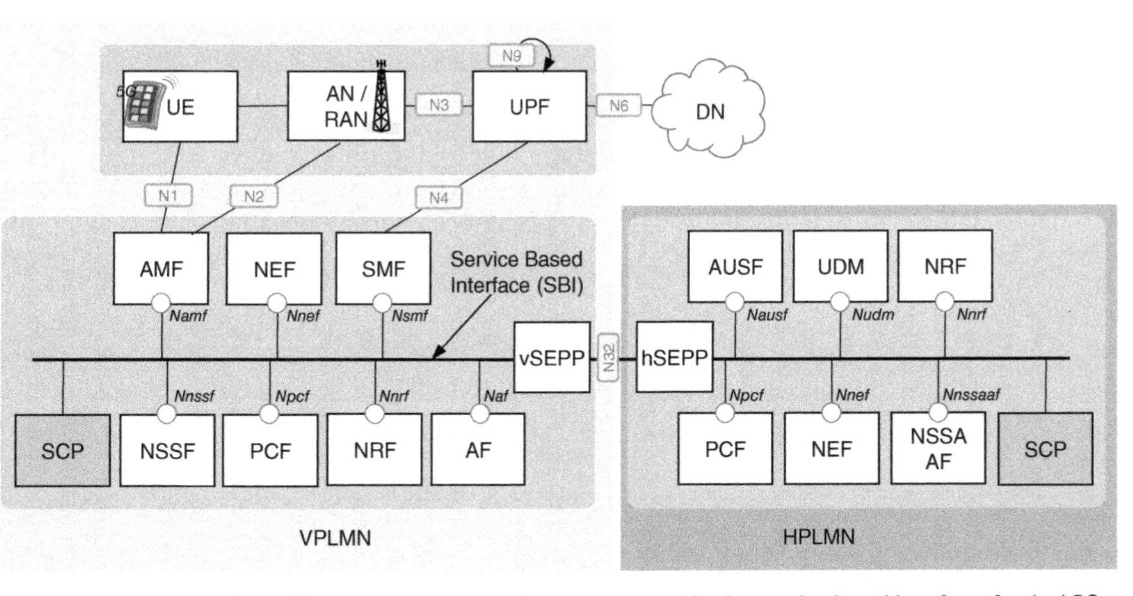

Figure 3.6 An example of the 5G architecture in a roaming case, presented using service-based interfaces for the LBO scenario. The basic roaming scenario is defined in Release 15, whereas Release 16 adds new components such as the highlighted SCP.

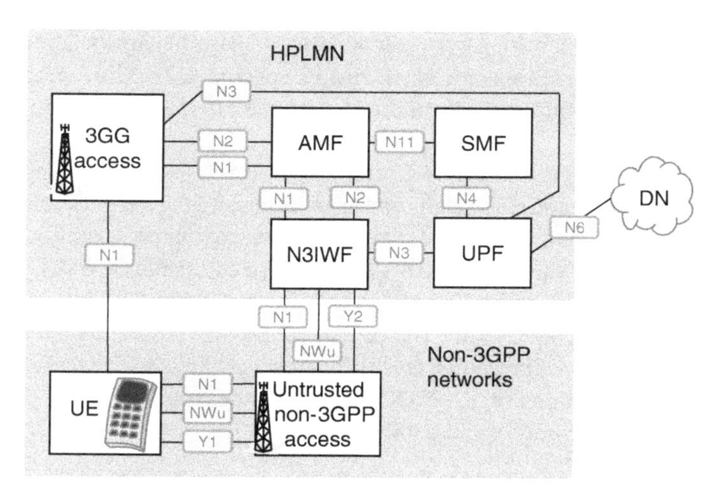

Figure 3.7 5G core network architecture for non-roaming via non-3GPP access. Please note that this figure exposes only direct connections for the ANs.

3.3.4 Interworking with Non-3GPP Networks

As an example of the non-3GPP access architecture scenarios, Figure 3.7 presents the non-roaming architecture for a 5G core network via the non-3GPP AN. The UE can be connected simultaneously to the 5GC via 3GPP and non-3GPP ANs in which case the single AMF is communicating via two *N2* interfaces.

Nevertheless, the two *N3* instances presented in Figure 3.7 can apply to a single or different UPF elements when separate PDU sessions are active based on 3GPP and non-3GPP ANs.

There is also a 5G architecture scenario for interworking with the 4G core network, i.e., EPC in a non-roaming environment depicted in Figure 3.8.

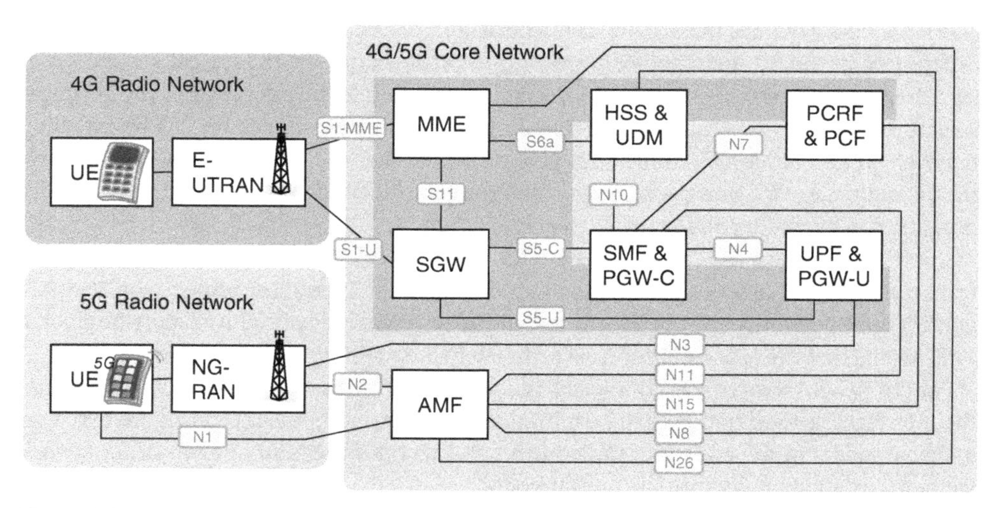

Figure 3.8 Non-roaming architecture for interworking between 5G system (5G NR and 5GC) and 4G system (EPC and E-UTRAN). The 4G-specific elements are highlighted by a darker gray color.

In Figure 3.8, the interworking between 4G and 5G core networks takes place via the optional *N26* interface via 4G-specific MME and 5G-specific AMF. The *N26* interface is able to support the essential functionalities of the 4G-specific *S10* interface for making the interworking possible.

The PCF and PCRF, SMF, and P-GW-C, and UPF and P-GW-U are optional elements for interworking between 4G and 5G systems and are able to serve the networks and UE based on their supported set of capabilities. The UE that are not able to support 4G and 5G interworking are served by elements of the respective 4G or 5G native components, which are P-GW and PCRF for 4G, and SMF, UPF, and PCF for 5G. As dictated by 3GPP TS 33.501, another UPF can be deployed between the NG-RAN and the UPF and P-GW-U via additional interface *N9*.

In addition to the figures presented in this chapter, 5G architectural scenarios contain various other cases as defined in 3GPP TS 23.501 and other relevant technical 3GPP specifications. These cases include, e.g., the following:

- Home-routed roaming architecture for non-3GPP accesses, N3IWF in the same PLMN as the 3GPP access;
- Home-routed roaming architecture for non-3GPP accesses, N3IWF in a different PLMN from the 3GPP access;
- Interworking between 5GC via non-3GPP access and E-UTRAN connected to EPC;
- LBO roaming architecture for non-3GPP accesses, N3IWF in a different PLMN from the 3GPP access;
- LBO roaming architecture for non-3GPP accesses, N3IWF in the same PLMN as the 3GPP access;
- Network analytics architecture.

Please refer to 3GPP TS 23.501 for more details on these scenarios.

3.3.5 5G User and Control Plane

3GPP TS 38.401 defines the 5G NR architecture and respective interfaces. One of the main differences in the 5G architecture compared to previous ones is the logical separation of signaling and data transport networks. In fact, in addition to the cases within the 5G infrastructure itself, this separation of the signaling and user data connections can be extended to cover 5G and other ANs of previous generations so that the user data could go through the 5G radio network, and the signaling load could be diverted via the 4G radio. This functionality can be understood as an extension to CA.

Another key aspect in 5G is that the 5G RAN and 5G core NFs are completely separated from the transport functions. Thus, as an example, the addressing scheme of the 5G RAN and 5G core networks are not tied to the addressing schemes of the transport functions. This means that the NG-RAN controls completely the mobility for an RRC connection.

For the NG-RAN interfaces, the specifications provide a fluent functional division across the interfaces as their options can be minimized. Furthermore, the interfaces are based on a logical model of the entity controlled through this interface, and one physical network element can implement multiple logical nodes.

This division of the user data and signaling traffic also means that the *Uu* and *NG* interfaces include both UP and CP protocols. The UP protocols implement the PDU session service carrying user data via the Access Stratum (AS), while the CP protocols control the

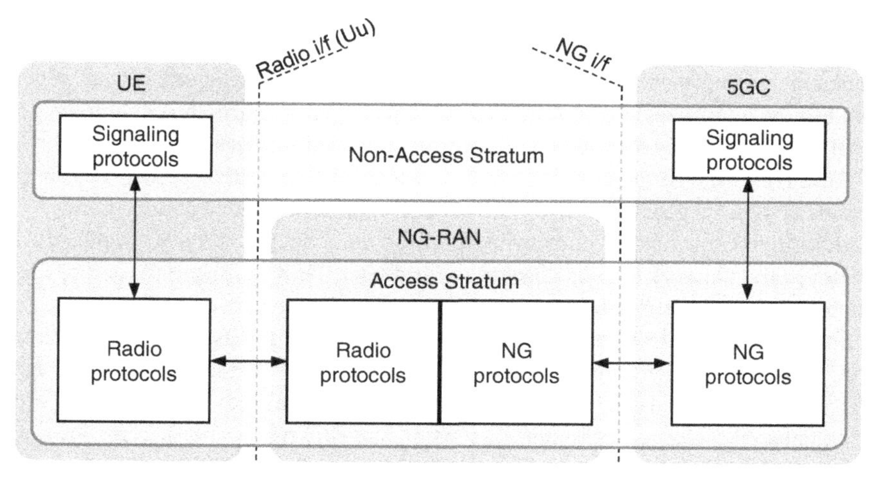

Figure 3.9 The principle of the 5G NG and *Uu* UP [16].

PDU sessions and the connection between the UE and the network, such as service requests, transmission resource control, and handover [16].

The 5G UP is designed for transferring data via the PDU session resource service from one Service Access Point (SAS) to another [16]. The high-level involvement of 5G protocols is depicted in Figure 3.9. The respective RAN and core network protocols provide the PDU session resource service. The 5G radio protocols are defined in the 3GPP TS 38.2*xx* and 38.3*xx* series, whereas the 5G core protocols are found in the TS 38.41*x* series.

Figure 3.10 depicts the high-level idea of the CP, i.e., the principle of the division of the signaling protocol stacks of the 5G radio and core networks. For the CP, the protocols are defined in the same TS series 38.2*xx*, 38.3*xx*, and 38.41*x* as for the UP. The radio and NG protocols contain a mechanism to transparently transfer NAS messages. The CP protocols shown in Figure 3.10 are presented as examples, where CM refers to Connection Management and SM to Session Management.

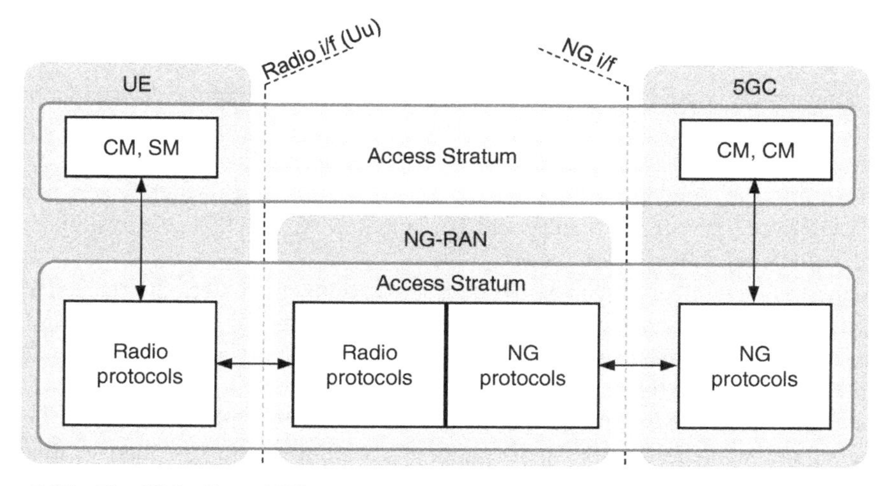

Figure 3.10 The CP for *Uu* and NG.

3.3.6 Edge Computing

5G relies increasingly on the cloud concept. Clouds enable intelligent service awareness, which, in turn, optimizes connectivity, latency, and other performance characteristics. 5G networks are also highly scalable and they offer advanced services thanks to the virtualization of the NFs. One of the consequences of this evolution is that 5G will rely largely on data centers. The 5G ecosystem has centralized, regional, and edge data centers.

In addition to centralized clouds residing in the main data centers, 5G applications may be located at the network's edge, closer to the end-user. These applications can be hosted in the mobile edge computing nodes, referred to as cloudlets.

5G thus requires high-capacity cloud processing power to fulfill its demanding performance criteria. As 5G deployments evolve, there will be the need for further data storage, increased server capacity, additional cooling for the equipment, and more space for housing the respective racks. Oftentimes, the same data center will serve 5G as well as many other systems. Preparedness can be seen already in practice as data centers are being deployed actively over wide areas.

Data centers will be changing from the previous decentralized mobile network model to better serve the centralized processing of the 5G network functionalities. This means that instead of using SA equipment for each function of the network, 5G will rely on cloud-based solutions both in main data centers and edge regions.

The 5G NFs run thus on virtualized software environments instead of SA hardware elements. This principle could potentially evolve further so that a major part of base station processing can take place in the cloud, whereas the base station could occasionally process more, e.g., when cloud's own load threshold reaches a predefined limit. Please refer to Chapter 5 for more details on the respective options for the functional RAN division.

The edge cloud can bring both the contents and data processing closer to the end-users. One of the clearest benefits of the edge cloud is the reduced latency. The closer to the mobile device the contents can be located, the lower the latency, thanks to the shorter transmission path. The edge can be located either in the cloud core network or cloud RAN, or even at the base station site.

Another benefit of the cloud edge is that part of the data processing of the device such as a smartphone can be offloaded to it. 5G will make this scenario reasonable in many cases thanks to the high data speeds and low latency. A practical example of this is an AR/VR device, which typically requires heavy processing power to provide a fluent user experience.

5G connectivity could be used for the AR/VR device embedding the device in a headset display. The challenge is that the mobile phone is not capable of processing data as efficiently as the SA hardware is. To overcome this issue, processing of the contents could be offloaded into the cloud edge while the 5G device serves the user merely as a relatively simple interface between the display and 5G connectivity, which is able to support high-speed, low-latency transmission modes.

The increasing number of such demanding applications could take advantage of 5G connectivity and cloud computing. At the same time, the terminal cost could be optimized because it would not need so much processing power, as long as it supports eMBB and URLLC connectivity to the edge cloud.

Business-wise, edge computing and data centers open new opportunities for both established and new stakeholders. Some operators may want to deploy and manage their edge infrastructure, but it could be equally feasible to buy or lease cloud processing and capacity based on the cloud-as-a-service model.

Virtualization of the 5G architecture opens this opportunity on a completely new level compared to earlier generations. The older networks are more isolated environments where operators take care of their own SA network components, i.e., machines with dedicated hardware and software running on top, whereas the functions are instance based and hardware agnostic in 5G.

The infrastructure model can also be hybrid consisting of owned and purchased services depending on the deployment phase, area of operation, and other factors. This mode makes it possible to optimize the Return on Investment (RoI). Especially for smaller 5G operators, it might not make sense to build a large data center at least at the beginning of the deployment.

Outsourcing cloud-based 5G NFs and other tasks generates new business models. There might also be a need to take into account the assurance of service availability. There is typically a Service Level Agreement (SLA) negotiated between an operator and cloud service provider to set the expectations for service up-time. In the critical environment, it can be the so-called "five nines," i.e., the service is guaranteed to stay up 99.999% of the time. This may require an active/active pairing of parallel servers in geo-redundant configuration, which provides the highest assurance level, but with an increased cost. For less critical applications, less expensive passive/passive configuration and non-geo-redundancy may be an adequate solution.

For more information on cloud computing, please refer to ITU-T Y.3515 [18].

3.4 Dual Connectivity

The 5G system allows DC within the native 5G network, or relies on MR access technologies as defined in Release 16 3GPP TS 37.340 V16.2.0. This states that MR-DC is a generalization of the intra-E-UTRA DC that is described in 3GPP TS 36.300. This allows multiple Rx/Tx-capable UE to utilize resources of two separate nodes so that one provides NR access whereas the other one provides either E-UTRA or NR access.

In this scenario, one node assumes a role of Master Node (MN), whereas the other acts as the Secondary Node (SN). There is interconnection for MN and SN, or both MN and SN connect to the core network. The MN and SN may use also a shared spectrum for access, and DC is applicable also for the Mobile Terminating Integrated Access and Backhaul (IAB-MT) use case relying on the UE functions. Nevertheless, the backhauling traffic over the E-UTRA is not supported in the EN-DC scenario.

Along with 5G, the "base station" can be referred to as gNB, ng-eNB, or en-gNB. Furthermore, the en-gNB houses MeNB, SgNB, and EN-DC. Table 3.2 clarifies these terms.

3.4.1 Multi-radio Dual Connectivity with EPC

E-UTRAN supports MR-DC via EN-DC. In this scenario, the UE connects to an eNB that acts as an MN and to one en-gNB that acts as an SN. The eNB connects to the EPC via the *S1* interface and to the en-gNB via the *X2* interface. The en-gNB can also connect to the EPC via the *S1-U* interface and to other en-gNB elements via the *X2-U* interface. Figure 3.11 depicts the EN-DC architecture.

Table 3.2 Terminology for 5G NodeB variants as per Release 16 3GPP TS 38.300 V16.2.0 [17] and TS 37.340 V16.2.0 [19].

5G NodeB type	Description
gNB	5G Next Generation NodeB is a logical 5G radio node comparable with the 4G eNB of LTE, and 3G NB of UMTS/HSPA. In this scenario, the UE uses 5G *Uu* to connect to gNB, and the gNB uses the NG interface to connect to 5GC.
ng-eNB	Intermediate 5G Next Generation NodeB is an enhanced 4G eNodeB. It connects to the 5C core network via the NG interfaces using 4G LTE radio interface to connect the 5G UE. In this scenario, the UE uses 4G *Uu* to connect to ng-eNB, and the ng-eNB uses the NG interface to connect to 5GC. The gNB and ng-eNB interconnect via the *Xn* interface. The ng-eNB facilitates use of the already existing 4G radio network, and it can thus coexist with the gNBs. The ng-eNB can serve 5G UE in the outage area of 5G.
en-gNB	Intermediate 5G-RAN node for the EN-DC option. This option provides fast intermediate deployment to connect 5G gNB access to the 4G LTE core network. The 4G eNB acts as an MN (or MeNB) that is in control of the radio connection with the UE and the en-gNB is used as an SN (or SgNB). In this architecture we can provide very high bit rates to the UE that support DC without too much impact on the core infrastructure.
MN (MeNB)	4G eNB serves as an MN (or MeNB) in the EN-DC option. The MeNB is in control of the radio connectivity.
SN (SgNB)	en-gNB is used as an SN (or SgNB) in the EN-DC option.

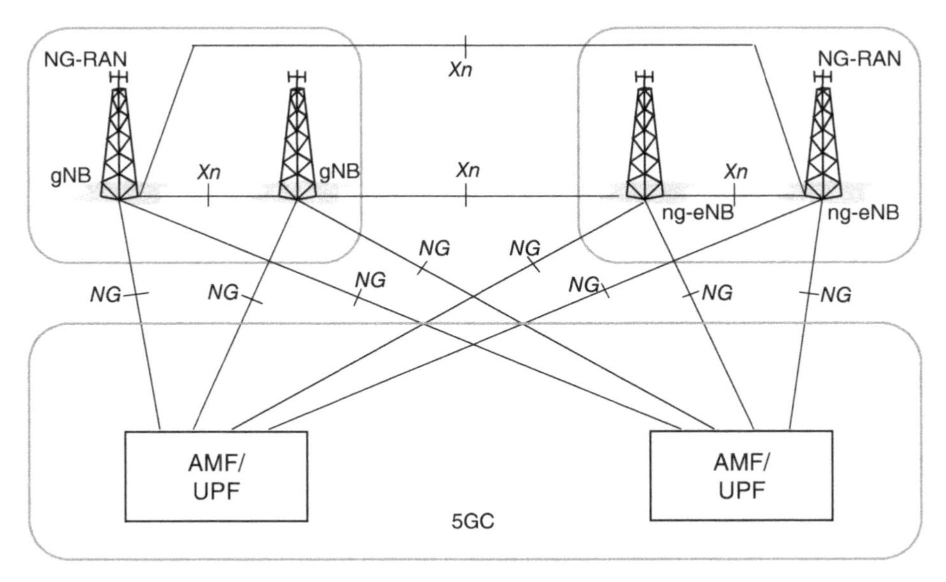

Figure 3.11 The high-level EN-DC architecture.

3.4.2 Multi-radio Dual Connectivity with 5GC

Figure 3.12 depicts the scenario of a device that is simultaneously connected to multiple NG-RAN nodes. The MR-DC involving 5GC has three scenarios: EN-DC, NR-E-UTRA DC, and NR-NR DC [19].

- In EN-DC, NG-RAN supports NG-RAN EN-DC (NGEN-DC, sometimes referred to as NE-DC). In this scenario, the UE connects to an ng-eNB that is an MN (sometimes referred to as MeNB) and to a gNB that is an SN (sometimes referred to as SgNB). The ng-eNB connects to the 5GC, while the gNB connects to the ng-eNB via the *Xn* interface.
- In EN-DC, NG-RAN supports NE-DC. In this scenario, the UE connects to a gNB that is an MN and to an ng-eNB that is an SN. The gNB connects to 5GC and the ng-eNB connects to the gNB via the *Xn* interface.
- NG-RAN supports NR-NR DC (sometimes referred to as NR-DC). In this scenario, the UE connects to a gNB that is an MN and yet to another gNB that is an SN. Furthermore, the NR-DC is also applicable for a UE connected to two gNB-DUs so that one is serving the MCG and the other is serving the SCG. The gNB connects to 5GC and the gNB elements interconnect via the *Xn* interface.

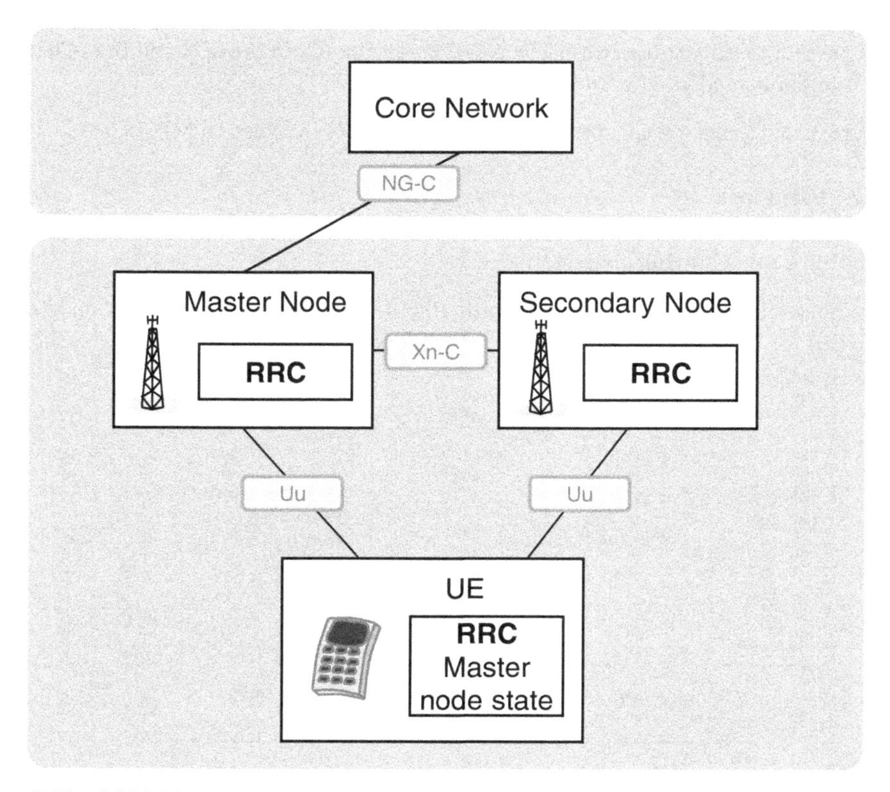

Figure 3.12 5G Multi-radio access technology DC.

Release 15 defines two cases for the DC, i.e., NGEN-DC and NE-DC, whereas Release 16 brings the third option, NR-DC. For more details on DC in Release 16, please refer to the latest release of 3GPP TS 37.340 version 16.

3.4.3 Dual Connectivity Network Interfaces

Section 4.3 of 3GPP TS 37.340 defines CP and UP interfaces for DC [19].

3.4.3.1 Control Plane

Multi-radio access technology DC requires an interface between the MN and the SN for CP signaling and coordination. For each MR-DC UE, there is also one CP connection between the MN and a corresponding core network entity. The MN and SN involved in MR-DC for a certain UE control their radio resources and are primarily responsible for allocating radio resources of their cells.

DC signaling involves two scenarios:

- **MR-DC with EPC**: In MR-DC with EPC (EN-DC), the involved core network entity is the MME. The *S1-MME* is terminated in MN and the MN and SN are interconnected via the *X2-C*.
- **MR-DC with 5GC**: In MR-DC with 5GC (NGEN-DC, NE-DC, and NR-DC), the involved core network entity is the AMF. The *NG-C* is terminated in the MN and the MN and SN are interconnected via the *Xn-C*.

Figure 3.13 depicts the CP connectivity of MN and SN involved in MR-DC.

3.4.3.2 User Plane

Figure 3.14 depicts UP connectivity options of the MN and SN involved in MR-DC. These options have the following characteristics:

- For the scenario involving MR-DC with EPC (EN-DC), the UP is based on the *X2-U* between the MN and SN, whereas the *S1-U* serves the UP between the MN, the SN or both and the S-GW.

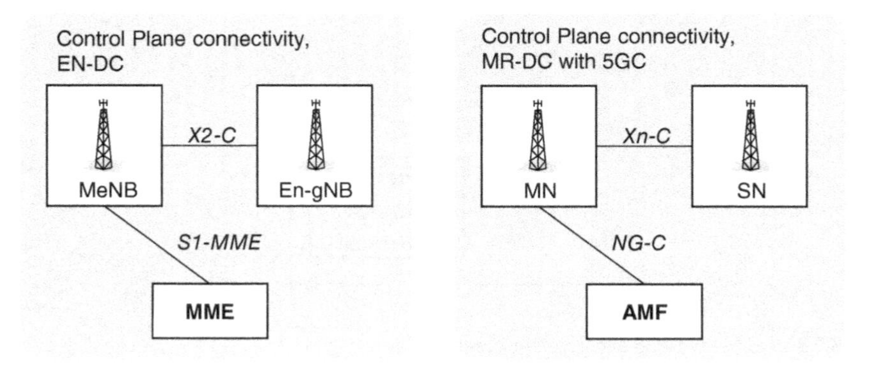

Figure 3.13 CP connectivity for EN-DC and MR-DC with 5GC.

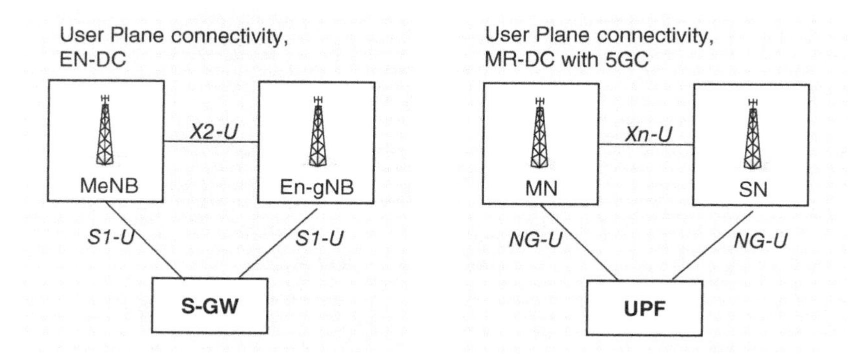

Figure 3.14 UP connectivity for EN-DC and MR-DC with 5GC.

- For the scenario involving MR-DC with 5GC (NGEN-DC, NE-DC, and inter-gNB NR-DC), the *Xn-U* serves the UP between the MN and SN, whereas the *NG-U* serves the UP between the MN, the SN or both and the UPF.

UP connectivity depends on the bearer configuration. For MN-terminated bearers, UP connection to the core network entity is terminated in the MN, whereas for SN-terminated bearers, the UP connection to the core network entity is terminated in the SN.

The transport of UP data over the *Uu* involves MCG or SCG radio resources or both. In this scenario, for MCG bearers, only MCG radio resources are involved, whereas for SCG bearers, only SCG radio resources are involved. For split bearers, both MCG and SCG radio resources are involved.

3.5 NG-RAN Architecture

3.5.1 Interfaces

The 5G network architecture can be described by using reference point representation or service-based interface representation. Figure 3.15 depicts the overall architecture and interfaces of the 3GPP 5G network in the traditional reference point format. As can be seen in this figure, the new terminology for the 5G core network is 5GC, whereas the NR system is referred to as NG-RAN.

The logical interfaces of the 5G architecture are referred to as *NG*, *Xn*, and *F1*. A complete description of these interfaces can be found in the following specifications:

- 3GPP TS 38.410 defines the general aspects and principles of the *NG* interface;
- 3GPP TS 38.420 defines the general aspects and principles of the *Xn* interface;
- 3GPP TS 38.470 defines the general aspects and principles of the *F1* interface.

The 5GC includes the same type of network elements as defined in the EPC of the LTE prior to Release 15, i.e., MME for signaling and S-GW, P-GW, and supporting elements for the policy rules, added by various functional 5G elements.

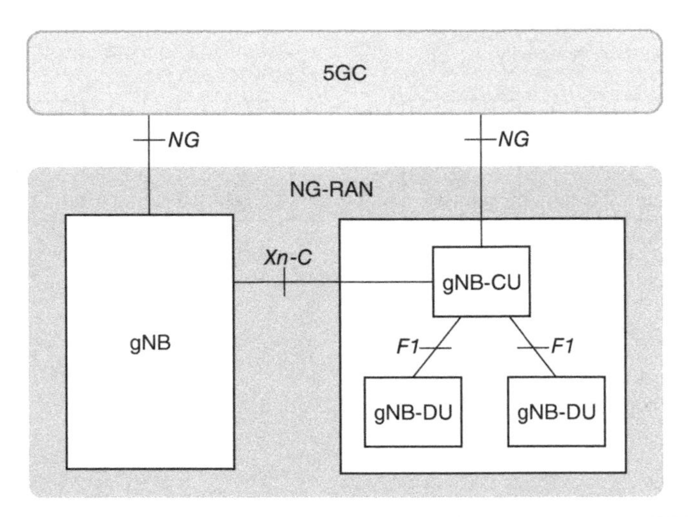

Figure 3.15 The overall 5G architecture in Releases 15 and 16 as defined in TS 38.401, V16.2.0 [16].

For the 5G radio system, the base station elements, cells, are referred to as gNB. The *NG* interface interconnects the gNB elements of the NG-RAN and the 5GC. The *Xn* network interface is used between NG-RAN nodes.

In 5G, the gNB elements can support Frequency Division Duplex (FDD) and Time Division Duplex (TDD) modes as well as dual mode operation. The gNB elements are interconnected via the *Xn-C* interface.

The NG-RAN is defined via the Radio Network Layer (RNL) and Transport Network Layer (TNL). The NG-RAN interfaces are defined as part of the RNL, whereas the TNL provides services for UP transport and signaling transport. TNL protocol and functionality are specified for the NG-RAN interfaces.

Each gNB may consist of a single gNB-CU and a set of gNB-DU. The connection within a gNB, between the gNB-CU and gNB-DU, takes place via the *F1* interface either so that a single gNB-DU is connected to only one gNB-CU, or, depending on the implementation, a single gNB-DU may also be connected to multiple gNB-CU elements. The connections from outside the gNB element, from other gNB elements, and from the 5GC terminate always at the gNB-CU element.

Also, in the case of the EN-DC, the *S1-U* and *X2-C* interfaces for a gNB that has the gNB-CU and gNB-DU elements, terminate in the gNB-CU element. Nevertheless, the set of gNB-CU and one or more gNB-DU elements per gNB are only visible to other gNB elements and the 5GC as a gNB. Furthermore, each gNB is connected to all AMF elements within an AMF region in an NG-Flex configuration. The AMF region is defined in 3GPP TS 23.501 [15]. It consists of one or more AMF sets. The AMF set, in turn, consists of (some) AMF elements that serve a given area and network slice. Multiple AMF sets may be defined per AMF region and network slice or slices.

3.5.2 Functions of gNB and ng-eNB

The functions of the native 5G gNB and evolved 4G ng-eNB supporting connectivity to the 5G system are defined in 3GPP TS 38.300. It states that both elements host the following functions:

- Connection setup and release;
- Distribution function for NAS messages;
- DC;
- Functionality related to network slicing;
- IP header compression, data encryption, and integrity protection;
- Measurement and respective reporting configuration to be used for the tasks related to mobility and scheduling;
- QoS flow management and mapping to data radio bearers;
- RAN sharing;
- Radio Resource Management (RRM). This function includes radio bearer and admission control, connection mobility control, as well as scheduling, which refers to the dynamic allocation of UL and DL resources for devices;
- Routing of UP data towards UPF and CP information towards AMF;
- Scheduling and transmission of paging messages and system broadcast information provided by the AMF or Operations Administration and Maintenance (OAM);
- Selection of an AMF during the UE attachment procedure in such a case when no routing to an AMF can be determined from the information provided by the UE;
- Session Management (SM);
- Support of devices in the RRC_INACTIVE state;
- Tight interworking between NR and E-UTRA;
- UL packet marking at the UL transport level.

3.6 5G Interfaces and Reference Points

3.6.1 Service-Based Interfaces

As interpreted from TS 23.501 V16.5.2, Table 3.3 summarizes the Release 16 service-based interfaces of the 5G system.

3.6.2 Reference Points

As interpreted from TS 23.501 V16.5.2, Table 3.4 summarizes the Release 16 reference points.

In addition to the reference points summarized in Table 3.4 and presented in 3GPP TS 23.501, there are other interfaces and reference points, such as between the SMF and the CHF. Furthermore, 3GPP TS 23.501 presents the reference points of the SMS over the NAS in Section 4.4.2.2, and 3GPP TS 23.273 presents the location services-related reference points.

Table 3.3 Release 16 service-based interfaces of the 5G system. Please note that 3GPP TS 32.240 defines the service-based interface for the CHF [3].

Interface	Description
Namf	Service-based interface exhibited by AMF
Nsmf	Service-based interface exhibited by SMF
Nnef	Service-based interface exhibited by NEF
Npcf	Service-based interface exhibited by PCF
Nudm	Service-based interface exhibited by UDM
Naf	Service-based interface exhibited by AF
Nnrf	Service-based interface exhibited by NRF
Nnssaaf	Service-based interface exhibited by NSSAAF
Nnssf	Service-based interface exhibited by NSSF
Nausf	Service-based interface exhibited by AUSF
Nudr	Service-based interface exhibited by UDR
Nudsf	Service-based interface exhibited by UDSF
N5g-eir	Service-based interface exhibited by 5G-EIR
Nnwdaf	Service-based interface exhibited by NWDAF
Nchf	Service-based interface exhibited by CHF
Nucmf	Service-based interface exhibited by UCMF

Table 3.4 Release 16 service-based reference points of the 5G system.

Reference point	Description	Note/reference
N1	Reference point between the UE and the AMF	
N2	Reference point between the AN or RAN and the AMF	
N3	Reference point between the AN or RAN and the UPF	
N4	Reference point between the SMF and the UPF	
N5	Reference point between the PCF and an AF	TS 23.228
N6	Reference point between the UPF and a DN	
N7	Reference point between the SMF and the PCF	
N8	Reference point between the UDM and the AMF	
N9	Reference point between two UPFs	
N10	Reference point between the UDM and the SMF	
N11	Reference point between the AMF and the SMF	
N12	Reference point between the AMF and the AUSF	
N13	Reference point between the UDM and the AUSF	
N14	Reference point between two AMFs	
N15	Reference point between the PCF and the AMF in the case of a non-roaming scenario, PCF in the visited network and the AMF in the case of a roaming scenario	

Table 3.4 (*Continued*)

Reference point	Description	Note/reference
N16	Reference point between two SMFs within the 5G network or between the SMF in the visited network and the SMF in the home network	
N16a	Reference point between the SMF and the I-SMF	
N17	Reference point between the AMF and the 5G-EIR	
N18	Reference point between any NF and the UDSF	
N19	Reference point between two PSA UPFs for 5G LAN-type service	
N22	Reference point between the AMF and the NSSF	
N23	Reference point between the PCF and the NWDAF	
N24	Reference point between the PCF in the visited network and the PCF in the home network	
N27	Reference point between the NRF in the visited network and the NRF in the home network	
N28	Reference point between the PCF and the CHF	TS 23.503
N29	Reference point between the NEF and the SMF	TS 23.503
N30	Reference point between the PCF and the NEF	TS 23.503
N31	Reference point between the NSSF in the visited network and the NSSF in the home network	TS 32.240
N32	Reference point between the SEPP in the visited network and the SEPP in the home network	TS 33.501
N33	Reference point between the NEF and the AF	
N34	Reference point between the NSSF and the NWDAF	
N35	Reference point between the UDM and the UDR	
N36	Reference point between the PCF and the UDR	
N37	Reference point between the NEF and the UDR	
N38	Reference point between I-SMFs	
N40	Reference point between the SMF and the CHF (please note that N40–N49 are reserved for future definition in TS 23.503)	TS 23.503
N50	Reference point between the AMF and the CBCF, related to a public warning system	TS 23.041
N51	Reference point between the AMF and the NEF	
N52	Reference point between the NEF and the UDM	
N55	Reference point between the AMF and the UCMF	
N56	Reference point between the NEF and the UCMF	
N57	Reference point between the AF and the UCMF	
N58	Reference point between the AMF and the NSSAAF	
N59	Reference point between the UDM and the NSSAAF	
N70	Reference point to support the SBA in the IMS (together with N5 and N71)	TS 23.228
N71	Reference point to support the SBA in the IMS (together with N5 and N70)	TS 23.228

3.7 IMS in 5G

As 3GPP TS 23.501 states, IMS support for 5GC is defined in 3GPP TS 23.228. To enable the IMS service, the 5G system architecture includes *N5* and *Rx* interfaces between the PCF and Proxy Call Session Control Function (P-CSCF) as defined in 3GPP TS 23.228, TS 23.503, and TS 23.203. *N5* is the native 5G interface, whereas *Rx* provides backwards compatibility for diameter, which can be used in early 5G deployments between IMS and 5GC functions.

The P-CSCF executes the functions of a trusted AF in the 5GC when PCF and P-CSCF use service-based interfaces within the same PLMN.

3GPP TS 23.501, Chapter 5.16.3, contains overall aspects of the 5G IMS, and TS 23.228 presents IP connectivity AN-specific concepts when using 5GS to access the IMS. Based on these sources, 5GS supports the IMS with the following aspects:

- Network informs the UE about IMS voice over PS session support;
- System supports domain selection for UE originating sessions;
- System supports NRF-based HSS discovery;
- System supports NRF-based P-CSCF discovery without the need for UE to know about P-CSCF IP address discovery;
- System supports the P-CSCF restoration procedure;
- System supports terminating domain selection for IMS voice;
- The IMS has capability to transport the P-CSCF addresses to the UE;
- TS 23.167 defines the IMS emergency service;
- TS 23.228 defines paging policy differentiation for the IMS.

3GPP TS 23.501 and TS 23.502 present further details about the IMS-related procedures, such as registration over 3GPP access, and aspects related to, e.g., policy, UE capabilities, HPLMN, IP address preservation, NG-RAN to UTRAN SRVCC support, extended NG-RAN coverage, and the voice support match indicator from the NG-RAN.

The 5G system architecture supports the *N5* interface between the PCF and P-CSCF. It also supports the *Rx* interface between the PCF and P-CSCF to enable the IMS service as per 3GPP TS 23.228, TS 23.503, and TS 23.203. Please note that *Rx* support between the PCF and P-CSCF is for backwards compatibility for early deployments using diameter between the IMS and 5GC functions. In addition, when service-based interfaces are present between the PCF and P-CSCF in the same PLMN, the P-CSCF performs the functions of a trusted AF in the 5GC.

Annex AA of 3GPP TS 23.228 describes support for SBA for IMS nodes in conjunction with the 5GC. Figure 3.16 shows the architecture to support SBA interactions between IMS entities, and Figure 3.17 shows the architecture using reference point representation.

Table 3.5 summarizes the reference points and service-based interfaces to support the IMS.

The above-mentioned service-based interface services provide equivalent functionality to *Rx* and *Cx*/*Sh* reference points of the diameter. In practice, the service-based interface-enabled IMS nodes may have both service-based and non-service-based interfaces to support the coexistence of IMS nodes that support SBA services and IMS nodes that do not support SBA services.

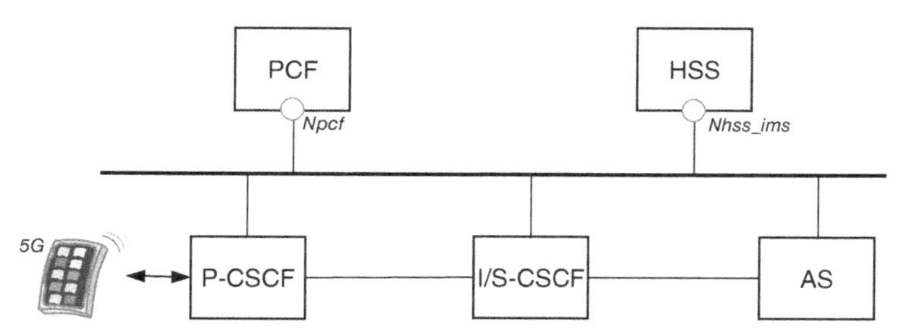

Figure 3.16 System architecture to support SBA in the IMS.

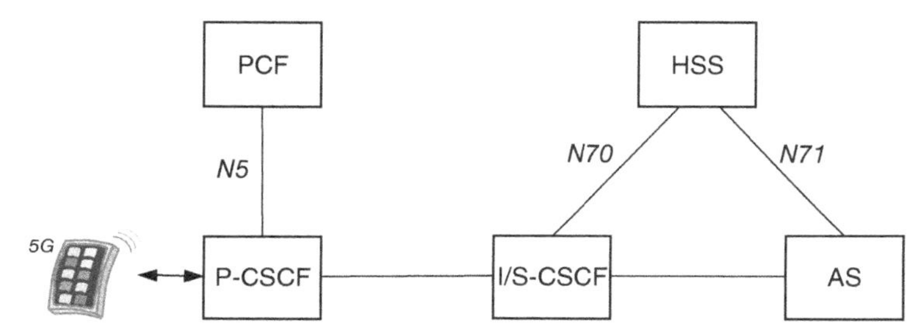

Figure 3.17 System architecture to support SBA in the IMS in reference point representation.

Interpreting 3GPP TS 23.228, Figure 3.18 depicts the architecture to support IMS services for roaming scenarios, including voice over IMS in deployments without IMS-level roaming interfaces. In this scenario, the home network's P-CSCF identifies the serving VPLMN where the UE is located. This takes place by using the procedures of 3GPP TS 23.228. For the IMS services with roaming-level interfaces, the P-CSCF and UPF are located in the VPLMN as per the LBO model; please see Sections 4.2.3 and M.1 of the specification for more details on this option.

For the roaming architecture involving voice over IMS with LBO, Section 4.15a of the same specification details the procedure, and for the special case of a loopback, please refer

Table 3.5 Reference points and SBA interfaces for IMS support.

Interface	Description
N5	Reference point between the PCF and the AF. The P-CSCF acts as an AF from the perspective of the PCF. 3GPP TS 23.501 defines the *N5* reference point.
N70	Reference point between a service-based interface-capable I/S-CSCF and a service-based interface-capable HSS
N71	Reference point between a service-based interface-capable IMS AS and a service-based interface-capable HSS
Npcf	Service-based interface exhibited by the PCF
Nhss	Service-based interface exhibited by a service-based interface-capable HSS

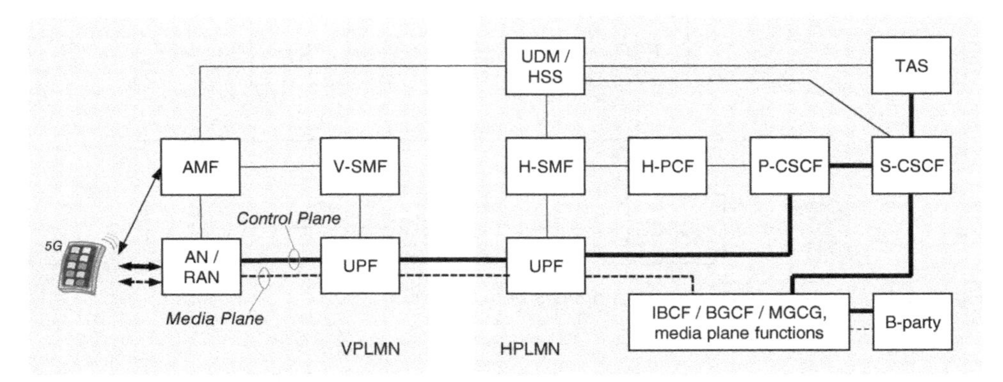

Figure 3.18 The IMS call scenario in visited and home 5G network use cases as interpreted from 3GPP TS 23.228, Y.9.2.

to Section M.3 for more details. Further details can also be found in 3GPP TS 23.501, Section 4.4.3, and 3GPP TS 23.502, Section 4.13.6, as well as from the white paper of the GSMA titled "Road to 5G: Introduction and Migration."

For more detailed functioning of the IMS, VoLTE, and VoNR, please see Chapter 5.

References

1 3GPP, "Release 16," 16 July 2018. [Online]. Available: https://www.3gpp.org/release-16.
2 3GPP, "RAN Adjusts Schedule for 2nd Wave of 5G Specifications," 14 December 2018. [Online]. Available: https://www.3gpp.org/news-events/2005-ran_r16_schedule.
3 3GPP, "TS 32.240 V16.1.0, Telecommunication Management, Charging Management, Charging Architecture and Principles, Release 16," 3GPP, December 2019.
4 3GPP, "TS 23.501 V16.5.2, System Architecture for the 5G System (5GS), Stage 2," 3GPP, August 2020.
5 3GPP, "3GPP TR 37.876 V15.0.0, Study on eNB(s) Architecture Evolution for E-UTRAN and NG-RAN," 3GPP, 2018.
6 3GPP, "TS 32.298 V16.5.0, Telecommunication Management, Charging Management, Charging Data Record (CDR) Parameter Description, Release 16," 3GPP, July 2020.
7 3GPP, "3GPP TS 32.291 V16.4.2, Telecommunication Management, Charging Management, 5G System, Charging Service, Stage 3, Release 16," 3GPP, July 2020.
8 Broadband Forum, "MR-427, 5G Fixed-Mobile Convergence," Broadband Forum, July 2018.
9 Wilke, J., "5G Network Architecture and FMC Joe Wilke," ITU, 2017.
10 3GPP, "Control and User Plane Separation of EPC Nodes (CUPS)," 3GPP, 3 July 2017. [Online]. Available: http://www.3gpp.org/cups. [Accessed 29 July 2018].
11 "What Is REST?" [Online]. Available: https://restfulapi.net. [Accessed 7 July 2019].
12 Oracle, "Ava Telephony API (JTAPI)," Oracle. [Online]. Available: https://www.oracle.com/technetwork/java/jtapi-136088.html?printOnly=1. [Accessed 7 July 2019].

13 ITU-T, "ITU-T Focus Group IMT-2020 Deliverables," 2017. [Online]. Available: https://www.itu.int/dms_pub/itu-t/opb/tut/T-TUT-IMT-2017-2020-PDF-E.pdf.
14 5GXcast, "Broadcast and Multicast Communication Enablers for the Fifth-Generation of Wireless Systems," EU, 22 July 2019. [Online]. Available: http://5g-xcast.eu/wp-content/uploads/2019/08/5G-Xcast_D2.4_v2.0_web.pdf. [Accessed 28 September 2020].
15 3GPP, "TS 23.501. System Architecture for the 5G System, Release 15, V. 15.1.0," 3GPP, 2018.
16 3GPP, "TS 38.401, V16.2.0, NG-RAN Architecture Description," 3GPP, June 2020.
17 3GPP, "TS 38.300 V16.2.0, NR; NR and NG-RAN Overall Description, Stage 2," 3GPP, July 2020.
18 ITU, "Y.3515: Cloud Computing – Functional Architecture of Network as a Service," ITU, July 2017.
19 3GPP, "TS 37.340 V16.2.1, Evolved Universal Terrestrial Radio Access (E-UTRA) and NR, Multi-Connectivity, Stage 2, Release 16," 3GPP, July 2020.

4

Phase 2 Radio Network and User Equipment

4.1 Overview

4.1.1 Key Specifications

This chapter refers to a multitude of 3rd Generation Partnership Project (3GPP) Technical Specifications (TS) and Technical Reports (TR) on New Radio (NR) access. The 3GPP defines 5G radio technology in the 38-series, under the term NR, whereas the name for the 5G core network is Next Generation Core (NGC). Table 4.1 summarizes the fundamental technical specifications that form 5G radio. A complete list of the 38-series is available on the Web in Ref. [1].

For additional frequency bands and their respective bandwidth variants that are specified for 5G, some of the respective key 3GPP TR are summarized in Table 4.2.

4.1.2 Summary of Key Release 16 Enhancements

The key enhancements to the 5G radio network now include the following aspects [2]: Cellular Vehicle-to-Everything (C-V2X), enhanced Ultra-Reliable Low Latency Communications (eURLLC), Integrated Access and Backhaul (IAB), massive Multiple In, Multiple Out (MIMO), Non-Public Networks (NPNs), positioning, power-saving functionality, Time-Sensitive Networking (TSN), and unlicensed spectrum.

4.1.2.1 Cellular Vehicle-to-Everything

3GPP Release 14 defines a Cellular Vehicle-to-Everything (C-V2X) mode, which can also serve as a base for traffic safety. Release 16 NR continues enhancing it and presents a Sidelink for advanced safety use cases. It is also the foundation for solutions forming autonomous driving. The Sidelink multicast uses Hybrid Automatic Repeat Request (HARQ) feedback. It also provides a means for forming distance-based ad-hoc multicast groups. The benefits of the Sidelink for consumers include optimized traveling and increased energy efficiency.

5G Second Phase Explained: The 3GPP Release 16 Enhancements, First Edition. Jyrki T.J. Penttinen.

Table 4.1 Some of the key technical specifications of the 3GPP NR interface.

TS	Title
23.501	System architecture for the 5G System, Stage 2, Release 16
38.101	User Equipment (UE) radio transmission and reception
38.104	Base Station (BS) radio transmission and reception
38.201	Physical layer; General description
38.211	Physical channels and modulation
38.300	NR overall description (Stage-2)
38.305	NG Radio Access Network (NG-RAN); Stage 2 functional specification of User Equipment (UE) positioning in NG-RAN
38.306	User Equipment (UE) radio access capabilities
38.321	Medium Access Control (MAC) protocol specification
38.322	Radio Link Control (RLC) protocol specification
38.323	Packet Data Convergence Protocol (PDCP) specification
38.331	Radio Resource Control (RRC); Protocol specification
38.401	NG-RAN; Architecture description
38.410	NG-RAN; NG general aspects and principles
38.801	Study on new radio access technology: Radio access architecture and interfaces

Table 4.2 The key 3GPP TR detailing 5G-specific Radio Frequency (RF) bands.

TR	Title
38.812	Study on Non-Orthogonal Multiple Access (NOMA) for NR
38.813	New frequency range for NR (3.3–4.2 GHz)
38.814	New frequency range for NR (4.4–4.99 GHz)
38.815	New frequency range for NR (24.25–29.5 GHz)
38.817-01	General aspects for UE RF for NR
38.817-02	General aspects for BS RF for NR
38.900	Study on channel model for frequency spectrum above 6 GHz
38.901	Study on channel model for frequencies from 0.5 to 100 GHz

4.1.2.2 eURLLC

3GPP Release 15 introduced the basis for the URLLC mode. Release 16 improves further the reliability in terms of availability, which can be as high as 99.999% in Release 16, as well as the latency that can be more often below 10 ms than in Long Term Evolution (LTE). Those aspects make 5G attractive to new verticals such as factory automation as their respective use cases benefit from the evolved connectivity. The two aspects, reliability and latency, have interdependency in terms of the resulting joint performance.

As an example, increasing the number of retransmission increments also increases the reliability but it does have a negative impact on latency value. Release 16 presents Coordinated Multi-Point (CoMP) that uses multi-Transmission and Reception Point (TRP) to create spatial diversity for redundant communication paths.

In addition to the increased redundancy, improved HARQ, and CoMP, the eURLLC of Release 16 is a result of increased interservice multiplexing, intradevice channel prioritization, and increasingly flexible scheduling.

4.1.2.3 Integrated Access and Backhaul

The IAB of 5G Release 16 improves the cost-efficiency of the system by allowing the reutilization of access-network mmWaves for transmission between the 5G NR and core subsystems. As such, the mmWave spectrum provides only small cell coverage for mobile use cases, while the respective transmission on fiber optics can have a big impact on the infrastructure cost, whereas the Pont-to-Point (PTP) wireless transmission on mmWave can serve extended areas, so the extension of the licensed access mmWaves needed to cover the backhaul data transmission provides additional justifications for the investment [3].

Along with Release 16 IAB, 5G Next Generation NodeB (gNB) is capable of handling access and backhaul connectivity simultaneously, offering operators an additional set of highly dynamic deployment options for fast extension of the network.

4.1.2.4 Massive MIMO

MIMO antenna technology has already been a continuous work item in previous 3GPP releases, and Release 16 continues the evolution. Release 16 MIMO enhancements consist of higher rank support of Multi-User MIMO (MU-MIMO), multi-TRP support, power efficiency optimization via Peak-to-Average Power Ratio (PAPR) reference signal improvements, and link reliability enhancement via evolved multibeam management. Release 16 also allows the full power Uplink (UL) transmission of MIMO devices for extending the respective cell-edge coverage.

4.1.2.5 Non-public Network

3GPP Release 16 includes support for the NPN. NPN refers to the private networks that use dedicated, independently manageable resources. Some benefits of the concept include security and privacy assurance as the communication takes place on-site, and ensures low latency for the local end-users. It is possible to deploy the NPN to a variety of locations such as industrial premises to serve the respective use cases.

4.1.2.6 Positioning

5G Release 16 enhances cellular positioning further, providing indoor accuracy of 3 m, while accuracy outdoors stays within 10 m. The new accuracy is a result of positioning techniques within single cells and combining multiple cells. Some examples of the solutions are Roundtrip Time (RTT), Angle of Arrival (AoA), Angle of Departure (AoD), and Time Difference of Arrival (TDOA). Many use cases benefit from the new accuracy, including indoor positioning, public safety, drone trail, and smart manufacturing.

4.1.2.7 Power Saving

3GPP Release 16 further optimizes the power consumption of devices. One of the drivers for this item is the ITU IMT-2020 requirement for long battery duration of Internet of Things (IoT) devices, up to 10 years. Release 16 power-saving features include optimized low-power settings, evolved power control mechanisms, overhead reduction, adaptive MIMO layer power reduction, enhanced cross-slot scheduling, and low-power carrier aggregation control. Release 16 also introduces a Wakeup Signal (WUS) to inform the device about potential pending transmissions. It optimizes the needed terminal's low-power Discontinuous Reception (DRX) monitoring periods reducing overall power consumption.

4.1.2.8 Time-Sensitive Networking

TSN supports the ongoing Industry 4.0 concept and respective use cases. Release 16 introduces support for TSN integration for optimal performance in such time and delay critical environments. Respective solutions include time synchronizing based on the Generalized Precision Timing Protocol (gPTP), TSN configuration information mapping with the 5G Quality of Service (QoS) framework, and header compression for optimized Ethernet frame transport.

4.1.2.9 Unlicensed Spectrum

3GPP Release 16 introduces the NR on Unlicensed (NR-U) spectrum with the goal of helping the ecosystem expand the reachability of 5G. It benefits the existing and new verticals that consider relying on the new generation. The NR-U offers two new operation modes: anchored NR-U and standalone NR-U.

Anchored NR-U combines unlicensed spectra as an add-on with other licensed or shared spectra. The combination of licensed spectra and unlicensed NR-U spectra increases the bandwidth by up to 100 MHz in UL and up to 400 MHz in Downlink (DL).

The standalone NR-U solely uses unlicensed spectra. Examples of the latter are the globally adopted 5 GHz band that has been popular for Wi-Fi and LTE Licensed Assisted Access (LAA), and the new defined band on 6 GHz, which adds UL and DL band capacity by 100 and 400 MHz, respectively, or up to 1.2 GHz in the United States [4].

4.1.2.10 Other Release 16 Enhancements and Additions

Release 16 enhances many existing functions and brings various new functions to comply with the ITU IMT-2020 requirements for the 5G system. In addition, Release 16 brings many LTE Advanced (LTE-A) solutions into the 5G platform. An example of this is Mobile IoT (MIoT) as both Enhanced Machine Type Communication (eMTC) and NB-IoT are now possible to integrate with 5G NR and 5G core networks. Other examples include mobility management, spectrum aggregation, interference management, and device capability signaling improvements, as well as two-step Random Access Channel (RACH), data collection, and Voice over New Radio (VoNR) circuit-switched fallback.

4.2 Radio Network

4.2.1 5G MIMO and Adaptive Antennas

MIMO antenna systems are an integral part of the 5G era. They provide more capacity thanks to the multiple radio paths. There will also be adaptive beamforming antennas, so instead of static cellular coverage, 5G will provide "beam coverage," a set of individual beams sweeping the area.

The MIMO concept is related closely to transmitter and receiver antenna beamforming to improve further the performance and to limit interference. Not only is beamforming useful for high frequencies, but it also forms an important base for many low-frequency scenarios to extend coverage and to provide higher data speeds.

MIMO helps spatially filter the interference level caused by the nearby transmission points. Spatial multiplexing also improves the network capacity as the MIMO antenna transfers multiple data streams between subscribers reusing time-frequency resource blocks. MIMO also helps lower electromagnetic emissions.

Referring to Ref. [5], the MIMO antenna system can increase channel capacity and transmission rate within a fixed communication bandwidth. Thus, multi-element MIMO antennas are becoming increasingly popular in mobile communication systems. As reasoned in Ref. [5], the 5G MIMO antenna array system requires more than six antenna units, which typically serve frequency bands below 6 GHz.

Nevertheless, the adequate design of 5G multi-element MIMO antennas is challenging in practice due to the limited space of mobile communication terminals. The solution is to shrink the size of the antenna elements while ensuring a sufficient level of isolation between the elements. The challenge arises because improved isolation can be achieved by increasing the distance between antenna elements, while the typical terminal models are rather small. Thus, device manufacturers need to consider additional decoupling methods.

The advanced design models thus allow higher-grade MIMO deployments in 5G devices. As an example, [6] studies demonstrate the feasibility of a 12 5G MIMO antenna setup embedded in a smartphone. The result indicates acceptable antenna performance and low envelope correlation coefficients for 5G communications. In this specific case, a combination of 2 × 2 LTE Low-Band (LB) MIMO, 4 × 4 LTE Mid-Band (MB)/High-Band (HB) MIMO, 12 × 12 5G MIMO, GPS, and 2 × 2 WiFi MIMO was considered. These antennas use coupling feed to ensure impedance matching and isolation of the elements. Especially, the most advanced 5G terminals can thus include such solutions in practice, too, along with Release 16.

3GPP RP-182863 [7] presents enhancements of the 3GPP RAN group on MIMO for NR. As a result, Release 16 includes enhanced beam handling and Channel-State Information (CSI) feedback, support for transmission to single User Equipment (UE) from multi-TRP, and full-power transmission from multiple UE antennas in the UL [8]. These enhancements increase throughput, reduce overhead, and/or provide additional robustness.

Additional mobility enhancements enable reduced handover delays, in particular when applied to beam management mechanisms used for deployments in mmWave bands.

The industry has developed 5G concepts and tested them actively. 3GPP Release 16 enhances radio and core functionalities and performance, and there are novelty technologies expected to appear during the evolution path of the systems after Release 16 specifications. The demanding performance requirements of 5G can indeed be satisfied with a set of many different types of technologies that are orchestrated for optimizing the user experience.

One of the enabling technologies that has already been under development for some time is the active antenna concept. In 5G, this technology is expected to be developed and deployed further.

4.2.2 5G Radio Access

The general term for the 5G RAN base station is NG-RAN node. It can be a gNB (which refers to the native 5G base station) or an ng-eNB (which is the intermediate radio base station based on the evolved 4G era). These gNB and ng-eNB elements are interconnected by the *Xn* interface within the radio network. Both elements are connected furthermore via *NG* interfaces to the 5G core network. This interface is divided into two parts: user and control interfaces.

The control interface is referred to as *NG-C* and connects the radio base stations to the Access and Mobility Management Function (AMF). The user interface is referred to as *NG-U* and connects the radio base stations to the User Plane Function (UPF). These interfaces are described in 3GPP TS 23.501, whereas the functional interface for the Control Plane (CP) and User Plane (UP) split as well as the 5G architecture are explained in 3GPP TS 38.401.

In 5G, further optimization of the radio resources takes place by separating user and control communications. This refers to the decoupling of user data and CPs. This provides a means to separate the scaling of UP capacity and control functionality. An example of this is delivery of the user data via a dense access node layer while the system information messages are delivered via an overlaying macro layer. So, 5G gives the possibility to optimize capacity via these different paths for signaling and data by applying Control and User Plane Separation (CUPS).

This separation also applies to multiple frequency bands and radio access technologies. In 5G, it provides the possibility to deliver the user data via a dense, high-capacity 5G layer on a higher frequency, whereas an overlaying LTE system provides reliable signaling for call control. This possibility, which enables mobile devices to connect both to LTE and 5G NR simultaneously, is referred to as Dual Connectivity (DC) E-UTRAN New Radio – Dual Connectivity (EN-DC).

An overview of the physical layer channels of 5G is presented in 3GPP TS 38.201 (general description of NR) and TS 38.202 (services provided by the physical layer of NR). These specifications also detail 5G protocol architecture and its functional split, and the overall radio protocol architectures that include Medium Access Control (MAC), Radio Link Control (RLC), Packet Data Convergence Protocol (PDCP), and Radio Resource Control (RRC).

The NR interface covers the interface between the UE and the network on layers 1, 2, and 3. The TS 38.200 series describe the layer 1 (physical layer) specifications. Layers 2 and 3 are described in the TS 38.300 series. TS 38.300 also defines the elements and functions for Release 15 and Release 16. Some of these include conditional and Dual Active Protocol Stack-based Handover (DAPS HO), early data forwarding, gNB, IAB-DU (Distributed Unit), IAB-MT (Mobile Terminating), ng-eNB, NG-RAN node, and IAB-node. 3GPP TS 38.300 also defines the IAB-donor, IAB-donor-CU (Centralized Unit) (which is the gNB-CU of an IAB-donor terminating the *F1* interface towards IAB-nodes and IAB-donor-DU), and IAB-donor-DU (which is the gNB-DU of an IAB-donor hosting the IAB Backhaul Adaptation Protocol (BAP) sublayer as defined in TS 38.340, and which provides wireless backhaul to IAB-nodes). 3GPP TS 37.340 also defines the conditional PS-cell change.

Some of the key components of the Release 16 radio network architecture that are defined in other radio specifications of the 3GPP include the following:

- **en-gNB**: defined as per 3GPP TS 37.340.
- **gNB-CU**: a logical node hosting RRC, SDAP, and PDCP protocols of the gNB or RRC and PDCP protocols of the en-gNB that controls the operation of one or more gNB-DUs. The gNB-CU terminates the *F1* interface connected with the gNB-DU.
- **gNB-DU**: a logical node hosting RLC, MAC, and physical layers of the gNB or en-gNB, and its operation is partly controlled by gNB-CU. One gNB-DU supports one cell or multiple cells. One cell is supported by only one gNB-DU. The gNB-DU terminates the *F1* interface connected with the gNB-CU.
- **gNB-CU-CP**: a logical node hosting the RRC and the CP part of the PDCP protocol of the gNB-CU for an en-gNB or a gNB. The gNB-CU-CP terminates the *E1* interface connected with the gNB-CU-UP and the *F1-C* interface connected with the gNB-DU.
- **gNB-CU-UP**: a logical node hosting the UP part of the PDCP protocol of the gNB-CU for an en-gNB, and the UP part of the PDCP protocol and the SDAP protocol of the gNB-CU for a gNB. The gNB-CU-UP terminates the *E1* interface connected with the gNB-CU-CP and the *F1-U* interface connected with the gNB-DU.
- **ng-eNB-CU**: defined in TS 37.470.
- **ng-eNB-DU**: defined in TS 37.470.
- **Packet Data Unit (PDU) session resource**: this term is used for specification of the *NG*, *Xn*, and *E1* interfaces. It denotes the *NG-RAN* interface and radio resources provided to support a PDU session.

For the concept of public networks and NPNs in 5G, the related public network integrated NPN and standalone NPN are defined in TS 23.501.

4.2.3 5G gNB Functions

The gNB is referred to as standalone 5G NB, whereas the ng-eNB represents a non-standalone 4G NB. They both host the following functions: Radio Resource Management (RRM), Internet Protocol (IP) header management, AMF management, routing, connection setup and release, scheduling, measurements, packet marking, Session Management (SM), network slicing, QoS, support of UE in the *RRC Inactive* state, distribution function for Non-Access Stratum (NAS) messages, RAN sharing, DC, and tight interworking between NR and E-UTRA.

Table 4.3 summarizes the key functionalities of 5G base station variants.

Table 4.3 Key functionalities of 5G gNB and ng-eNB as interpreted from Ref. [9]

Functionality	Description
Connection setup and release	Connection setup and release performed by 5G gNB relate to the procedures for initiating and terminating data sessions.
Connectivity to AMF	AMF management from the gNB side refers to the selection of an AMF upon UE attachment in those scenarios when the AMF cannot determine the routing information from the UE messaging.
Measurements	5G gNB can perform radio interface measurements and deliver respective reports that eases configuration for mobility and scheduling.
Network slicing	Network slicing refers to the forming of dynamic virtual subnetworks that have different capabilities and that can optimize network resource utilization via Network Functions Virtualization (NFV). Network slicing only consumes selected resources that are needed for the specific event.
Packet marking	Packet marking of 5G NB refers to the ability to mark the transport-level packets in the UL.
QoS	QoS functions refer to QoS flow management and mapping to data radio bearers.
RRM of gNB and ng-eNB	Includes RRM, radio bearer and radio admission control, connection and mobility control, and scheduling, which refers to the dynamic resource allocation to a set of UE in both UL and DL. Furthermore, 5G gNB IP header management provides data compression, encryption, and integrity protection.
Routing	Refers to the routing of UP data towards a set of UPF elements, and routing of CP information to the AMF.
Scheduling	The scheduling functionality of 5G gNB refers to the ability to schedule and transmit paging messages originating from the AMF, and schedule and transmit system broadcast information originating from the AMF or Operations Administration and Maintenance (OAM).
Security and radio configuration	Maintaining of security and radio configuration for UP Cellular IoT (C-IoT) 5GS optimization, as defined in TS 23.501 for an ng-eNB-only scenario.
SM	SM refers to the procedures and functionalities applied during active data connection.
Other functions	5G gNB and ng-eNB also support UE in the inactive state of the RRC, distribution function for NAS messages, RAN sharing, DC, and tight interworking between NR and E-UTRA.

4.2.3.1 AMF

The AMF contains the following key functions as referred to in 3GPP TS 23.501 and TS 38.300:

- NAS signaling termination;
- NAS signaling security;
- AS security control;
- Intercore network node signaling for mobility between 3GPP access networks;
- Idle mode UE reachability (including control and execution of paging retransmission);
- Registration area management;
- Support of intrasystem and intersystem mobility;
- Access authentication;
- Access authorization, including checking of roaming rights;
- Mobility management control (subscription and policies);
- Support of network slicing;
- Session Management Function (SMF) selection;
- Selection of C-IoT 5GS optimizations (Release 16 addition).

4.2.3.2 UPF

UPF can perform the following key functions as described in 3GPP TS 23.501 and TS 38.300, such as:

- Acts as an anchor point for intra- and inter-Radio Access Technology (RAT) mobility;
- Is an external PDU session point of interconnect to the data network;
- Performs packet routing and forwarding, packet inspection, and acts in the UP part of policy rule enforcement;
- Forms traffic usage reports;
- Is a UL classifier supporting routing traffic flows to a data network;
- Is a branching point supporting a multihomed PDU session;
- Manages QoS handling for the UP;
- Performs UL traffic verification, being Service Data Flow (SDF) to QoS flow mapping;
- Makes DL packet buffering and DL data notification triggering.

There have been no additions to the UPF since the Release 15 definitions.

4.2.3.3 SMF

SMF hosts the following main functions, as described in 3GPP TS 23.501 and TS 38.300 [9, 10]:

- Manages sessions;
- Allocates and manages UE IP addresses;
- Selects and controls UP function;
- Configures traffic steering at UPF routing traffic to the proper destination;
- Manages the control part of policy enforcement and QoS;
- Takes care of DL data notification.

4.2.4 3GPP RAN Interfaces

Table 4.4 summarizes the 3GPP-defined 5G RAN interfaces in Release 16 3GPP TS 38.401 V16.2.0 (2020-07) [11].

The protocol layers of *NG* and *Xn* for the CP and UP are depicted in Figure 4.1. The *Xn*-Application Protocol (AP) serves as the application layer signaling protocol. The transport layer protocol relies on the CP of the *Xn* in Stream Control Transmission Protocol (SCTP), while the UP relies on the GDP-U, respectively.

NG-AP is defined in 3GPP TS 38.423 and TS 38.424 (*Xn* data transport). It is a reliable transport layer protocol defined by the Internet Engineering Task Force (IETF). It operates on top of a connectionless packet network such as IP. IP is as defined by the IETF. The data link layer is for the data transport of the NR as defined in 3GPP TS 38.414. 3GPP TS 38.411 and 38.421 TS (*Xn* layer 1) define the next generation physical layer (Phy), i.e., the physical aspects of the NR.

4.2.5 The Split Architecture of RAN

Figure 4.2 depicts the high-level split architecture of gNB. The 5G architecture model allows the division of gNB functions applying CU for UP and CP, while the UP and CP are common for the DU and Remote Radio Head (RRH) – also referred to as Remote Radio Unit (RRU) – of the gNB [11].

Table 4.4 5G NR-RAN interfaces of Release 16 as per 3GPP TS 38.401 [12]

Interface	Description	Source
NG	*NG* refers to new generation interface. It is the renewed radio interface between the UE and gNB. It is based on Orthogonal Frequency Division Multiplexing (OFDM) in both DL and UL.	TS 38.410...38.414, *NG* interface general aspects and principles
Xn	The *Xn* interface is between NG-RAN nodes gNB and ng-eNB, whereas the interface between LTE eNB elements is *X2*.	TS 38.420...38.424, *Xn* interface general aspects and principles
E1	*E1* is the point-to-point interface between gNB-CU-CP and gNB-CU-UP.	TS 38.460...38.463
F1	*F1* is the interface between gNB-CU and gNB DU elements.	TS 38.470...38.475, *F1* interface general aspects and principles
F2	*F2* is the interface between the lower and upper parts of the 5G NR physical layer; includes *F2-C* and *F2-U*.	TS 38.401
Iuant	The *Iuant* interface forms part of the NG-RAN for the control of Remote Electrical Tilt (RET) antennas or Tower-Mounted Amplifiers (TMAs).	TS 38.401

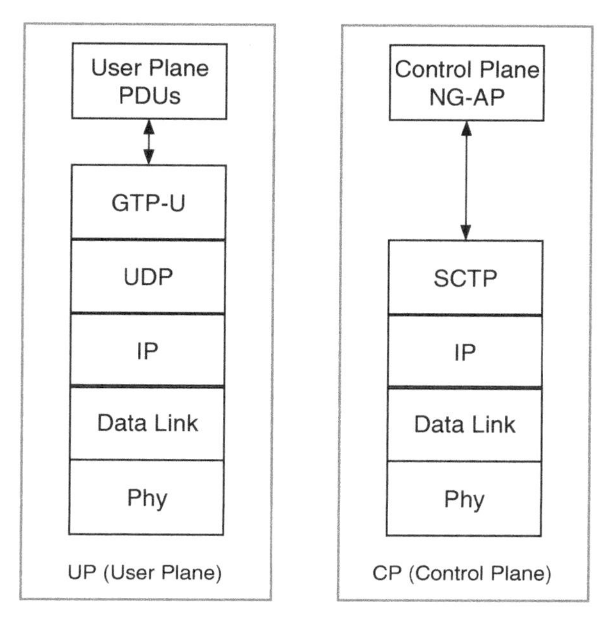

Figure 4.1 The protocol stacks for *NG* and *Xn* interfaces.

There is thus a set of gNB elements in an NG-RAN connected to the 5G Core (5GC) network via the *NG* interface, whereas the gNG elements can be interconnected via the *Xn* interface.

A single gNB element may consist of a gNB-CU and one or more gNB-DU components. The gNB-CU and gNB-DU interconnect via the *F1* interface. One gNB-DU connects to only one gNB-CU.

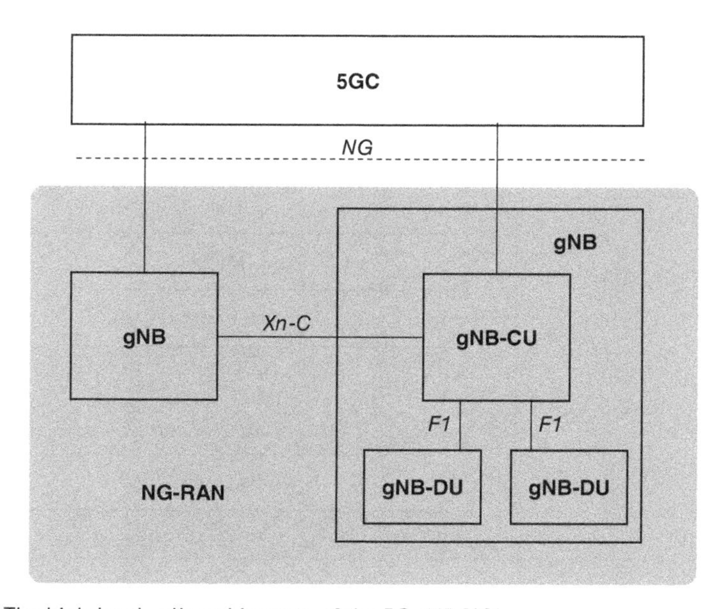

Figure 4.2 The high-level split architecture of the 5G gNB [12].

3GPP TS 38.300, Release 16, also allows a set of ng-eNB elements in the NG-RAN, and each ng-eNB may consist of an ng-eNB-CU and one or more ng-eNB-DU components. The *W1* interface connects the ng-eNB-CU and ng-eNB-DU to the 5GC.

Figure 4.3 presents the 5G RAN architecture in more detail, exposing the protocol layers and the principle of the higher and lower split.

4.2.6 IAB

3GPP Release 16 introduces IAB to enhance the cost-efficiency of the frequencies originally meant for radio access. Reutilization of the same frequencies for transmission is especially efficient on the mmWave bands as they can complement, or replace, the traditional radio links. Figure 4.4 depicts the overall Release 16 architecture of the IAB.

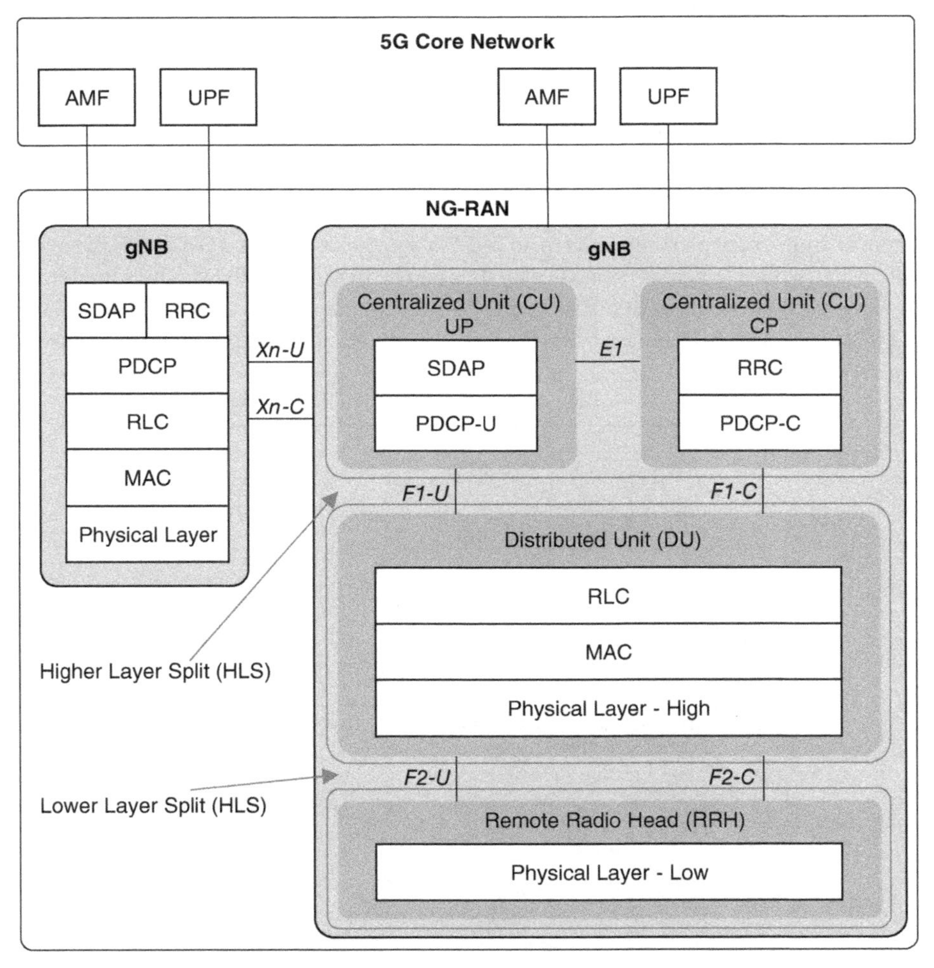

Figure 4.3 The functional split of UP and CP in gNB based on DU, as per the 3GPP specifications [12].

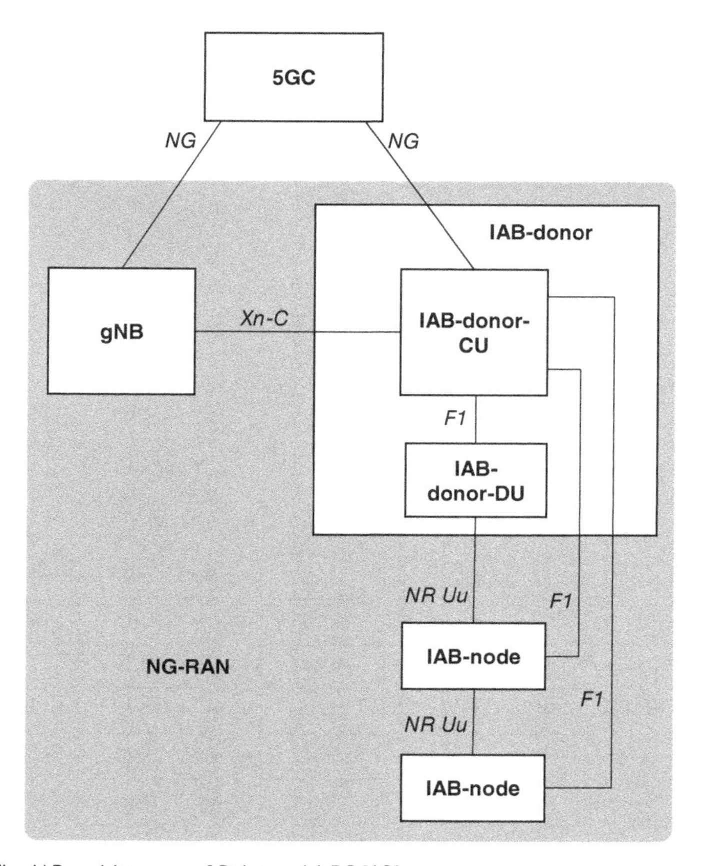

Figure 4.4 The IAB architecture of Release 16 5G [12].

The 5G RAN can interconnect the IAB via the IAB-nodes. They connect wirelessly to the gNB via an IAB-donor. To do so, the gNB needs to support IAB functionality.

As can be seen in Figure 4.4, the IAB-donor consists of two parts: IAB-donor-CU and IAB-donor-DU. The concept supports one or more IAB-donor-DU components. For the separated gNB-CU-CP and gNB-CU-UP, the IAB-donor can house IAB-donor-CU-CP, one or more IAB-donor-CU-UP components, and one or more IAB-donor-DU components [12]. All NG-RAN gNB-DU and gNB-CU functions are typically also applicable for the IAB-DU/IAB-donor-DU and IAB-donor-CU, respectively. As Ref. [12] states, the following applies to the IAB concept:

- The IAB architecture relies on a subset of UE functions on the NR *Uu* interface, which in the IAB context is referred to as "IAB-MT function of IAB-node." The IAB-nodes use this interface in their interconnection or connectivity to the IAB-donor-DU.
- The IAB-node serves as a wireless backhaul to the downstream IAB-nodes and UE by relying on the network functionalities of the NR *Uu* interface. In this context, the interface is referred to as "IAB-DU function of IAB-node."

- The *F1-C* signaling traffic is backhauled between the IAB-node and IAB-donor-CU. This happens via the IAB-donor-DU and a set of optional intermediate hop IAB-nodes.
- The *F1-U* data traffic is backhauled between the IAB-node and IAB-donor-CU. This takes place via the IAB-donor-DU and the optional set of intermediate hop IAB-nodes.

4.2.7 5G Network Layers

The 5G protocol stack consists of layers 1, 2, and 3. Layer 1 refers to the physical layer. Layer 2 includes MAC, RLC, and PDPC, while layer 3 refers to the RRC layer [13]. Figure 4.5 summarizes the high-level 5G protocol stacks on UP and CP.

Figure 4.6 presents the frame structure of the RLC and PDCP.

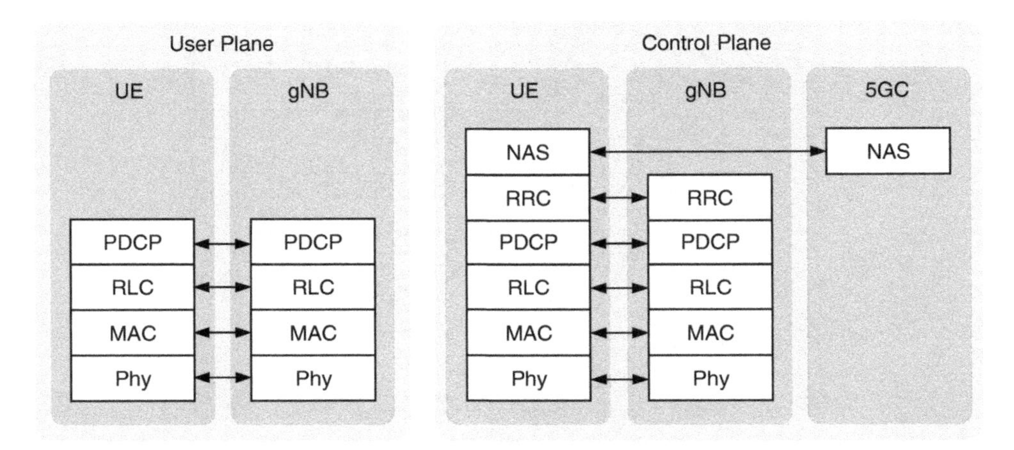

Figure 4.5 The high-level 5G UP and CP protocols.

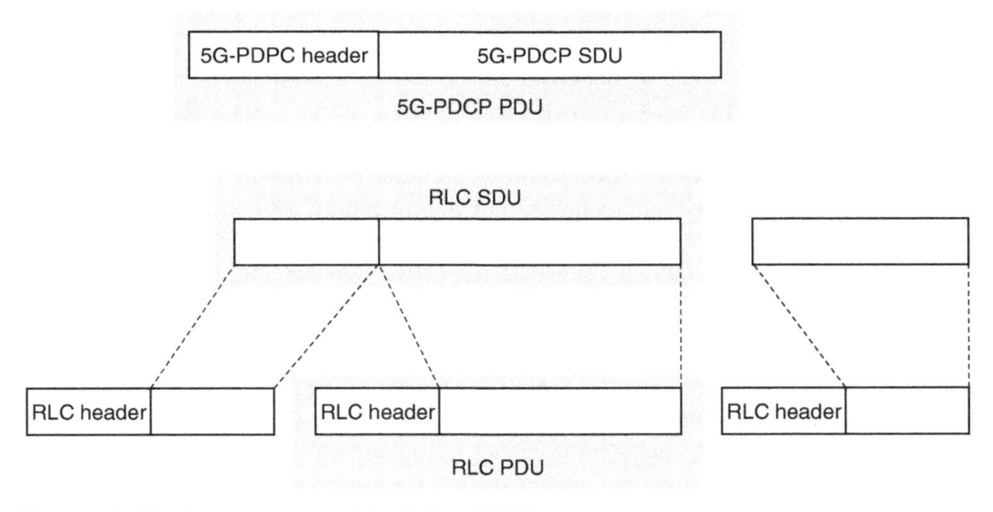

Figure 4.6 The frame structure of the RLC and PDCP.

4.2.7.1 Layer 1

5G layer 1, i.e., the physical layer, includes the following main functionalities:

- Coded transport and physical channel rate matching and mapping;
- MIMO antenna processing, beamforming, and transmit diversity management;
- Physical channel power weighting, modulation, and demodulation;
- RF processing, radio measurements, and reporting of radio characteristics;
- Soft-combining of HARQ;
- Synchronization in frequency and time domains;
- Transport channel error detection, Forward Error Coding (FEC), and decoding.

4.2.7.2 Layer 2

Layer 2 refers to 5G MAC, RLC, and PDPC functionalities. For the MAC, these include:

- 5G MAC Service Data Unit[1] (SDU) multiplexing and demultiplexing;
- Antenna beam management;
- HARQ-based error correction;
- Logical and transport channel mapping;
- MAC SDU concatenation from logical channel into the Transport Block (TB);
- Padding;
- Random access procedures;
- Reporting of scheduling information;
- Transport format selection;
- UE logical channel priority management, and UE priority management by dynamic scheduling.

For the RLC, the functions of layer 2 include:

- 5G RLC re-establishment;
- Acknowledged Mode (AM) data transfer's ARQ-based error correction, protocol error detection, and resegmentation;
- Unacknowledged Mode (UM) and AM data transfer's 5G RLC PDU reordering, duplicate packet detection, 5G RLC SDU discard, and segmentation;
- Upper layer PDU transfer.

For PDCP, the functions of layer 2 include:

- 5G PDCP SDU retransmission in the connected 5G RLC mobility mode;
- Ciphering and deciphering based on mandatory Advanced Encryption Standard (AES);
- CP data transfer;
- UL SDU discard based on timer value;
- Upper layer PDU in-sequence delivery at the 5G PDCP re-establishment procedure, and lower layer SDU duplicate detection, for 5G RLC AM;
- User data transfer;
- UP ciphering and integrity protection.

4.2.7.3 Layer 3

5G layer 3, i.e., the RRC layer, includes the following main functionalities:

- Key management and other security procedures;
- Message transfer between NAS and UE;
- Mobility functions;
- PTP radio bearer establishment, configuration, maintenance, and release;
- Radio interface measurements and reporting from the UE;
- RRC connection establishment, maintenance, and release;
- System information broadcasting for NAS and AS.

4.2.8 IAB Protocol Stacks

Figure 4.7 shows the protocol stack for the *F1-U* between IAB-DU and IAB-donor-CU-UP, and Figure 4.8 shows the protocol stack for the *F1-C* between IAB-DU and IAB-donor-CU-CP. In these example figures, the *F1-U* and *F1-C* traffic is carried over two backhaul hops. Please note that the *F1* needs to be security protected as described in TS 33.501.

The LTE system contains DC to provide users with higher data throughput, robustness of mobility, and an evolved means for load balancing. If the LTE UE supports DC, it is capable of connecting simultaneously to two eNBs, which are a Master eNB (MeNB) and a Secondary eNB (SeNB). The MeNB and SeNB operate on different carrier frequencies and they interconnect via *X2* backhaul links.

Figure 4.9 depicts the scenario for *F1-C* involving IAB-DU and IAB-donor-CU-CP, the *F1-C* traffic passing through the MeNB.

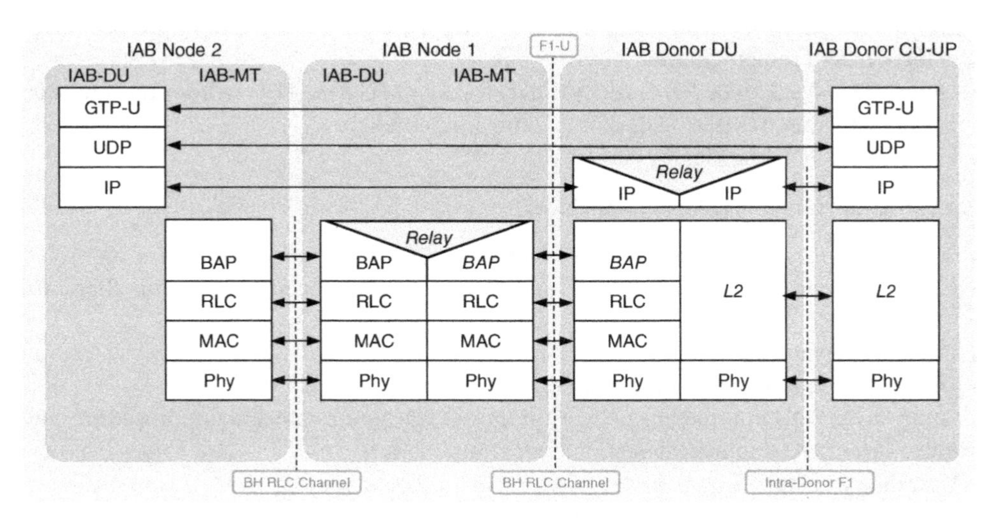

Figure 4.7 Protocol stack for *F1-U* of IAB.

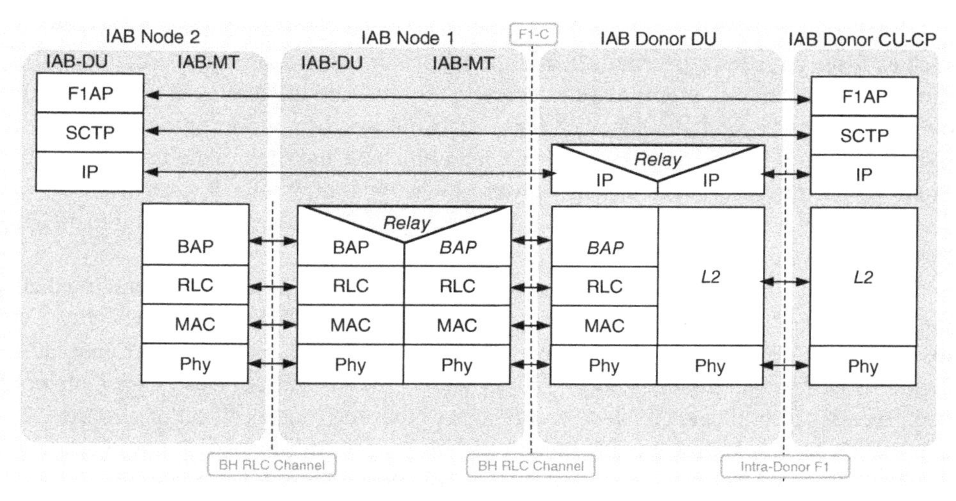

Figure 4.8 Protocol stack for *F1-C* of IAB.

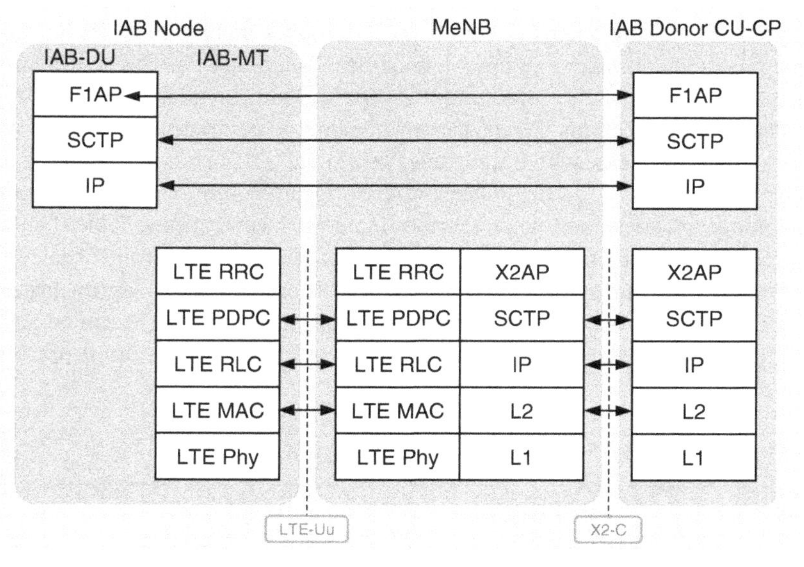

Figure 4.9 Protocol stack for IAB *F1-C* traffic delivered via the MeNB.

4.3 User Equipment

4.3.1 Background

Release 16 5G devices can support new features as per device vendor strategies. It can be assumed that there will be a variety of devices from the most basic models up to the high-end, very rich set of complex features. An example of a Release 16 addition is the direct communications mode that can benefit many verticals such as critical communications.

At the same time, the overall technologies beyond the 3GPP specifications evolve, too, such as battery-related performance, and the models may have advanced accessories, sensors, high-performance memories and processors, advanced displays, and other accompanying solutions beyond the specifications. It also is worth mentioning that overall CO_2 reduction also depends on the optimized materials and logistics related to the devices. There will be many familiar-looking as well as new types of mobile devices in the 5G era. These devices will support a variety of usage types, including smart devices and fixed wireless access via Customer Premise Equipment (CPE).

There will also be IoT devices such as Machine-to-Machine (M2M) equipment, intelligent sensors, and modules for automotive and industry, as well as new, advanced devices for virtual reality. 3GPP TS 38.306 defines the 5G UE radio access capabilities as of Release 15, including new terminal aspects, local connectivity, evolved battery, display, sensor, memory, and processor technologies. Table 4.5 summarizes other relevant specifications for the UE.

3GPP TS 38.300 defines the UE capabilities for 5G NR. In practice, the network determines the UL and DL data speed of UE based on the supported band combinations and baseband capabilities, i.e., modulation scheme, MIMO antenna layers, and other characteristics on the radio path.

According to the Global Mobile Suppliers Association (GSA), there were early-stage device types in the consumer markets from 19 vendors in July 2019, including phones, hotspots, customer premise equipment, modules, "snap-on" dongles and adapters, and USB terminals, while the 5G chipset modules are becoming available [14].

In critical communications, 5G devices will be able to communicate directly with other devices – similar to the case with the current Push-to-Talk (PTT) service. This is beneficial for special circumstances as, e.g., the police and fire brigade require reliable radio communications in emergency areas even when there is no network coverage available. The respective security requirements include mutual authentication of a single or a group of devices without the need to negotiate and agree keys, as well as data encryption and integrity protection.

Only authorized users may access the 5G device. This requirement can be activated by verifying a PIN or biometric characteristic of the user. In addition, theft prevention is a

Table 4.5 Some of the key 5G specifications describing the UE

3GPP TS	Description
38.300	UE key functionalities and capabilities
38.101–1	NR UE radio transmission and reception, part 1, range 1 standalone
38.101–2	NR UE radio transmission and reception, part 2, range 2 standalone
38.101–3	NR UE radio transmission and reception, part 3, range 1 and range 2 interworking operation with other radios
38.101–4	NR UE radio transmission and reception, part 4, performance requirements

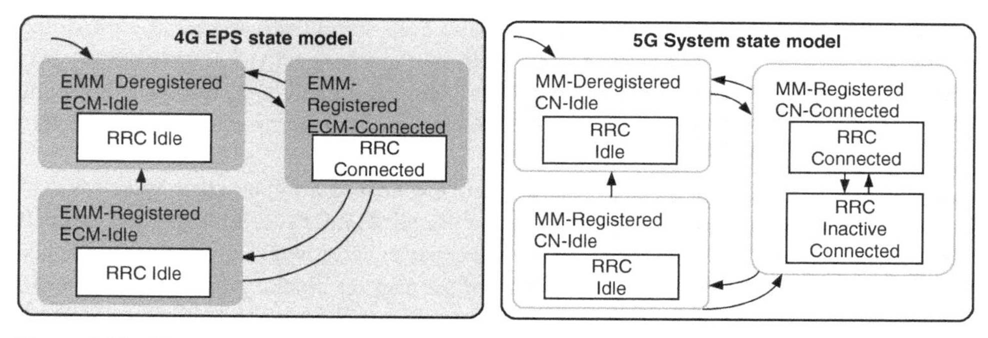

Figure 4.10 The state model for 4G and 5G systems as interpreted from 3GPP TS 23.502 [15].

potential issue that requires means to ensure only authorized entities may be able to disable or re-enable stolen devices. Also, users may want to store and process private and confidential data on the device, leveraging the secure entity within the device for this purpose.

4.3.2 Terminal States

When the user switches on the 5G device, it performs an initial network attach procedure and registration. Other signaling flows include call establishment and mobility management procedures in the idle and connected modes, PDU session establishment (the actual 5G data connection), session termination, and network deregistration. 5G performs a detach procedure when the user switches off the equipment, or it runs out of battery. Figure 4.10 depicts the state diagram of 4G and 5G as interpreted from 3GPP TS 23.502 (procedures for the 5G system).

The registration procedure happens when the UE performs initial registration to the 5G system, or when it executes a mobility registration update procedure whenever the Tracking Area (TA) changes. TA is a set of gNBs. When the 5G customer is about to receive a connection request, the network sends paging signaling to the UE within the 5G TA where it was last registered via all the gNB elements.

If the TA is large, it triggers a large amount of radio signaling. The UE needs to send registration signaling to the network each time it moves between the TA areas. If the TA areas are small, this increases signaling and consumes the UE's battery, so it is important to optimize the TA size.

Registration is also done to update the UE capabilities, and in a periodic registration update. In the 5G deregistration procedure, the UE informs the network to stop accessing the 5GC, and the 5GC informs the UE about the released access.

4.4 Cloud RAN

4.4.1 Introduction

Cloud RAN (C-RAN) is a centralized RAN architecture model based on cloud computing for RANs. Equally, the core network can be based on cloud. Both support current mobile communication systems and future wireless standards.

Historically, each mobile communication radio base station serves users by providing coverage areas individually and relying on their own local capacity. The traditional radio networks are dimensioned based on a busy-hour blocking probability. The busy hour of the network refers typically to the most occupied 60 minutes of a complete 24-hour period of time.

The operators dimension the offered capacity of each base station based on this peak traffic profile so that the maximum blocking probability does not exceed the desired value. It is inevitable that a small number of customers experience service degradation or congestion during the busy hour. The dimensioning of the average blocking probability is one of the many optimization tasks of operators to balance investments and adequate user experiences. The average load is typically much lower, which results in unused capacity reserve during off-peak hours.

In the traditional architecture model, the processor resources of base stations cannot be shared with others. The processing of the baseband radio signals happens within the same site to which the power amplifier was located. An additional challenge was the relatively long coaxial cables that were deployed from the transmitter (residing at the physical base station shelter) to the antennas (installed on walls, rooftops, or masts), resulting in cable losses – which, in turn, impacted negatively on the radio link budget and costs.

Ever since, radio base stations have been optimized by applying more sophisticated solutions. In the distributed base station model, the base stations rely on separation of the RRH and Baseband Unit (BBU), which are connected by fiber. These elements can be located far away from each other as the fiber optics has minimal signal loss compared to coaxial cable. If the RRH is next to the antenna, the respective coaxial cable loss is insignificant. Also, as remotely located BBU processing can be shared among more than one single base station, the overall capacity can be maximized. These solutions are already popular in 3G deployments.

C-RAN is the next step in this evolution. It can be based on an up-to-date or evolved Common Public Radio Interface (CPRI) standard, which serves as a protocol in the respective digital baseband data streaming. Coarse/Dense Wavelength Division Multiplexing (CWDM/DWDM) is used in the transmission on mmWave bands for the baseband signal over long distances wirelessly from the base station to the core network.

C-RAN is a cost-efficient, centralized base station deployment model relying on data center infrastructure. This model also provides 5G operators with wide bandwidth access to the cloud BBU pool with high reliability and low latency. The concept includes fluent dynamic resource sharing in multivendor, multitechnology environments [16].

5G thus benefits from the C-RAN concept, and it can be deployed via fronthaul transmission interfaces as depicted in Figure 4.11 [17]. In addition, if the operator so decides, 5G can still use the more traditional distributed and centralized RAN architecture models.

Both radio and core as well as the transport networks of 5G can be deployed on cloud, facilitated by Software-Defined Networking (SDN) and NFV. Cloud supports a variety of diversified services. In addition, cloud is the foundation for network slicing, which provides fast and efficient deployment of many new 5G services.

C-RAN can be connected via 3GPP RAN or other, non-3GPP access network sites. Examples of the former are 3GPP base stations (gNB of 5G, and base stations of any earlier generations), whereas Wi-Fi access points represent the latter type. In addition to access

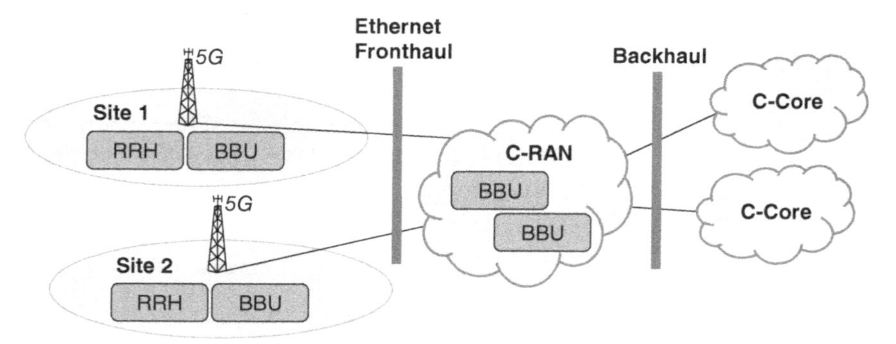

Figure 4.11 Cloud-based 5G RAN architecture model.

connectivity, the C-RAN architecture has Mobile Cloud Engines (MCE), which coordinate services of Real-Time (RT) and non-RT resources.

The C-RAN connects to the service-oriented Cloud Core (C-Core) network as depicted in Figure 4.12. The C-Core network has various tasks such as control function that manages mobility, services, policies, security, and user data. It also takes care of dynamic policy control of the supported services, and manages the storing of data in the unified database.

Furthermore, the C-RAN has a CP with a set of components for performing network functions, and a programmable UP fulfilling a variety of service requirements. The concept provides the possibility to orchestrate the network functions to select CP and UP functions.

As depicted in Figure 4.12, the SDN controller is in the transmission section between the C-RAN and C-Core. The SDN decouples the CP and data plane, and centralizes the network control functions, which optimizes the network management.

One of the key benefits of SDN-enabled C-RAN deployment is adaptive reconfiguration. In this mode, the SDN controller receives the status changes of the RAN in RT,

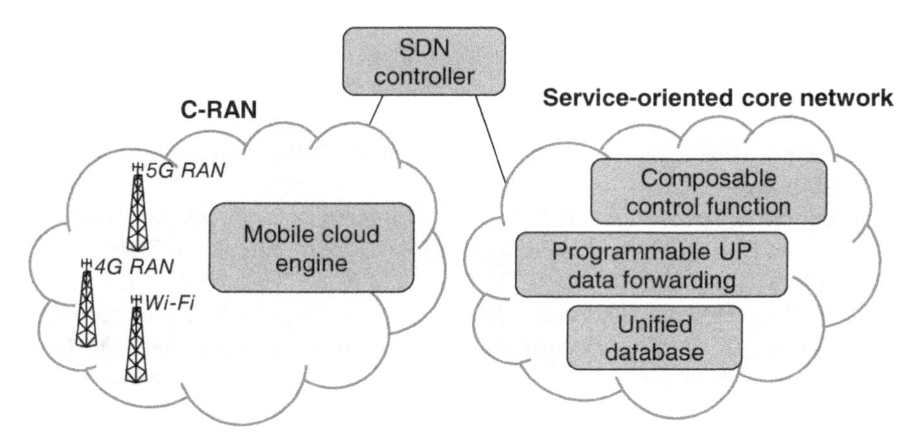

Figure 4.12 The principle of C-RAN and C-Core of 5G.

which enhances the performance of mobility management and load balancing of the BBUs when 5G users move within the service area and generate varying data transmission patterns.

For further details on the cloud concept, please refer to ETSI White Paper 23, "Cloud RAN and MEC: A Perfect Pairing" [18], and ITU Technical Report, GSTR-TN5G, "Transport Network Support of IMT-2020/5G" [19].

4.4.2 Open RAN Terminology

"Open RAN," sometimes presented in its short form of "O-RAN," refers to the overall movement of the telecom industry to disaggregate hardware and software, and create respective open interfaces in between.

The term "Open RAN" refers to one or another subgroup of the Telecom Infra Project [20]. There are two subgroups: the Open RAN project group, which defines and builds 2G, 3G, and 4G RAN solutions based on vendor-neutral hardware and software-defined technology, and the Open RAN 5G NR project group with the focus on 5G NR.

Instead, "O-RAN" refers to the O-RAN Alliance. It publishes RAN specifications, releases open software for the RAN, and supports O-RAN Alliance members in integration and implementation testing. O-RAN Alliance works on open, interoperable interfaces, RAN virtualization, and big data-enabled RAN intelligence [20, 21].

4.4.3 Open RAN Alliance

AT&T, China Mobile, Deutsche Telekom, NTT DoCoMo, and Orange founded the O-RAN Alliance in February 2018. The organization includes an operating board, Technical Steering Committee (TSC), and the actual technical workgroups. The technical workgroups report to the TSC. The following summarizes the focus areas of the workgroups:

- WG1 defines use cases and overall architecture of the O-RAN.
- WG2 defines the non-RT RAN Intelligent Controller (RIC) and the *A1* interface. It optimizes higher-layer procedures, policy in RAN, and provides Artificial Intelligence/Machine Learning (AI/ML) models to near-RT RIC.
- WG3 defines the near-RT RIC and the *E2* interface and optimizes RAN elements and resources.
- WG4 defines the open fronthaul interfaces to ensure multivendor DU-RRU interoperability.
- WG5 defines the open *F1*, *E1*, *W1*, *X2*, and *Xn* interfaces to provide operators with an interoperable multivendor environment that is still compliant with the 3GPP specifications.
- WG6 considers the cloudification and orchestration of the O-RAN model to decouple the RAN software from the respective hardware, facilitating Commercial Off-The-Shelf (COTS)-based deployments.
- WG7 considers the white-box hardware by releasing decoupled software and hardware reference design, which is aimed at reducing 5G deployment costs.

- WG8 considers stack reference design with the goal of developing software architecture and a design plan for the O-RAN CU and O-RAN DU, based on the 3GPP NR specifications.
- WG9 considers open X-haul transport with the focus on transport equipment, respective physical media, as well as control and management protocols.

4.4.4 Open RAN Reference Architecture

The focus areas of the O-RAN Alliance include software-defined, AI-enabled RIC, RAN virtualization, open interfaces, white-box hardware, and open source software. As a result, the O-RAN Alliance has designed a reference architecture for Open RAN as depicted in Figure 4.13 [21].

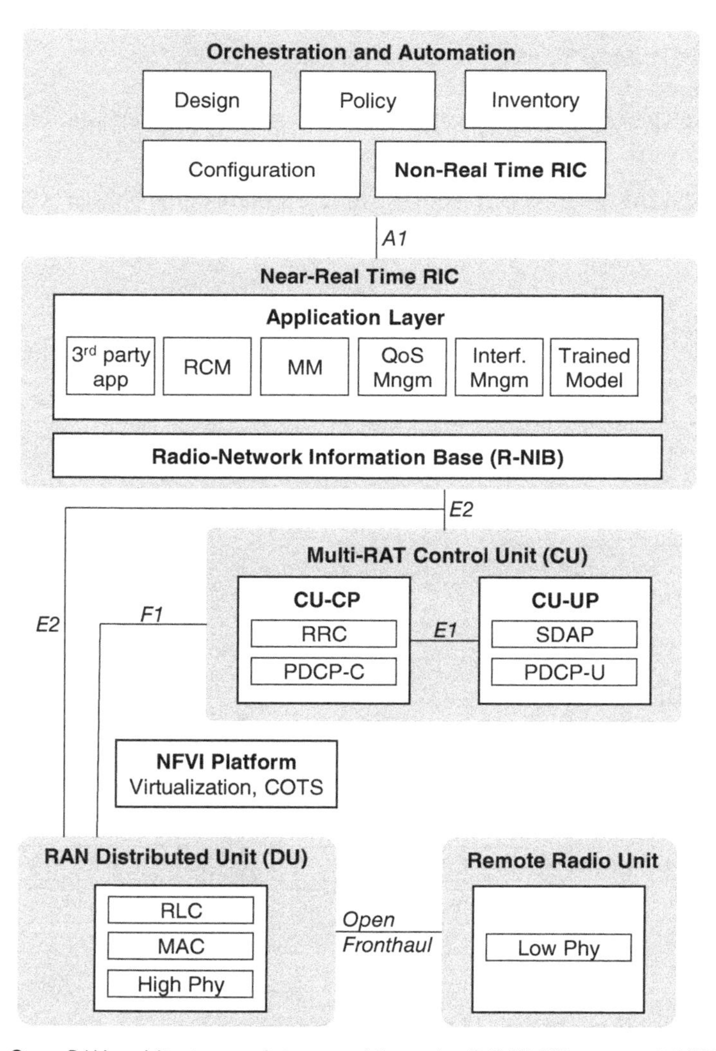

Figure 4.13 Open RAN architecture as interpreted from the O-RAN Alliance model [21].

The O-RAN Alliance reference architecture consists of the following layers:

- RAN RRU, which refers to the physical RF and antennas;
- RAN DU, which houses RLC, MAC, and a higher physical layer protocol stack;
- Network Functions Virtualization Infrastructure (NFVI) platform;
- Multi-RAT CU, which houses CU-CP and CU-UP protocol stacks;
- Near-RT RIC, which houses an application layer and the database for radio network information – the application layer includes third party applications;
- Orchestration and automation layer, which contains design, policy, inventory, configuration, and non-RT RIC functions.

For more information on the CU-UP and CU-CP, please refer to Section 4.2.5. The following subsections summarize the O-RAN architecture blocks based on Refs. [21, 22].

4.4.4.1 O-RAN Protocol Structure

Figure 4.14 depicts the 5G gNB protocol stack [23].

5G O-RAN gNB communicating with 5G UE consists of three elements and their respective protocol layers:

- O-RAN Radio Unit (O-RU) is responsible for the section of the physical (Phy) functions. Some examples of these include beamforming, Fast Fourier Transform (FFT), and inverse

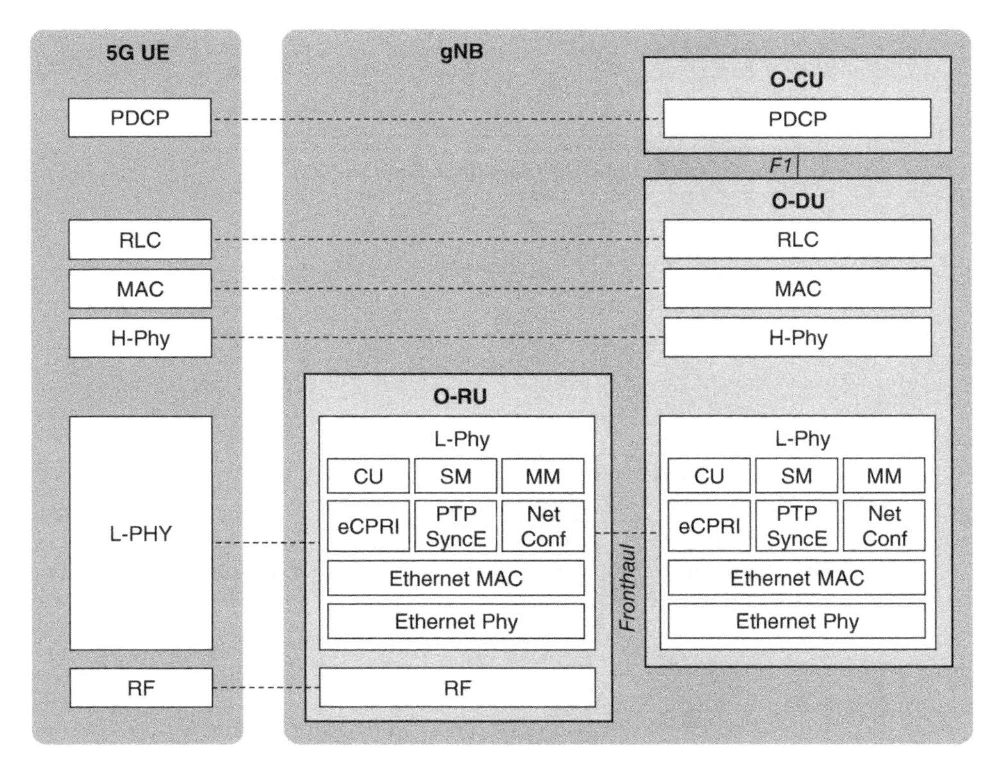

Figure 4.14 The 5G gNB protocol stack of the O-RAN model.

FFT (iFFT). This layer consists of the digital and analogue functions of the 5G transceiver. It is possible to virtualize the O-RU relying on non-proprietary components.
- O-RAN Distributed Unit (O-DU) manages the RLC, MAC, and the upper segment of the physical layer. Some of the key functions include baseband processing and scheduling. It is possible to virtualize the O-DU using hardware accelerators like Field-Programmable Gate Array (FPGA) or Graphics Processing Unit (GPU).
- O-RAN Central Unit (O-CU) is a virtualized section of the RAN, which handles the functions of the PDCP layer. It connects the backhaul network to the core network.

4.4.4.2 Orchestration Based on Non-RT RIC

The non-RT RAN intelligent control refers to functions with delays of more than 1 second, while the decoupled near-RT control functions can handle delays up to 1 second. The non-RT functions include service and policy management, RAN analytics, and model training. The non-RT RIC produces trained models and RT control functions, and distributes them via the standardized *A1* interface to the near-RT RIC for runtime execution. The CU and DU send data to the non-RT RIC via the *A1* interface, too. The policies rely on AI- and ML-based models based on messages of non-RT RIC and near-RT RIC. The non-RT RIC core algorithms facilitate the RAN behavior modification. Network operators own and deploy the core algorithm to optimize the individual network deployments.

4.4.4.3 Near-RT RIC Functions

The O-RAN reference architecture houses the evolved RRM function with its embedded intelligence. The architecture also facilitates the utilization of optional legacy RRM that is located to the near-RT RIC layer. It receives an AI model from the non-RT RIC and executes it to modify the behavior of the network interface reference points *A1* and *E2*.

The near-RT RIC provides the possibility to introduce new functions and enhance the existing functions such as load balancing based on UE control, radio bearer management, interference detection, and mitigation. The near-RT RIC has the following capabilities:

- Delivers embedded intelligence that enhances QoS management, Connectivity Management (CM), and seamless handover control;
- Provides a robust, secure, and scalable platform for third party control application onboarding;
- Leverages a Radio-Network Information Base (R-NIB) to monitor the state of the underlying network;
- Collects RAN measurements data to facilitate RRM;
- Initiates configuration commands to the CU and DU.

Established telecom equipment manufacturers and new stakeholders alike can provide near-RT RIC.

4.4.4.4 Multi-RAT CU

The multi-RAT protocol stack supports, among others, 4G and 5G protocols. The CU executes functions such as handovers based on near-RT RIC control commands.

Multi-RAT CU works on the virtualization platform, which is the execution environment for the CU and near-RT RIC. The platform can allocate virtual resources across multiple network elements by accelerator resource encapsulation ensuring security isolation.

The current architecture relies on the 3GPP-defined interfaces *F1*, *E1*, *X2*, and *Xn* that can be further adjusted for the multivendor environment, while telecom equipment manufacturers can provide a CU as a regional anchor for CP and UP of the DUs.

4.4.4.5 DU and RRU

The DU executes layer 2 functions while the RRU takes care of the base band and RF processing. The interface in between provides the DU-RRU lower layer split via the open fronthaul interface, while the CU-DU higher layer split provides an interoperable *F1* interface for serving multiple telecom equipment manufacturers.

4.4.5 Logical Architecture of O-RAN

Figure 4.15 depicts the logical architecture of the Open RAN model as interpreted from Ref. [22].

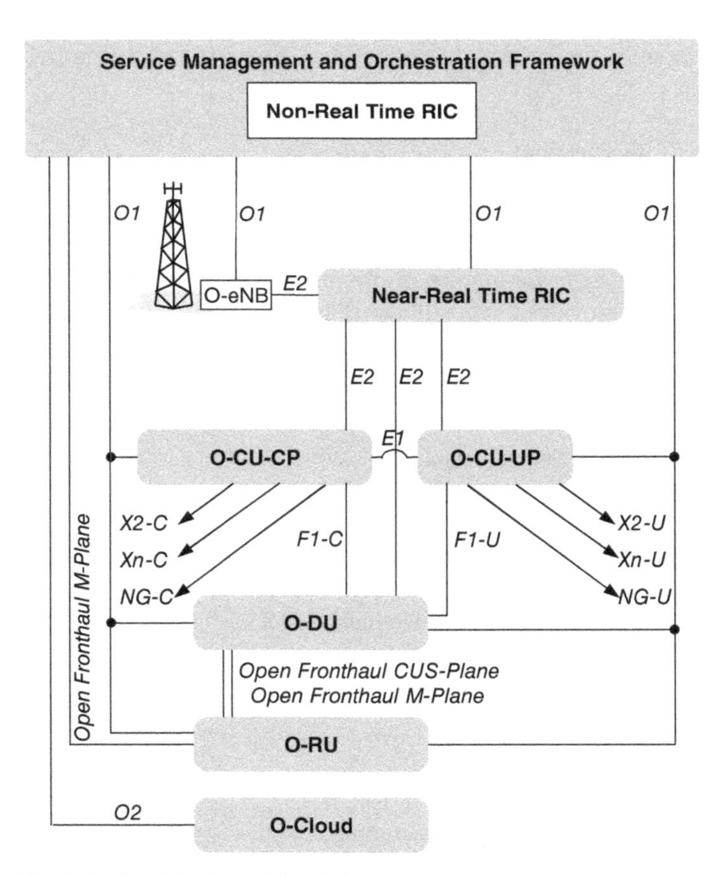

Figure 4.15 The logical architecture of the O-RAN.

The O-RAN builds upon the ETSI NFV reference architecture, and relies on the COTS hardware and virtualization software that enables the abstraction. The Virtual Machines (VM) and containers of the model provide multiple hierarchical cloud setup options and their configurations for the geophysical deployment of the near-RT RIC, O-CU, and O-DU, e.g., the regional and edge clouds.

The O-RAN model provides a means to automate the deployment and provisioning of O-RAN-based RANs, whereas the O-cloud is the actual cloud-based computing platform. It includes physical nodes for hosting O-RAN functions, software components, management, and orchestration functions.

4.4.5.1 Open Interfaces

The objective of the work of the O-RAN Stack Reference Design and Open Interfaces is to provide fully operable multivendor profile specifications, compliant with 3GPP specifications for the *F1*, *E1*, *W1*, *X2*, and *Xn* interfaces. This working group will also propose 3GPP specification enhancements as appropriate.

3GPP 5G specifications define the split model for the CU and DU, referred to as High Layer Split (HLS), as depicted in Figure 4.16. The operator can thus decide the physical deployment location of the DU, CU-UP, and CU-CP independently from each other.

As defined in 3GPP Release 15, the HLS provides an *F1* interface between the gNB of DU and CU functions. Within the gNB-CU, the *E1* interface is defined between the CU-UP and CU-CP functions. Release 15 also specifies the *W1* interface between the eNB-DU and eNB-CU. The *Xn* interface connects the gNB-CU functions, and the *X2* interface serves the connection between the eNB-CU and gNB-CU.

The 3GPP has defined the above-mentioned interfaces aiming to provide operators with a sufficient number of multivendor deployment options and adequate level of interoperability, whereas the O-RAN Alliance has further enhanced the respective technical specifications to ensure the openness of the *F1*, *E1*, *W1*, *X2*, and *Xn* interfaces at user, control, and management planes. Another development area of the O-RAN Alliance is the framework for the open-sourced protocol stack to provide respective

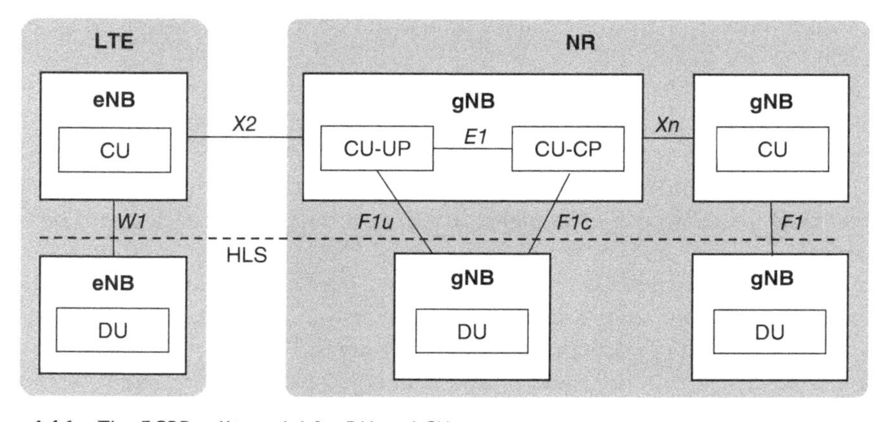

Figure 4.16 The 3GPP split model for DU and CU.

functions and procedures of the CU. As a result, equipment vendors can develop protocol stack algorithms as open modules [24].

The O-RAN reference architecture thus enables new RAN infrastructures based on openness and AI-powered radio control, and it works as a foundation for the virtualized RAN deployment on open hardware. The 3GPP specifications back the functions, while the O-RAN definitions enhance them further.

4.5 5G Spectrum

4.5.1 Advances of 5G Frequencies

The elemental enhancement of 5G is the ability to handle much faster data rates and provide higher capacity for simultaneously communicating consumers and machines. To cope with these demands and requirements, 5G networks will provide radio equipment with extended support of bands and bandwidths.

Among other advances in the RF, there will be new frequency bands below and above the 7 GHz band. The globally agreed frequencies are decided at the ITU-R, while the country-specific deployments depend on each area's regulatory bodies. The current discussion includes many variants up to about the 100 GHz band. In practice, the most favorable frequency strategies support as many and big chunks of contiguous bands as possible. The final decision of each region and country is based on the complete picture of the entities requiring and needing the frequencies.

As there is typically more demand for the frequencies, there might be the need for their optimized utilization. One of the ways to utilize the bands as efficiently as possible is the white space approach. This refers to shared bands that different stakeholders can utilize when needed. In addition to the traditional modes for business models (MNO purchasing rights for licensed frequency utilization) and these novel ideas for more optimal performance via capacity sharing, there are also potential options for further, new spectrum chunks such as satellite communication and radio location. Some examples of these sharing modes are seen via Licensed Shared Access (LSA) that is currently at the planning stage in Europe at the 2.3 GHz band, as well as the Citizens Band Radio Service (CBRS) in the United States that would rely on the 3.5 GHz band [25].

As the need for 5G frequency bands increases along with the expected, much higher utilization of 5G services as before within the previous mobile generations – partially due to the huge increase in M2M-type communications – the propagation characteristics of the radio waves are the bottleneck. Thus, for the largest coverage areas per radio cell, the frequencies need to be low, while the highest capacity coverage areas need to rely on higher frequencies.

The most concrete radio frequencies for 5G range from about 1 GHz up to 30 GHz, but the technically functional, very near-range cells may rely on solutions up to about 100 GHz, while the lowest bands provide the widest coverage areas. The highest frequencies provide much-needed capacity for the limited locations and the lowest frequencies ensure the basic functioning of 5G services within the widest areas.

4.5.2 ITU-R WRC-19 Results

The ITU-R WRC-19 discussed 5G frequency strategies and the respective decisions work as a basis for practical 5G deployments. The ITU-R WRC-15 identified the first set of frequencies to be studied for the feasibility of 5G. The then-identified frequencies included 24.25–27.50, 37.00–40.50, 42.50–43.50, 45.50–47.00, 47.20–50.20, 50.40–52.60, 66.00–76.00, and 81.00–86.00 GHz. These frequencies have been undergoing further detailed studies for the primary use of mobile services. There have also been frequency bands under study requiring possibly additional allocations to mobile services on a primary basis, and these bands include 31.80–33.40, 40.50–42.50, and 47.00–47.20 GHz.

As a most concrete step in this evolution, the WRC-19 decided the use of the frequency bands. As stated in Ref. [26], the ultra-low latency and very high bit-rate applications of IMT will require large contiguous blocks of spectrum. Furthermore, the harmonized worldwide IMT bands are beneficial for global roaming. The WRC-19 thus identified additional bands to enable 5G deployment such as 24.25–27.5, 37.00–43.50, 45.50–47.00, 47.20–48.20, and 66.00–71.00 GHz.

In practice, the first frequencies were auctioned for the initial 5G deployments during the first half of 2019. As an example, US operators have purchased capacity on 2.4, 3.5, 24, and 28 GHz bands. There are also other regulators that are investigating options for the mobile industry's preferred bands above 30 GHz.

4.5.3 RF Bands

4.5.3.1 Principles

5G brings many new frequency bands, including the mmWave spectrum, which offers more capacity for users. 5G has also many enhanced and NR-related solutions. The capacity of the radio interface can be increased, widening the RF bandwidth. As the low-band and mid-band frequency ranges are already rather occupied, the high bands will bring highly needed additional capacity for 5G users thanks to the considerably wider bandwidths of the mmWave regions. The downside of higher-frequency bands is the more limited radio propagation. The coverage area of the mmWave bands might be up to some hundreds of feet, while the lowest bands are adequate for large, suburban and rural areas – at the cost of less data throughput, though.

5G is capable of selecting automatically the most adequate modulation scheme from Quadrature Phase Shift Keying (QPSK) and Quadrature Amplitude Modulation of 16-QAM, 64-QAM, and 254-QAM. QPSK is the most robust mode, which provides the largest coverage, but the lowest data speeds. 256-QAM is the most sensitive to interference and thus works only in very limited areas where it provides the highest bit rates.

Both the UL and DL of 5G are based on OFDM, whereas LTE uses OFDM in the DL and Single Carrier Time Division Multiple Access (SC-TDMA) in the UL. The 5G UL has better PAPR performance than LTE, benefiting especially low-power IoT devices [27].

4.5.3.2 LTE Release 16 Bands

In the initial phase of each 5G operator, radio coverage is typically still fragmented and rather limited because the deployment of the 5G base stations takes time. Especially during the Non-Standalone (NSA) deployment architectures, 4G radio coverage still plays a major role as customers rely largely on the existing LTE bands.

The larger the country and the higher the population, the more challenging is 5G base station deployment at a fast pace to ensure adequate coverage and capacity for all subscribers, which means that the LTE and 5G dual mode devices are in reality still a rather long way off. Therefore, not only is the 5G coverage area as such an important aspect in offering new, advanced, and high-performance services, but the service level and data rates within those combined 4G and 5G coverage areas can also vary vastly depending on the deployed capacity and the band types. The dynamic use of channel coding and modulation schemes becomes more complex balancing 4G and 5G radio coverage and performance, including shared frequency strategies based on, e.g., dynamic spectrum sharing of the 4G and 5G radio spectrum [28, 29].

3GPP TS 36.104 lists the frequency bands for LTE base station radio transmission and reception, whereas 3GPP TS 38.104 defines the allowed bands for 5G NR [30]. As Tables 4.6 and 4.7 indicate, there are various common bands in LTE and NR.

The number of both LTE and NR bands has increased steadily along with new Releases of 3GPP TS. Table 4.6 summarizes the LTE bands presented in 3GPP TS 36.104 version 16.4.0, dated June 2020. For up-to-date information, please refer to the latest version because the 3GPP will update this specification regularly.

Compared to Release 15, Release 16 brings bands 53, 87, and 88. Release 16 also divides unlicensed band 46 further into four subsegments.

The bandwidth of the LTE can be 1.4, 5, 10, 15, or 20 MHz, depending on the band number. Carrier aggregation provides a further means to combine these bands up to five components to achieve wider total bandwidth per single user, up to 100 MHz.

Table 4.6 The 3GPP frequency bands and frequency ranges for the LTE as interpreted from 3GPP TS 36.104 V16.4.0. The channels presented in transparencies are not applicable at present. CA refers to the use of carrier aggregation, and CA* refers to partial restrictions for the use of carrier aggregation (please refer to TS 36.104 for details). Please note that band 46 is for unlicensed operations and consists of sub-bands. The bands in italic represent the additional Release 16 bands.

CH	$f_{UL, low}$ MHz	$f_{UL, high}$ MHz	$f_{DL, low}$ MHz	$f_{DL, high}$ MHz	Mode
1	1920.0	1980.0	2110.0	2170.0	FDD
2	1850.0	1910.0	1930.0	1990.0	FDD
3	1710.0	1785.0	1805.0	1880.0	FDD
4	1710.0	1755.0	2110.0	2155.0	FDD
5	824.0	849.0	869.0	894.0	FDD
(6)	830.0	840.0	875.0	885.0	FDD
7	2500.0	2570.0	2620.0	2690.0	FDD
8	880.0	915.0	925.0	960.0	FDD
9	1749.9	1784.9	1844.9	1879.9	FDD
10	1710.0	1770.0	2110.0	2170.0	FDD

Table 4.6 (*Continued*)

CH	$f_{UL,\ low}$ MHz	$f_{UL,\ high}$ MHz	$f_{DL,\ low}$ MHz	$f_{DL,\ high}$ MHz	Mode
11	1427.9	1447.9	1475.9	1495.9	FDD
12	699.0	716.0	729.0	746.0	FDD
13	777.0	787.0	746.0	756.0	FDD
14	788.0	798.0	758.0	768.0	FDD
15	N/A	N/A	N/A	N/A	FDD
16	N/A	N/A	N/A	N/A	FDD
17	704.0	716.0	734.0	746.0	FDD
18	815.0	830.0	860.0	875.0	FDD
19	830.0	845.0	875.0	890.0	FDD
20	832.0	862.0	791.0	821.0	FDD
21	1447.9	1462.9	1495.9	1510.9	FDD
22	3410.0	3490.0	3510.0	3590.0	FDD
(23)	2000.0	2020.0	2180.0	2200.0	FDD
24	1626.5	1660.5	1525.0	1559.0	FDD
25	1850.0	1915.0	1930.0	1995.0	FDD
26	814.0	849.0	859.0	894.0	FDD
27	807.0	824.0	852.0	869.0	FDD
28	703.0	748.0	758.0	803.0	FDD
29	N/A	N/A	717.0	728.0	FDD (CA)
30	2305.0	2315.0	2350.0	2360.0	FDD
31	452.5	457.5	462.5	467.5	FDD
32	N/A	N/A	1452.0	1496.0	FDD (CA)
33	1900.0	1920.0	1900.0	1920.0	TDD
34	2010.0	2025.0	2010.0	2025.0	TDD
35	1850.0	1910.0	1850.0	1910.0	TDD
36	1930.0	1990.0	1930.0	1990.0	TDD
37	1910.0	1930.0	1910.0	1930.0	TDD
38	2570.0	2620.0	2570.0	2620.0	TDD
39	1880.0	1920.0	1880.0	1920.0	TDD
40	2300.0	2400.0	2300.0	2400.0	TDD
41	2496.0	2690.0	2496.0	2690.0	TDD

(*Continued*)

Table 4.6 *(Continued)*

CH	$f_{UL, low}$ MHz	$f_{UL, high}$ MHz	$f_{DL, low}$ MHz	$f_{DL, high}$ MHz	Mode
42	3400.0	3600.0	3400.0	3600.0	TDD
43	3600.0	3800.0	3600.0	3800.0	TDD
44	703.0	803.0	703.0	803.0	TDD
45	1447.0	1467.0	1447.0	1467.0	TDD
46	5150.0	5925.0	5150.0	5925.0	TDD
46a	*5150.0*	*5250.0*	*5150.0*	*5250.0*	*Unlicensed*
46b	*5250.0*	*5350.0*	*5250.0*	*5350.0*	*Unlicensed*
46c	*5470.0*	*5725.0*	*5470.0*	*5725.0*	*Unlicensed*
46d	*5725.0*	*5925.0*	*5725.0*	*5925.0*	*Unlicensed*
47	5855.0	5925.0	5855.0	5925.0	TDD
48	3550.0	3700.0	3550.0	3700.0	TDD
49	3550.0	3700.0	3550.0	3700.0	TDD LAA
50	1432.0	1517.0	1432.0	1517.0	TDD
51	1427.0	1432.0	1427.0	1432.0	TDD
52	3300.0	3400.0	3300.0	3400.0	TDD
53	*2483.5*	*2495.0*	*2483.5*	*2495.0*	*TDD*
65	1920.0	2010.0	2110.0	2200.0	FDD
66	1710.0	1780.0	2110.0	2200.0	FDD (CA*)
67	N/A	N/A	738.0	758.0	FDD (CA)
68	698.0	728.0	753.0	783.0	FDD
69	N/A	N/A	2570.0	2620.0	FDD
70	1695.0	1710.0	1995.0	2020.0	FDD
71	663.0	698.0	617.0	652.0	FDD (CA*)
72	451.0	456.0	461.0	466.0	FDD
73	450.0	455.0	460.0	465.0	FDD
74	1427.0	1470.0	1475.0	1518.0	FDD
75	N/A	N/A	1432.0	1517.0	FDD (CA)
76	N/A	N/A	1427.0	1432.0	FDD (CA)
85	698.0	716.0	728.0	746.0	FDD
87	*410.0*	*415.0*	*420.0*	*425.0*	*FDD*
88	*412.0*	*417.0*	*422.0*	*427.0*	*FDD*

4.5.3.3 5G NR Bands in Release 16

The current 5G bands, and their bandwidths as per the global specifications, are presented in Figures 4.17 and 4.18. The first phase 5G networks rely on a small number of 5G base stations, while the already existing 4G base stations and 4G core network support the connectivity in a parallel fashion. Thus, the 5G device can provide somewhat faster data speeds within fragmented radio coverage areas, whereas it functions still in 4G mode elsewhere.

Due to practical economic reasons, mobile device manufacturers typically want to limit the supported frequency bands in their offered devices. This is a reality because each band adds to the complexity and cost of the device. As an example, the more bands the device contains, the more components it needs to house such as antennas. This, in turn, increases the size and weight of the device. So, device manufacturers tend to seek an optimal set of supported frequencies and features for their market areas.

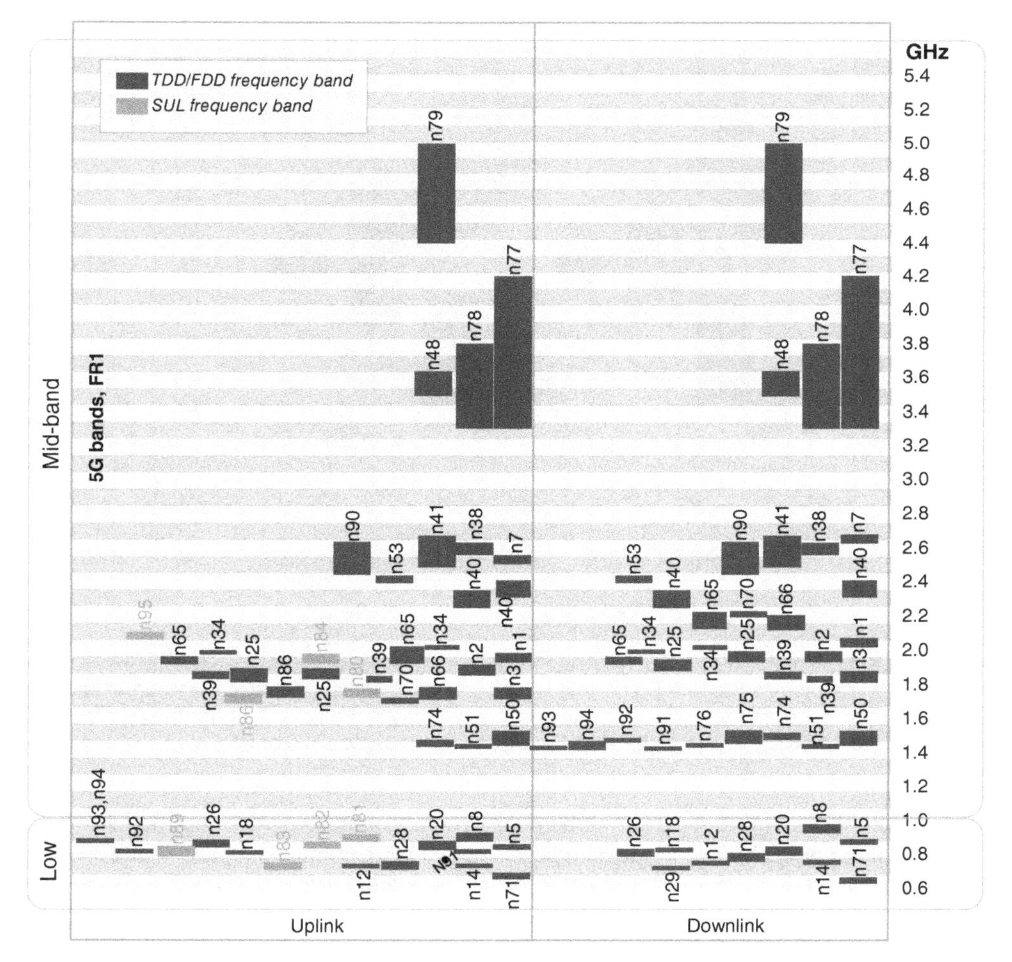

Figure 4.17 The low and mid-bands of 5G and their bandwidths. The wider the bandwidth, the more capacity and data speeds it is capable of offering. As can be seen in this figure, the bandwidths of the low-band area are very limited, while some of the allocations in the mid-band offer significantly more capacity.

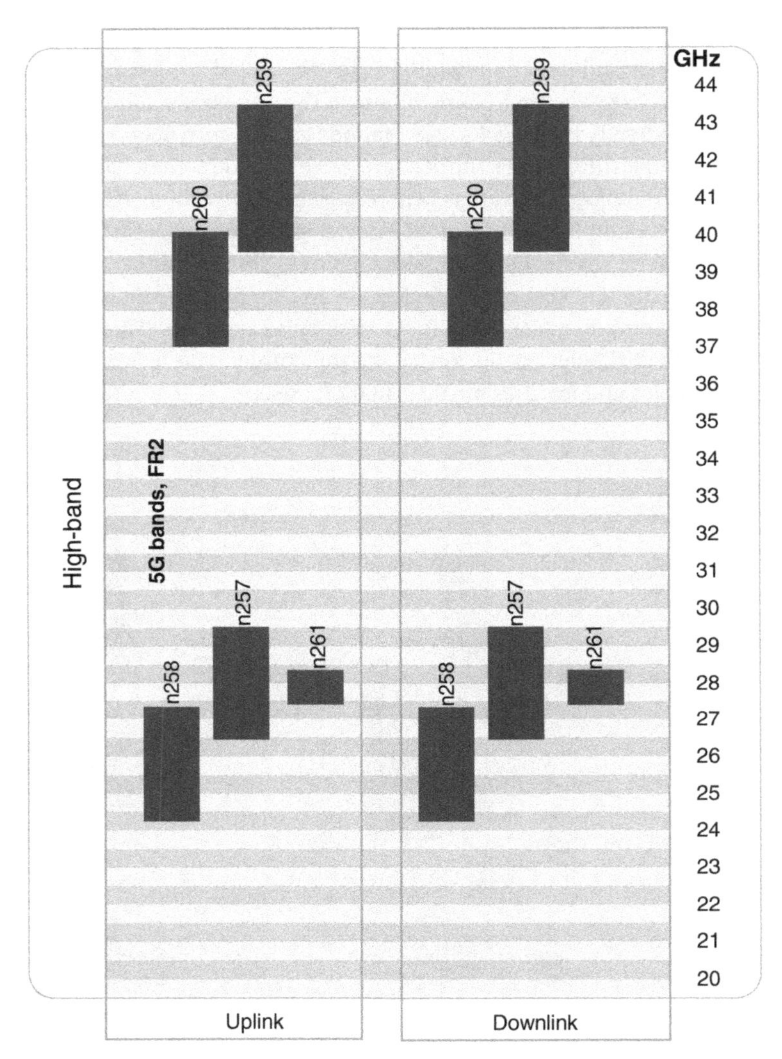

Figure 4.18 The high bands of 5G. These are the new mmWave blocks designed for 5G. They contain much more bandwidth than mid- and low bands do, and their ability to deliver high capacity is superior. High bands are essential for taking full advantage of the new performance of the 5G networks. Please note that the 3GPP will keep extending the list of 5G bands.

As a practical example of this principle, GSMArena summarizes the supported 5G bands of some of the typical 5G smart devices based on available public data [31]. The task of the consumer is to ensure that the desired device supports at least some of the common frequency bands with the mobile network operator.

3GPP TS 38.104 lists the 5G bands. Currently, TS 38.104 presents 51 5G NR band options of which five are on the mmWave band. As can be seen, the number of mmWave bands is still limited at this time. The 3GPP will add new bands to the forthcoming versions.

Table 4.7 summarizes the 5G NR RFs and bands interpreted from the above-mentioned source, version 16.4.0, dated June 2020. As can be seen by comparing Tables 4.5 and 4.6,

Table 4.7 The NR bands and frequency ranges as interpreted from 3GPP TS 38.104 V16.4.0. The bands in italic indicate the additions along with Release 16. Please note that for the bands n91–n94, the UL and DL ranges are independent from each other. Please also note the band n95 is applicable only in China.

CH	$f_{UL,\,low}$ MHz	$f_{UL,\,high}$ MHz	$f_{DL,\,low}$ MHz	$f_{DL,\,high}$ MHz	Mode
n1	1920.0	1980.0	2110.0	2170.0	FDD
n2	1850.0	1910.0	1930.0	1990.0	FDD
n3	1710.0	1785.0	1805.0	1880.0	FDD
n5	824.0	849.0	869.0	894.0	FDD
n7	2500.0	2570.0	2620.0	2690.0	FDD
n8	880.0	915.0	925.0	960.0	FDD
n12	*699.0*	*716.0*	*729.0*	*746.0*	*FDD*
n14	*788.0*	*798.0*	*758.0*	*768.0*	*FDD*
n18	*815.0*	*830.0*	*860.0*	*875.0*	*FDD*
n20	832.0	862.0	791.0	821.0	FDD
n25	*1850.0*	*1915.0*	*1930.0*	*1995.0*	*FDD*
n26	*814.0*	*849.0*	*859.0*	*894.0*	*FDD*
n28	703.0	748.0	758.0	803.0	FDD
n29	*N/A*	*N/A*	*717.0*	*728.0*	*SDL*
n34	*2010.0*	*2025.0*	*2010.0*	*2025.0*	*TDD*
n38	2570.0	2620.0	2570.0	2620.0	TDD
n39	*1880.0*	*1920.0*	*1880.0*	*1920.0*	*TDD*
n40	*2300.0*	*2400.0*	*2300.0*	*2400.0*	*TDD*
n41	2496.0	2690.0	2496.0	2690.0	TDD
n48	*3550.0*	*3700.0*	*3550.0*	*3700.0*	*TDD*
n50	1432.0	1517.0	1432.0	1517.0	TDD
n51	1427.0	1432.0	1427.0	1432.0	TDD
n53	*2483.5*	*2495.0*	*2483.5*	*2495.0*	*TDD*
n65	*1920.0*	*2010.0*	*2110.0*	*2200.0*	*FDD*
n66	1710.0	1780.0	2110.0	2200.0	FDD
n70	1695.0	1710.0	1995.0	2020.0	FDD
n71	663.0	698.0	617.0	652.0	FDD
n74	1427.0	1470.0	1475.0	1518.0	FDD
n75	N/A	N/A	1432.0	1517.0	SDL
n76	N/A	N/A	1427.0	1432.0	SDL

(Continued)

Table 4.7 *(Continued)*

CH	$f_{UL, low}$ MHz	$f_{UL, high}$ MHz	$f_{DL, low}$ MHz	$f_{DL, high}$ MHz	Mode
n77	3300.0	4200.0	3300.0	4200.0	TDD
n78	3300.0	3800.0	3300.0	800.0	TDD
n79	4400.0	5000.0	4400.0	5000.0	TDD
n80	1710.0	1785.0	N/A	N/A	SUL
n81	880.0	915.0	N/A	N/A	SUL
n82	832.0	862.0	N/A	N/A	SUL
n83	703.0	748.0	N/A	N/A	SUL
n84	1920.0	1980.0	N/A	N/A	SUL
n86	*1710.0*	*1780.0*	*N/A*	*N/A*	*SUL*
n89	*824.0*	*849.0*	*N/A*	*N/A*	*SUL*
n90	*2496.0*	*2690.0*	*2496.0*	*2690.0*	*TDD*
n91	*832.0*	*862.0*	*1427.0*	*1432.0*	*FDD*
n92	*832.0*	*862.0*	*1432.0*	*1517.0*	*FDD*
n93	*880.0*	*915.0*	*1427.0*	*1432.0*	*FDD*
n94	*880.0*	*915.0*	*1432.0*	*1517.0*	*FDD*
n95	*2010.0*	*2025.0*	*N/A*	*N/A*	*SUL*
n257	26 500.0	29 500.0	26 500.0	29 500.0	TDD
n258	24 250.0	27 500.0	24 250.0	27 500.0	TDD
n259	*39 500.0*	*43 500.0*	*39 500.0*	*43 500.0*	*TDD*
n260	37 000.0	40 000.0	37 000.0	40 000.0	TDD
n261	*27 500.0*	*28 350.0*	*27 500.0*	*28 350.0*	*TDD*

part of the LTE and 5G bands are common, while some 5G NR bands are unique such as those of the mmWave spectrum, n257–n261.

The WRC-19 identified a total of 17.25 GHz of the 5G spectrum, while the 5G spectrum prior to the conference was 1.9 GHz. Of this number, 14.75 GHz of spectrum has been harmonized worldwide. The ITU-R WRC-19 was thus an important milestone in this evolution as it allocated new bands for 5G as an opportunity for further formalization at the standardization and national frequency auctions. As Ref. [26] points out, the WRC-19 also took measures to ensure an appropriate protection of the Earth Exploration Satellite Services, including meteorological and other passive services in adjacent bands.

As a next step, the WRC-19 formed a set of studies to identify frequencies for new components of 5G such as High Altitude IMT Base Stations (HIBS). HIBS benefits especially underserved areas where ground-based IMT base stations are challenging to deploy.

For the current 5G frequency bands adopted by the 3GPP specifications in Release 16, the frequency range FR1 covers frequencies in the 410 MHz–7.125 GHz range, while FR2 refers to frequencies within 24.250–52.600 GHz. Please note that the FR1 range was limited to 450 MHz–6 GHz in Release 15. As an example, the bands n257–n261 presented in Table 4.7 belong to FR2, whereas the rest of the 5G bands are in in FR1.

Please also note that NB-IoT is designed to operate in the NR operating bands n1, n2, n3, n5, n7, n8, n12, n14, n18, n20, n25, n28, n41, n65, n66, n70, n71, n74, and n90.

4.6 5G Radio Aspects

The key specification for 5G NR is 3GPP 38.300 [32]. Among other aspects, it defines the protocol architecture and functional split, interfaces, protocols, channels and procedures, mobility and states, scheduling, UE functionalities and capabilities, QoS, security, self-configuration, and self-optimization.

The following sections summarize the key radio aspects of Release 16.

4.6.1 Bandwidth

5G NR is able to support a variety of UE channel bandwidths in a flexible way while it operates within the base station's channel bandwidth. As 3GPP 38.104 states, the base station can transmit to and/or receive from one or more UE bandwidth parts that are smaller than or equal to the number of carrier resource blocks on the RF carrier, in any part of the carrier resource blocks.

TS 38.104 specifies multiple transmission bandwidth configurations N_{RB} per base station's channel bandwidth and respective subcarrier spacing for FR1 and FR2. The FR1 transmission bandwidth configurations can have bandwidth values of 5, 10, 15, 20, 25, 30, 40, 50, 60, 70, 80, 90, and 100 MHz, while the subcarriers can vary between the values of 15, 30, and 60 kHz. For FR2 mode, the transmission bandwidth configuration can have bandwidth values of 50, 100, 200, and 400 MHz, while the subcarriers can be either 60 or 120 kHz.

Further requirements for RF channel utilization, including the guard bands, tolerance values for interfering bands, etc., can be found in 3GPP TS 36.104 and TS 38.104 for 4G LTE and 5G NR, respectively.

5G NR supports different UE channel bandwidths in a flexible way. The base station can transmit to and receive from one or more UE bandwidth sections in any part of the carrier resource blocks. Along with the results of the ongoing feasibility studies for the 5G frequency bands, the current concrete bands are listed in 3GPP TS 38.104. It specifies transmission bandwidth configurations for the frequency ranges FR1 and FR2 as summarized in Table 4.8.

4.6.2 Duplex

5G uses FDD and TDD frequency bands. 5G also introduces a Supplementary (SUL) concept. Ref. [32] depicts the principle of SUL, and it is detailed in 3GPP TS 38.300.

Table 4.8 3GPP Release 16 5G frequency range definitions. Please note that Release 16 extends the FR1 range, and along with the identified additional frequency bands for 5G by the WRC-19, the FR2 is expected to widen, too, in future 3GPP releases

Frequency range	Release 15 band	Release 16 band
FR1	450–6000 MHz	410–7125 MHz
FR2	24 250–52 600 MHz	24 250–52 600 MHz

4.6.3 SUL

The idea of SUL in 5G is to compensate the cell radio coverage in the UL direction. The UE's transmitted power level in the UL is typically very low compared to the 5G gNB power levels. This may lead to considerable performance degradation closer to the cell edge area causing unbalance between the UL and DL directions. This can be seen, e.g., by failing two-way mobile video conferencing, especially if there is insufficient overlapping in the cell edge area.

As the lower frequencies propagate wider distances, the idea is to use such a secondary supplementary frequency band to enhance the UL performance of the UE. In 5G, the concept is adaptive, so the default UL band is used as such in normal conditions. As soon as the quality decreases below a defined threshold value on a cell edge area, the 5G network orders the UE to use a better-propagating SUL band for its UL transmission instead.

The currently defined 5G SUL bands are n81, n82, and n83, which are located at the sub-1 GHz spectrum to ensure a large coverage area. The bands n80, n84, and n86 all are below the 2 GHz spectrum, which still provides adequate coverage. In practical deployments, the potential interference with the overlapping LTE frequencies needs to be researched case by case based on 3GPP TR 37.872 and TR 37.716. Release 16 enhances the SUL further, e.g., by introducing extended support for interband aggregation via UL carrier aggregation, SUL without LTE/NR DC, and LTE/NR DC without SUL [33].

4.6.4 Dynamic Spectrum Sharing

Dynamic Spectrum Sharing (DSS) refers to the ability of the 5G RAN to take advantage of part of the 4G radio access. This combined resource utilization for a single session augments the performance accordingly [34–37]. As an example, AT&T has deployed in the United States 5G NR. AT&T has allocated their DSS on the Band n5/Band 5 spectrum with a 2 × 10 MHz configuration.

Along with the deployments during Release 15, and continuing as Release 16 expands the set of 5G services, DSS is thus a commercially viable 5G NR feature. It provides operators with the possibility of rapidly expanding their 5G NR footprint without having to refarm their LTE spectrum.

4.6.5 5G Antennas

Along with 5G deployments, it is expected that more advanced antenna technologies will be used, too. 3GPP TR 37.842 V2.1.0 defines RF requirements for an Active Antenna System (AAS) on the base station, as of Release 13. This specification acts as a base also for the 5G era with respect to the AAS.

4.6.6 Radio Performance

The key capabilities of Release 16 5G include high data rates and low latency values. Release 16 also enhances ultra-high reliability, energy efficiency, and extreme device density.

Among other advances, the frame structure of 5G will also be evolving. Table 4.9 summarizes the key statements relevant to the frame structure of the Release 16 radio interface, and compares the values with those of 4G.

Table 4.9 The key characteristics of the 4G and 5G radio interface.

Characteristics	4G LTE	5G NR Release 15
Radio frame duration	10 ms	10 ms
Subframe duration	1 ms	1 ms
Slot duration	0.5 ms	0.5 ms
Slot format	Predefined	Configurable in a dynamic and semi-statistical way
Channel coding for data	Turbo	LDPC
Channel coding for control	Tail Bit Convolution Code (TBCC)	Polar
Modulation scheme for UL	Single-Carrier Frequency-Division Multiplexing (SC-FDMA)	Discrete Fourier Transform spread Orthogonal Frequency Division Multiplexing (DFT-s-OFDM); OFDM (optional)
Modulation scheme for DL	OFDM	OFDM
Bandwidth (MHz)	1.4, 3, 5, 10, 15, 20	5, ..., 100 (sub-6 GHz); 50, ..., 400 (above 6 GHz)
Subcarrier spacing (kHz)	15 (unicast, Multimedia Broadcast Multicast Service (MBMS)); 7.5/1.25 (MBMS-dedicated carrier)	30, 60, 120; 240 (not for data)

(Continued)

Table 4.7 (*Continued*)

Characteristics	4G LTE	5G NR Release 15
Maximum carrier aggregation Component Carrier (CC)	32	16
Maximum MIMO antenna ports	8 (Single User MIMO (SU-MIMO)); 2 (Multi User MIMO (MU-MIMO))	8 (SU-MIMO); 16 (MU-MIMO)
HARQ transmission/retransmission	TB (Transport Block)	TB, code block group

4.6.7 OFDM in Release 16

3GPP Release 16 uses Cyclic Prefix OFDM (CP-OFDM) for radio interface, continuing the Release 15 solution. The following summarizes the key justifications for CP-OFDM [38].

- OFDM is spectral efficient in UL and DL, and provides fast data speeds.
- OFDM can rely fluently on MIMO for high spectral efficiency via both SU-MIMO and MU-MIMO.
- Release 16 adjustments of OFDM enhance further the PAPR. The PAPR of OFDM in 5G can be lowered by applying PAPR reduction techniques with minor impact in performance [39].
- OFDM supports high-speed use cases.
- The baseband complexity of an OFDM receiver is low.
- OFDM is well localized in the time domain, which is relevant in the support of latency-critical URLLC and dynamic TDD.
- The OFDM CP makes it robust to timing synchronization errors.
- OFDM is a flexible waveform so it supports a variety of use cases and services over a wide range of frequencies adjusting the subcarrier spacing accordingly. In 5G NR, the bandwidth can vary from 5 up to 100 or 400 MHz depending on the scenario.
- A guard interval, CP, is added at the beginning of each OFDM symbol to preserve orthogonality between subcarriers and eliminate Inter-Symbol Interference (ISI) and Inter-Carrier Interference (ICI).

4.6.8 Modulation (LTE)

LTE can use QPSK, 16-QAM, and 64-QAM modulation schemes, whereas 5G has the option to utilize up to 254-QAM. The channel estimation of OFDM is usually done with the aid of pilot symbols. The channel type for each individual OFDM subcarrier corresponds to flat fading. Pilot symbol-assisted modulation on flat fading channels involves the sparse insertion of known pilot symbols in a stream of data symbols.

QPSK modulation provides the largest coverage areas but with the lowest capacity per bandwidth. 64-QAM results in a smaller coverage, but it offers more capacity.

4.6.9 Coding

LTE uses turbo coding or convolutional coding, the former being more modern providing in general about 3 dB gain over the older and less effective, but at the same time more robust, convolutional coding. 5G further optimizes the coding by introducing new channel coding schemes, i.e., LDPC for data channels and polar codes for control channels [40].

4.6.10 OFDM

5G uses CP-OFDM and DFT-s-OFDM in UL communications. As stated in 3GPP TS 38.300, the DL transmission waveform of the 5G radio interface is conventional OFDM using a CP, which is referred to as CP-OFDM. It is the very same as in LTE and LTE-A, as described in the previous section. The UL transmission waveform of 5G, in turn, is conventional OFDM using a CP with a transform precoding function performing DFT spreading that can be disabled or enabled. The latter is referred to as DFT-s-OFDM.

The difference between 5G and LTE/LTE-A multiplexing is thus for the UL; instead of previously utilized SC-TDMA, 5G is based on OFDM in both the DL and UL. The DFT spreading that can be applied in 5G UL optimizes the PAPR performance, which was the reason to select the SC-TDMA in the first place in LTE/LTE-A because the sole conventional OFDM is not optimal for it, especially in use cases requiring low battery consumption such as low-power IoT devices.

In the DL, 5G supports closed loop Demodulation Reference Signal (DMRS)-based spatial multiplexing for the Physical Downlink Shared Channel (PDSCH). Furthermore, up to 8 and 12 orthogonal DL DMRS ports are supported for type 1 and type 2 DMRS, respectively. Up to eight orthogonal DL DMRS ports for each UE are supported for SU-MIMO and up to four orthogonal DL DMRS ports for each UE are supported for MU-MIMO, as defined in 3GPP TS 38.300.

A more detailed description of the Downlink Shared Channel (DL-SCH) physical layer model can be found in 3GPP TS 38.202, and the Physical Broadcast Channel (PBCH) physical layer model is described in 3GPP TS 38.202. The requirements of the ODFM in 5G have been defined in 3GPP TS 38.300, Section 5.1. It summarizes the flow of waveform generation for both DL and UL.

4.6.11 Modulation

In 5G, there is a set of modulations that the radio interface supports. For proper selection of the modulation scheme, a modulation mapper is applied. It receives binary values (0 or 1) as input and produces complex-valued modulation symbols as output. These output modulation symbols can be $\pi/2$-BPSK, BPSK, QPSK, 16-QAM, 64-QAM, and 256-QAM. The modulation schemes supported are listed in Table 4.10.

4.6.12 Frame Structure

The frame structure of 5G is defined in 3GPP TS 38.211 (NR; physical channels and modulation) [41]. In 5G, DL and UL transmissions are organized into frames, which are derived from the OFDM structure. A single frame has a duration of

Table 4.10 The modulation schemes of 5G.

	BPSK	QPSK	16-QAM	64-QAM	256-QAM
DL		✓	✓	✓	✓
UL, OFDM, and CP		✓	✓	✓	✓
UL, DFT-s-OFDM, and CP	✓	✓	✓	✓	✓

$$T_f = (\Delta f_{\max} N_f / 100) T_c = 10\,\text{ms}$$

Each frame has 10 subframes. An individual subframe has a duration of

$$T_f = (\Delta f_{\max} N_f / 1000) T_c = 1\,\text{ms}$$

The number of consecutive OFDM symbols per subframe varies depending on the number of the symbol and slot. Furthermore, each frame is divided into two half-frames of five subframes each with half-frame 0 consisting of subframes 0–4 and half-frame 1 consisting of subframes 5–9. A more detailed description of the frame structure and frequency band allocation can be found in 3GPP TS 38.133 and TS 38.213.

4.6.13 5G Channels

4.6.13.1 Channel Mapping

Figure 4.19 depicts the mapping of logical and transport channels of the 5G system.

The following sections summarize the 5G UL and DL channels.

4.6.13.2 Uplink

As defined in [41], an *UL physical channel* refers to a set of resource elements carrying information that originates from higher layers. 3GPP has defined the following *UL physical channels* for 5G:

- Physical Uplink Shared Channel (PUSCH);
- Physical Uplink Control Channel (PUCCH);
- Physical Random-Access Channel (PRACH).

The physical layer uses *UL physical signals* that do not carry information originating from higher layers. These UL physical signals are the following:

- Demodulation Reference Signals (DM-RS);
- Phase-Tracking Reference Signals (PT-RS);
- Sounding Reference Signal (SRS).

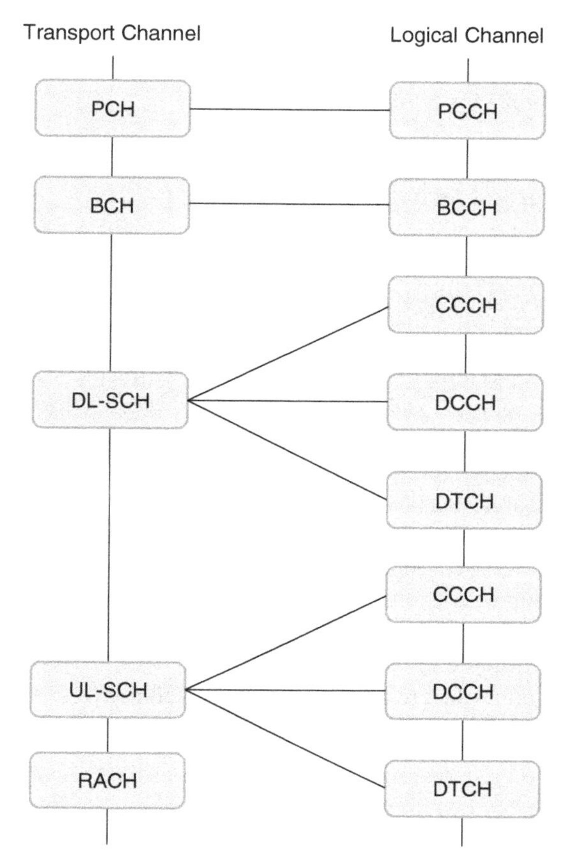

Figure 4.19 Logical and transport channel mapping in 5G [42].

The UE uses a frame structure and physical resources as dictated by Ref. [41] when it is transmitting in UL. Furthermore, Ref. [41] defines a set of antenna ports that are applied in UL.

4.6.13.3 Downlink

As defined in Ref. [41], a *DL physical channel* refers to a set of resource elements that carry information arriving from higher layers. 5G defines the following DL physical channels:

- Physical Downlink Shared Channel (PDSCH);
- Physical Broadcast Channel (PBCH);
- Physical Downlink Control Channel (PDCCH).

A *DL physical signal* refers to a set of resource elements that the physical layer uses, yet these do not carry information originating from higher layers. The DL physical signals are the following:

- DM-RS for PDSCH and PBCH;
- PT-RS;

- Channel-State Information Reference Signal (CSI-RS);
- Primary Synchronization Signal (PSS);
- Secondary Synchronization Signal (SSS).

4.6.14 General Protocol Architecture

A physical layer channels overview is presented in TS 38.201 (NR general description), and TS 38.202 (NR; services provided by the physical layer) [43]. These specifications also detail 5G protocol architecture and its functional split, and the overall and radio protocol architectures, MAC, RLC, PDCP, and RRC.

The 3GPP specifications describe the NR interface covering the interface between the UE and the network on layers 1, 2, and 3. TS 38.200 series describes the layer 1 (physical layer) specifications. Layers 2 and 3 are described in the TS 38.300 series.

Figure 4.20 depicts the NR interface protocol architecture related to layer 1, which interfaces the MAC layer 2 and RRC layer 3. The connectivity between layers is done by utilizing Service Access Points (SAP).

The transport channels between the physical radio layer and the MAC layer answer the question: How is the information transferred over the radio interface. In the upper layer, between the MAC and RRC layers, MAC provides logical channels. They answer the question: What type of information is transferred?

The physical layer offers data transport services to higher layers. Access to these services is gained by using the transport channel through the MAC layer.

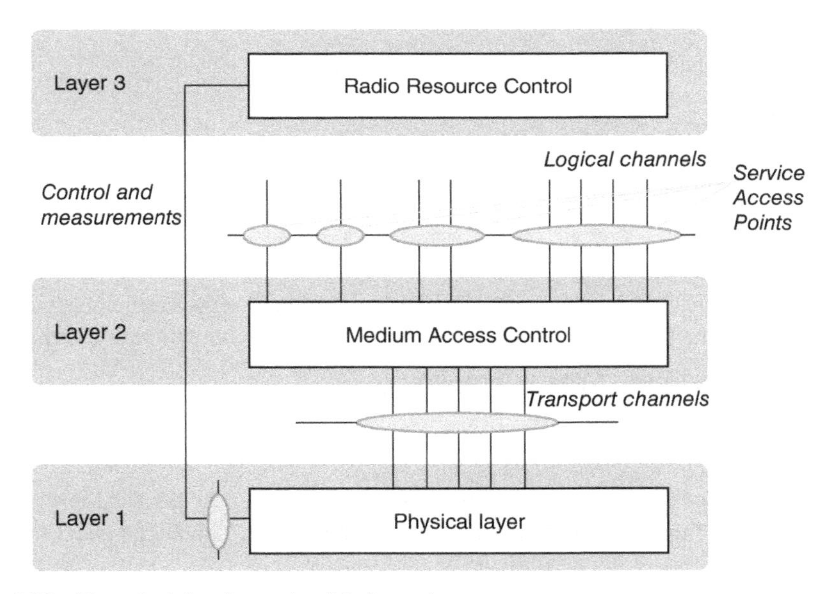

Figure 4.20 The principle of mapping 5G channels.

4.6.15 Physical Layer Procedures

Physical layer procedures include cell search, power control, UL synchronization, UL timing control, random access and HARQ procedures, antenna beam management, and CSI-related procedures. NR provides support for interference coordination via physical layer resource control in frequency, time, and power domains.

4.6.16 Physical Layer Measurements

Radio characteristics are measured by the UE and the network and reported to higher layers. These include, e.g., measurements for intra- and interfrequency handover, inter-RAT handover, timing measurements, and measurements for RRM. Measurements for inter-RAT handover are defined in support of handover to E-UTRA.

4.6.17 Quality of Service

5G aspects of the QoS are defined in 3GPP TS 38.300.

References

1 3GPP, "3GPP Specification Series 38," 3GPP, 2018. [Online]. Available: http://www.3gpp.org/DynaReport/38-series.htm. [Accessed 26 July 2018].

2 Casaccia, L., "Propelling 5G Forward: A Closer Look at 3GPP Release 16," Qualcomm, 7 July 2020. [Online]. Available: https://www.qualcomm.com/news/onq/2020/07/07/propelling-5g-forward-closer-look-3gpp-release-16. [Accessed 4 August 2020].

3 FierceWireless, "Verizon Aims to Deploy Small Cells in 3.5 GHz When Practical," FierceWireless, 2018. [Online]. Available: https://www.fiercewireless.com/tech/verizon-aims-to-deploy-small-cells-3-5-ghz-when-practical. [Accessed 29 July 2018].

4 Qualcomm, "Private LTE Networks. White Paper," Qualcomm, 2017.

5 Li, R., Mo, Z., Sun, H. et al., "A Low-Profile and High-isolated MIMO Antenna for 5G Mobile Terminal," *Micromachines* 12. 27 March 2020.

6 You, C., Jung, D., Song, M. et al., "Advanced 12 × 12 MIMO Antennas for Next Generation 5G," *2019 IEEE International Symposium on Antennas and Propagation and USNC-URSI Radio Science Meeting*, 7–12 July 2019.

7 "3GPP TSG RAN Meeting 82 Documents," 12–13 December 2018. [Online]. Available: https://www.3gpp.org/ftp/TSG_RAN/TSG_RAN/TSGR_82/Docs. [Accessed 4 August 2020].

8 "5G Evolution: 3GPP Releases 16 & 17 Overview," Ericsson, 9 March 2020. [Online]. Available: https://www.ericsson.com/en/reports-and-papers/ericsson-technology-review/articles/5g-nr-evolution. [Accessed 4 August 2020].

9 3GPP, "TS 38.300 V16.2.0, NR; NR and NG-RAN Overall Description, Stage 2," 3GPP, July 2020.

10 3GPP, "TS 23.501 V16.5.1, System Architecture for the 5G System (5GS), Stage 2," 3GPP, August 2020.

11 3GPP, "TS 38.301 V16.2.0, Architecture Description, Release 16," 3GPP, July 2020.
12 3GPP, "TS 38.401 V16.2.0," 3GPP, July 2020.
13 RF Wireless World, "5G Protocol Stack; 5G Layer 1, 5G Layer 2, 5G Layer 3," RF Wireless World, 1 July 2018. [Online]. Available: http://www.rfwireless-world.com/Terminology/5G-Protocol-Stack-Layer-1-Layer-2-and-Layer-3.html. [Accessed 1 July2018].
14 GSA, "5G Device Ecosystem Report March 2019," GSA. [Online]. Available: https://gsacom.com/paper/5g-device-ecosystem-report-march-2019. [Accessed 4 July 2019].
15 3GPP, "TS 23.502 V16.5.1 (2020-08), Procedures for the 5G System (5GS), Stage 2, Release 16," 3GPP, August 2020.
16 Murphy, K., "Centralized RAN and Fronthaul," Ericsson. [Online]. Available: https://www.isemag.com/wp-content/uploads/2016/01/C-RAN_and_Fronthaul_White_Paper.pdf. [Accessed 6 July 2019].
17 Bougioukos, M., "Preparing Microwave Transport Network for the 5G World," Nokia, 2017.
18 ETSI, "TSI White Paper No. 23: Cloud RAN and MEC: A Perfect Pairing," ETSI, February 2018. [Online]. Available: https://www.etsi.org/images/files/ETSIWhitePapers/etsi_wp23_MEC_and_CRAN_ed1_FINAL.pdf. [Accessed 30 August 2020].
19 ITU-T, "GSTR-TN5G: Transport Network Support of IMT-2020/5G," ITU, February 2018. [Online]. Available: https://www.itu.int/dms_pub/itu-t/opb/tut/T-TUT-HOME-2018-PDF-E.pdf. [Accessed 30 August 2020].
20 Shelton, M., "Open RAN Terminology," Parallel Wireless, 20 April 2020. [Online]. Available: https://www.parallelwireless.com/open-ran-terminology-understanding-the-difference-between-open-ran-openran-oran-and-more. [Accessed 29 July 2020].
21 O-RAN Alliance, "O-RAN: Towards an Open and Smart RAN," O-RAN Alliance, October 2018.
22 O-RAN Alliance, "O-RAN Use Cases and Deployment Scenarios," O-RAN Alliance, February 2020.
23 KeySight, "O-RAN Next-Generation Fronthaul Conformance Testing (Whitepaper)," KeySight, USA, 22 July 2020.
24 Techplayon, "Open RAN (O-RAN) Reference Architecture," Techplayon, 28 October 2018. [Online]. Available: http://www.techplayon.com/open-ran-o-ran-reference-architecture. [Accessed 10 August 2020].
25 Thinksmallcell, "Europe's Plans for CBRS," Thinksmallcell, 2018. [Online]. Available: https://www.thinksmallcell.com/Technology/europe-plans-for-cbrs-style-shared-spectrum-in-2-3ghz-band.html. [Accessed 29 July 2018].
26 ITU, "WRC-19 Identifies Additional Frequency Bands for 5G," ITU, 22 November 2019. [Online]. Available: https://news.itu.int/wrc-19-agrees-to-identify-new-frequency-bands-for-5g. [Accessed 19 August 2020].
27 IoT For All, "Unlicensed LTE Explained – LTE-U Vs. LAA Vs. LWA Vs. Multefire," IoT For All, 2018. [Online]. Available: https://www.iotforall.com/unlicensed-lte-lte-u-vs-laa-vs-lwa-vs-multefire. [Accessed 29 July 2018].
28 Atis, "Multi Operator Core Network," Atis, 2018. [Online]. Available: https://access.atis.org/apps/group_public/download.php/31137/ATIS-I-0000052.pdf. [Accessed 29 July 2018].
29 FCC, "The FCC Decisions for the 3.5GHz Band," 2018. [Online]. Available: https://www.fcc.gov/wireless/bureau-divisions/broadband-division/35-ghz-band/35-ghz-band-citizens-broadband-radio. [Accessed 29 July 2018].

30 3GPP, "TS 38.104: NR Base Station (BS) Radio Transmission and Reception, Release 16," 3GPP, June 2020.
31 GSMArena, "Phone Finder," 24 December 2019. [Online]. Available: https://www.gsmarena.com.
32 3GPP, "Ts 38.300 V15.0.0 (2017-12)," 3GPP, 2017.
33 Qualcomm, "Propelling 5G Forward: A Closer Look at 3GPP Release 16," Qualcomm, July 2020. [Online]. Available: https://www.qualcomm.com/media/documents/files/propelling-5g-forward-a-closer-look-at-release-16.pdf. [Accessed 30 August 2020].
34 5G Americas, "5G Americas, Shared Network Options," 5G Americas, 2018. [Online]. Available: http://www.5gamericas.org/files/4914/8193/1104/SCF191_Multi-operator_neutral_host_small_cells.pdf. [Accessed 29 July 2018].
35 CBRS Alliance, "CBRS Alliance," CBRS Alliance, 2018. [Online]. Available: www.cbrsalliance.org. [Accessed 29 July 2018].
36 Ericsson, 5G Radio Access, Ericsson, 2016.
37 Nokia, "Citizens Broadband Radio Service (CBRS): High-quality Services on Shared Spectrum. White Paper," Nokia, 2017.
38 Ericsson, "Waveform and Numerology to Support 5G Services and Requirements," Ericsson, 2017.
39 Lim, D.-W., Heo, S.-J., and No, J.-S., "An Overview of Peak-to-Average Power Ratio Reduction Schemes for OFDM Signals," *Journal of Communications and Networks* 11(3), 229–239, 2009.
40 Lagen, S., Wanuga, K., Elkotby, H. et al., "New Radio Physical Layer Abstraction for System-Level Simulations of 5G Networks," 28 January 2020. [Online]. Available: https://arxiv.org/pdf/2001.10309.pdf. [Accessed 30 August 2020].
41 3GPP, "TS 38.211; Physical Channels and Modulation," 3GPP, 2018.
42 Tabbane, S., "5G Networks and 3GPP Release 15: ITU PITA Workshop on Mobile Network Planning and Security," ITU, Nadi, Fiji Islands, 23–25 October 2019.
43 3GPP, "TS 38.201; NR; General Description," 3GPP, 2018.

5

Core and Transport Network

5.1 Overview

5.1.1 The 5G Pillars

The 5G system provides operators with optimal resources, security, capacity, and quality for a variety of use cases. To achieve this, the 3rd Generation Partnership Project (3GPP) has specified enhanced and new network technologies that form different service components, pillars, for servicing these use cases. The 5G dimensions are the enhanced Mobile Broadband (eMBB), Critical Communications (CriC) and Massive Internet of Things (mIoT), supported by vehicular communications (Vehicle-to-Vehicle (V2V) and its variants), and network operations.

5G must comply with a multitude of performance criteria such as area traffic capacity, peak data rate, user's experienced data rate, spectrum efficiency, mobility, latency, connection density, and energy efficiency. The needed and offered performance depends on the type of service dimension.

As an example, mIoT is capable of coping with very high simultaneously communicating device density and network efficiency due to the multitude of devices functioning for several years with the same battery. At the same time, e.g., Ultra-Reliable Low Latency Communications (URLLC)-type applications require the highest performance from a low-latency and high-mobility perspective, while it is not particularly demanding for any other requirement parameters. eMBB is demanding for all the criteria except for connection density and latency.

Low latency is the most challenging criteria for the core network to comply with, as it needs to be as low as 1 ms for the most critical use cases. There are a variety of tiers for 5G Core (5GC) deployment strategies. Some of these are:

- *Distributed data center footprint*. This refers to a model called Central Office Re-architected as Data Center (CORD). This, in turn, refers to the edge computing concept with physical assets residing closer to the core's edge, by transforming them as data centers. As network utilization increases, the distributed data centers can support increasingly Cloud Radio Access Network (Cloud-RAN) and virtual-RAN hub sites, as well as virtualized core Network Functions (NFs).

5G Second Phase Explained: The 3GPP Release 16 Enhancements, First Edition. Jyrki T.J. Penttinen.

- *Cloud-based 4G Evolved Packet Core (EPC) and 5GC.* This supports new techniques designed for the evolved 4G and 5G core networks such as separation of the User Plane (UP) and Control Plane (CP). The strict latency requirements can be met via distributed cloud infrastructure.
- *Software-Defined Networking (SDN).* This is needed to support the increasing edge-cloud locations for 5G, which need high performance and secure connectivity. The centralized data centers, edge clouds, and cell sites can thus be connected via wide-area SDN in an efficient way as it provides adequate performance and level of routing, security, redundancy, and orchestration [1].
- *Network slicing.* This concept optimizes 5G core services per use case. Network slicing refers to the concept where the Virtual Network Functions (VNFs) are formed by service, which results in optimized processing paths for data delivery over the whole network. Furthermore, the slices are isolated from each other, including the security aspects [2].
- *Network operation automation.* This concept makes it possible to take in new services efficiently and quickly. The impact can be monitored and analyzed, and it is possible to modify and scale of the services in a highly dynamic way.

The 5G core network is standardized to define a functional architecture in such a way that the implementation and replacement of further technologies is possible, and the 5G interfaces support much more fluently multivendor scenarios than in previous generations. Furthermore, the 5G architecture provides separate scaling of UP and CP functionality, which facilitates the deployment of the UP and CP independently. Furthermore, 5G supports authentication for both 3GPP and non-3GPP identities, which makes it possible to use it fluently, e.g., Wi-Fi networks as part of the communications. Also, 5G allows the utilization of different network configurations via respective network slices.

The most essential new core network concepts in 5G are thus the separation of UP and CP, network slicing, and Service-Based Architecture (SBA).

5.1.2 5G Core Network Services

There are many services defined in 3GPP releases prior to the 5G era. They keep evolving and are valid in the 5G era, too, in a parallel fashion. The specification 3GPP TS 23.501, Section 5.16, lists key services that are still valid in the 5G era. These include Public Warning System (PWS), Short Message Service (SMS) over Non-Access Stratum (NAS), IP Multimedia System (IMS), emergency services, multimedia priority services, eCall, and emergency services fallback.

To deliver these enhanced and new services, the 5G core network needs to be updated. A first logical step to support the advanced services is to deploy the fully end-to-end 5G architecture as per the Standalone (SA) mode.

5.2 Network Functions Virtualization

5.2.1 SDN

SDN refers to network architecture that has been designed to minimize the hardware constraints by abstraction of the low-level functions. Instead, SDN makes it possible to execute these functions in a software-based, centralized CP via Application Programming Interfaces (APIs). The benefit of this approach is that the network services are agnostic to the underlying hardware, and they can be offered and utilized regardless of the connected hardware elements.

The SDN concept can be used for 5G as a framework that makes it possible to provide 5G to function in the CP, thus optimizing the data transmission. As an example, instead of using dedicated network elements such as Policy Control Functions (PCFs), which was a standalone component in earlier networks, 5G enables the functionality of such a component to perform its tasks on a common hardware, which is shared among many other NFs.

Some of the benefits of the SDN concept are optimized bandwidth and enhanced latency performance. Furthermore, rerouting of the data flows takes place in practically real time via the SDN, which enhances considerably the prevention of network outages and thus contributes to the development of high-availability services, which are desired, e.g., in the CriC of 5G.

5.2.2 NFV

Together with the SDN concept, Network Functions Virtualization (NFV) has an important role in providing optimal 5G performance. The principal task of NFV is to decouple the software from the hardware via the deployment of Virtual Machines (VM) [3]. These VMs perform a multitude of NFs such as encryption of the communications.

The benefit of the VM concept is the highly dynamic characteristics, so whenever needed, the VM is created for the respective function automatically. This results in cost savings as the operator of such NFs does not need to invest in a standalone element, which typically is based on proprietary hardware architectures. Yet another benefit of the virtualization of NFs is the possibility of deploying them much faster than the dedicated hardware/software elements.

As for 5G, NFV enables the network slicing concept, which is an elemental part of the system. Network slicing enables the use of a multitude of virtual networks on top of a physical infrastructure, and it also provides a means for dividing a physical network into multiple virtual networks that can serve various RANs. Yet another benefit of NFV is the possibility to easily scale the functions based on different criteria for optimal performance, e.g., cost or energy consumption.

NFV simplifies the operation of the functions by decoupling them from the traditionally required standalone hardware components. The VNFs are deployed on high-volume servers or cloud infrastructure instead of specialized hardware. This model expedites cost-efficient commercial deployments.

NFV also enables the 5G network slicing that provides Mobile Network Operators (MNOs) with the possibility of customizing and optimizing resources for different types of verticals. In addition to network slicing as such, 5G NFV makes it possible to divide physical network into a set of virtual networks that may support multiple RANs.

The underlying NFV acts as a layer between hardware resources and the virtual environment such as virtual compute, storage, and networking. NFV is connected to the VNFs, which are controlled by an Element Management System (EMS). Furthermore, NFV management is connected to the MNO's Operations Support System (OSS) and Business

Support System (BSS) which, among other tasks, takes care of 5G network maintenance and billing.

The benefits of NFV also include the possibility to optimize further resource provisioning of the VNFs based on the cost or other criteria and scale the VNFs accordingly. 5G MNOs can use NFV to introduce new services when needed.

SDN is a network architecture model designed to overcome hardware limitations. SDN is related to NFV as it benefits in the abstraction of lower-level functions, and instead moves them to a normalized CP that manages network via the API as depicted in Figure 5.1.

SDN thus provides the MNOs with the possibility to offer 5G services via a centralized CP in a hardware-agnostic manner. It enhances the data flows via lower bandwidth and lower latency, and also network redundancy can be managed more efficiently.

SDN results in more flexible networks. The SDN network architecture supports the 5G ecosystem requirements and can be used to design, build, and manage 5G networks. As the CP and UP are separated, the CP is directly programmable, while the underlying infrastructure is abstracted for applications and network services to create various network hierarchies.

While control is distributed in traditional networks, SDN detaches the CP from the network hardware, enabling packet data flow control through a controller. The controller is

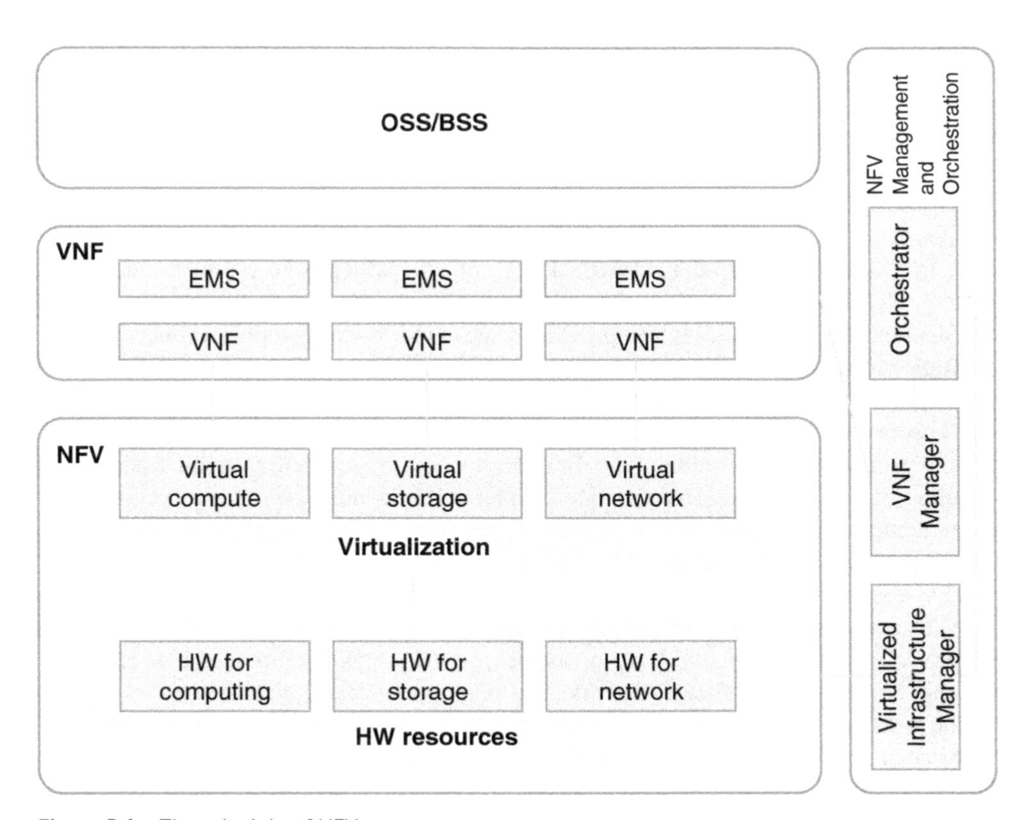

Figure 5.1 The principle of NFV.

located between network devices and applications as depicted in Figure 5.2. As a consequence, network control is programmable, facilitating the management of the 5G network as well as modification and addition of services [4].

For more details on network virtualization, please refer to the ETSI Network Function Virtualization Industry Specification Group documentation [5].

5.3 5G Cloud Architecture

5.3.1 Concept

The 5G system will change the philosophy of earlier mobile communication networks. One of the concepts indicating this evolution is the introduction of the cloud concept. This is a result of increasing the intelligence of the 5G network, which, in turn, enables user devices and network equipment to communicate efficiently via optimized transport and core functionalities.

In practice, this means that the cloud concept is utilized for enabling intelligent service awareness, which, in turn, optimizes the connectivity, latency, and other essential

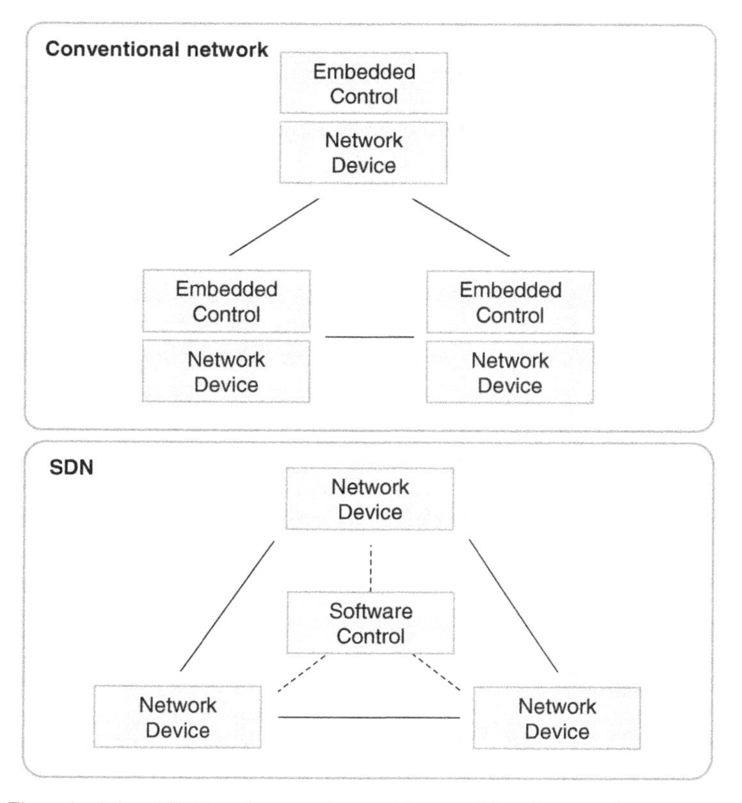

Figure 5.2 The principle of SDN and comparison with a traditional network.

performance characteristics. Thus, 5G networks will be flexible and scalable to offer advanced services to end-users, which can be done by virtualizing the NFs.

As a result, the role of data centers will increase considerably along with 5G network deployment. They enable highly dynamic, scalable, and flexible offering of capacity, whether the respective connected device is a tiny IoT sensor or advanced equipment capable of supporting high-speed data for Virtual Reality (VR) purposes.

In addition to centralized clouds hosted in data centers, 5G application can also reside in a network's edge, closer to the end-user. These applications can be hosted in mobile edge computing nodes, referred to as cloudlets.

5.3.2 Data Center as a Base for 5G Architecture

5G will require increased cloud computing to fulfill the strict performance criteria for advanced communications. Data centers must be upgraded to satisfy this need for 5G traffic. The elemental requirements are related to increased CPU performance and storage capacity, which requires an increased number of servers, enhanced cooling of the physical equipment, and more space for housing the respective racks.

Most remarkably, data centers require fundamental architectural renewal from the previous decentralized mobile network model to better serve the centralized processing of the network functionalities. As a result, instead of specific equipment for mobile system purposes, there will be cloud solutions for both centralized and edge regions.

The NFs are increasingly executed in these clouds in a virtualized software environment instead of local hardware processing. This principle could potentially evolve to such an extent that the major part of base station processing takes place in clouds and is handed back to the local processing of the original base stations merely in events of heavy cloud congestion.

There is a trend for relying on the Open Compute Project (OCP) in so-called hyperscale data center operators. This concept can be assumed to be adopted increasingly by telecommunication operations [6]. According to this source, various telecommunications companies investigated Proofs of Concept (PoCs) by applying Open Compute in their data centers. The significance of Open Compute has increased considerably to serve as a telecommunications cloud for OCP. In fact, telecommunications may turn out to be the driving force for such development of OCP, accompanied by the Telecom Infra Project (TIP).

According to Ref. [7], the trend of cloud business applications, including 5G and IoT in general, can lead into the Internet becoming increasingly distributed. This means that more data centers need to be located closer to users. The mayor benefits of these compact edge data centers include the ability to reduce costs and latency. This, in turn, leads into a hierarchical architecture consisting of a centralized data center and several connected edge clouds as can be interpreted from Refs. [7, 8].

Edge computing is thus an elemental part of the distributed data center infrastructure, ensuring that the computing and storage resources are optimally close to users [8].

The benefits of edge computing include ultra-low latency, reduction and offloading of core network (backhaul) traffic, and in-network data processing. Edge computing facilitates optimal performance so that latency-sensitive applications benefit from the concept. Some examples of these special environments are autonomous devices, including self-driving vehicles, drones, and robots, as well as applications offering immersive user experience such as VR and Augmented Reality (AR) solutions with highly interactive interfaces requiring practically real-time response.

The distribution of NFs can be divided, e.g., into common functions such as slice control, distributed data functions such as data forwarding and Quality of Service (QoS) control, and slice-specific control functions such as session/mobility control and policy management, while the transport network delivers the data and control information between the NFs.

The main reason for the utilization of edge data centers is the significantly lower latency and faster delivery of the content. Also, the cost of transport can be reduced significantly.

It can be assumed that much of the IoT traffic will be generated locally due to the characteristics of the devices; they typically are not too mobile but stay in their special environments such as factories, whereas the other extreme where mobility is considered may be mIoT adapted to the Vehicle-to-Everything (V2X) environment.

5.3.3 Network as a Service

One of the key documents for the cloud environment is ITU Recommendation ITU-T Y.3515, which provides Network as a Service (NaaS) functional architecture [9]. The Recommendation ITU-T Y.3500 defines NaaS as a cloud service category in which the capability provided to the cloud service customer is transport connectivity and related network capabilities. Concretely, the cloud service provider is the party that makes cloud services available for cloud service customers.

5.4 Network Functions Overview

5.4.1 5G Release 15 and 16 Network Functions

Virtualized 5G NFs replace the old mobile communications architectural model, which relied on dedicated hardware elements. Figure 5.3 depicts the 5G NFs in the new SBA model for roaming. In this scenario, User Equipment (UE), such as a smartphone or IoT device, roams the Visited Mobile Network (VPLMN). The UE establishes a connection to the Data Network (DN), while the Home Public Land Mobile Network (HPLMN), which is the user's home 5G network, enables its connectivity. The visited and home networks communicate with each other for the user's subscription information (via Unified Data Management (UDM)), subscriber authentication (via the Authentication Server Function (AUSF)), and policies (via PCF).

The interworking between HPLMN and VPLMN is protected by home Security Edge Protection Proxy (hSEPP) and visited network Security Edge Protection Proxy (vSEPP). In this example, the visited network provides functions for network slice selection (via the Network Slice Selection Function (NSSF)), network access control and mobility management (via the Access and Mobility Management Function (AMF)), data service management (via the Session Management Function (SMF)), and applications (via the Application Function (AF)). This example indicates the 5G network functions in Release 15 and 16, while Tables 5.1 and 5.2 summarize the 5G Release 15 and 16 functions.

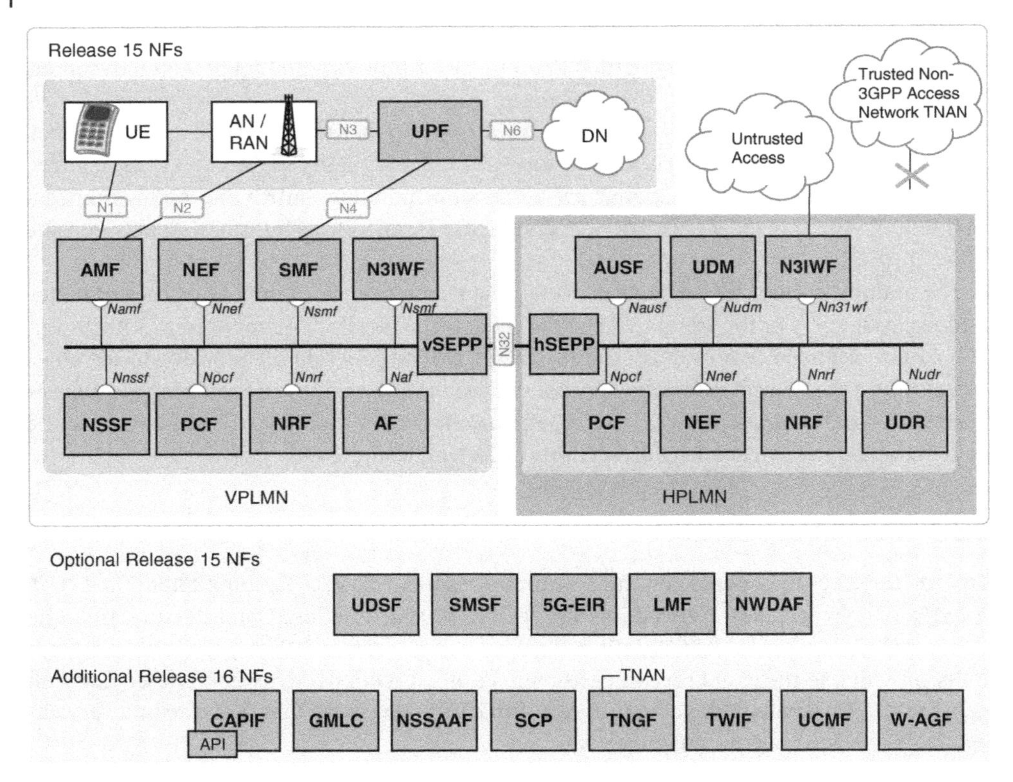

Figure 5.3 The 5G NFs in 3GPP Release 15 and 16 networks.

5.4.2 5G Core Network Aspects

The 5G core network architecture is service based, and it can take advantage of the SDN and NFV concepts. The 5G core network is also based on *network slicing*, which optimizes the offered service type per vertical [10]. Network slicing can be understood as "networks over a network." With it, a network operator can create a single network, or in typical cases, several parallel virtual networks within a physical network to serve a variety of very different types of verticals. As an example, a drone-controlling application benefits from ultra-low latency values to ensure the fastest possible response to remote commands. Meanwhile, an intelligent, permanently installed wireless remote sensor may need only a most basic and occasional data transfer service for a telemetric purpose. By setting up separate network slices, the MNO can optimize the network resources and provide fluent user experience for all customers and devices.

Because of the much more dynamic operations of the network slice-enabled network compared to traditional models, the 5G core network needs to manage slices efficiently and in real time. This takes place via the NSSF.

The 5G core network is based on the cloud concept, and the most efficient forms of the core will take advantage of network slicing and SBA. The deployment of these advanced

Table 5.1 The Release 15 NFs.

NF	Description
5G-EIR	The optional *5G Equipment Identity Register* is an evolved version of the Long Term Evolution (LTE) EIR. The task of the 5G-EIR is to check the status of the Permanent Equipment Identity (PEI) in case it has been blacklisted. The PEI is an evolved variant of the traditional International Mobile Equipment Identity (IMEI) used in the previous mobile networks.
AF	The *Application Function* is comparable with the LTE Application Server (AS) and the GSM Service Control Function (gsmSCF). It interacts within the core network to provide services such as application influence on traffic routing, access to the Network Exposure Function (NEF), and interaction with the policy framework.
AMF	The 5G *Access and Mobility Management Function* replaces the LTE Mobility Management Entity (MME). It takes care of 5G signaling. The AMF has many tasks such as access, authentication, and authorization, and transport for SMS. It embeds the *Security Anchor Functionality* (SEAF) and includes Location Services (LCS) management for regulatory services.
AUSF	The 5G *Authentication Server Function* replaces the LTE MME and Authentication, Authorization, and Accounting (AAA). It supports authentication for 3GPP access and untrusted non-3GPP access.
LMF	The *Location Management Function* determines the location of UE. It provides geodetic location determination for target UE. Please refer to Chapter 6 for more information on the LMF.
N3IWF	The *Non-3GPP Interworking Function* takes place when untrusted access such as Wi-Fi is used via the 5G infrastructure.
NEF	The 5G *Network Exposure Function* is an evolution of the Service Capability Exposure Function (SCEF) and API layer of LTE. In 5G, it assists in storing and retrieving exposed capabilities and events of the NFs together with the Unified Data Repository (UDR). This information is shared between NFs within the 5G network.
NRF	The 5G *Network Repository Function* is part of the evolution of the Dynamic Name Server (DNS) used in LTE. In 5G, it supports the service discovery function for providing information of the discovered NF instances to the requesting NF instance. One of the tasks of the NRF is to maintain the NF profile and respective services of available NF instances.
NSSF	The 5G *Network Slice Selection Function* supports the 5G-specific network slicing concept. The NSSF selects the needed set of network slice instances serving UE.
NWDAF	The *Network Data Analytics Function* is managed by the respective 5G MNO. The NWDAF provides slice-specific network data analytics to the NFs that are subscribed to it.
PCF	The 5G *Policy Control Function* is an evolution of the LTE Policy and Charging Enforcement Function (PCRF). In 5G, it supports the unified policy framework to govern network behavior. It provides policy rules to the CP functions to enforce them.
SEPP	The 5G *Security Edge Protection Proxy* interconnects 5G networks. The SEPP elements are located at the edge of the 5G core network (the hSEPP residing in home network and the vSEPP in visited network). The SEPP provisions confidentiality and integrity protection of the signaling between networks, and hides the topology of the network behind a firewall. The SEPP is a non-transparent proxy and supports message filtering and policing on the inter-PLMN CP interfaces.

(Continued)

Table 5.1 (Continued)

NF	Description
SMF	The 5G *Session Management Function* replaces, together with the 5G User Plane Function (UPF), the LTE Serving Gateway (S-GW) and PDN Gateway (P-GW). Among other tasks, the SMF handles Session Management (SM) for session establishment, modification, and release.
SMSF	The *Short Message Service Function* supports the SMS over a 5G NAS. It manages SMS subscription data and takes care of their delivery via signaling channels.
UDM	*Unified Data Management* generates 3GPP Authentication and Key Agreement (AKA) credentials for each user. UDM resides in the same HPLMN with the subscriber it serves. It performs user identification, including storage and management of the Subscription Permanent Identifier (SUPI) per each individual 5G subscriber. It also can de-conceal the Subscription Concealed Identifier (SUCI), which is the privacy-protected subscription identifier. Furthermore, UDM manages access authorizations such as roaming restrictions.
UDR	The 5G *Unified Data Repository* is an evolved version of LTE Structured Data Storage (SDS). It can store and retrieve subscription data, and it takes care of the storage and retrieval of policy data by the PCF, storage and retrieval of structured data for exposure, and application data by the NEF.
UDSF	The 5G *Unstructured Data Storage Function* is comparable with the LTE Structured Data Storage Function (SDSF). In 5G, the UDSF is an optional function for the storage and retrieval of information in a form of unstructured data by any NF.
UPF	The 5G *User Plane Function* replaces, together with the SMF, the LTE S-GW and P-GW. It takes care of the user data. The UPF acts as an anchor point for intra- and inter-Radio Access Technology (RAT) mobility. It is also the external PDU session point to interconnect DN and it takes care of packet routing and forwarding.

features enhances the end-user's perception of the quality of service as the network develops.

For the incumbent MNO, the first step for practical core deployment is to take advantage of the already existing 4G infrastructure. This happens by using virtual 4G core, which is prepared for the transition, i.e., the already existing infrastructure can be prepared to provide a base for the "5G-ready" networks. Then, relying on this base, it is possible to deploy new software functionalities such as network slicing, cloud-based VNFs, SDN, and distributed UP and automatization. The existing network can be optimized and scaled up to support increasing 5G-type use cases.

As is the case in Release 15, part of the new Release 16 elements are mandatory, while others are optional. Furthermore, some elements can be co-located physically into other elements. In 5G, the previous LTE elements and their functions have been reorganized, and new NFs are introduced. The SBA also reorganizes the CP so that instead of MME, S-GW, and P-GW as they have been defined in the EPC, their functionality has been divided into AMF and SMF. Along with the network slicing concept, which is new in 5G, a single UE can have one or more SMFs associated with it at a time, each representing a network slice.

The following sections summarize the key elements of the 3GPP Release 16 5G architecture and their functioning as interpreted from 3GPP Release 16 TS 23.501 [10].

5.5 NFs Enhanced in Release 16

The following sections summarize the NFs defined in Release 15. Release 16 has further shaped part of these original NFs.

5.5.1 5G-EIR

5G-EIR is an evolution of the LTE EIR. As has been the principle in previous generations, it also has optional functionality in the 5G network. The task of the 5G-EIR is to check the status of PEI whether it has been blacklisted or not. Functioning of the 5G-EIR is the same in Releases 15 and 16.

5.1.2 AF

The AF works in the application layer of 5G, and it is comparable with the LTE AS and gsmSCF. It interacts with the 3GPP core network to provide services such as application influence on traffic routing, access to the NEF, and interaction with the policy framework for policy function. As a new function in Release 16, it also manages the interactions of the IMS with 5GC. Depending on each deployment scenario, the AF, which the operator trusts, may interact with relevant NFs, while the non-trusted AF relies on an external exposure framework via the NEF for interaction with NFs.

5.5.3 AMF

The 5G AMF replaces the LTE MME. The AMF has a multitude of tasks and interfaces with various functional elements as depicted in Figure 5.4.

The AMF has the following functionality in Release 16:

- Termination of RAN CP interface *N2*.
- Termination of NAS interface *N1* for NAS ciphering and integrity protection.
- Registration, connection, reachability, and mobility management.
- Lawful Interception (LI) via the interface to the LI system.
- Transport for SM messages between UE and SMF, for SMS messages between UE and SMSF, and for LCS messages between UE/RAN and LMF.
- It is a transparent proxy for routing SM messages.
- Performs access authentication and authorization.
- Acts as an SEAF (TS 33.501).
- Manages LCS for regulatory services.
- In case of interworking, allocates EPS Bearer ID.
- Notifies UE mobility events.
- Supports CP and UP C-IoT 5GS optimization.
- Provisions external parameters such as "expected UE behavior" or "network configuration."
- Supports Network Slice-Specific Authentication and Authorization (NSSAA).

Table 5.2 The additional NFs presented in Release 16.

NF	Description
CAPIF	Common API Framework for 3GPP northbound APIs provides exposure of the NEF for external entities. It facilitates standardized integration of services with diverse service providers for interaction at the application layer.
CHF	The new Charging Function is specified in 3GPP TS 32.255. It considers various configurations and functionalities the SMF supports.
GMLC	The Gateway Mobile Location Centre extends the functionality of the LMF defined in Release 15, also adding roaming cases.
I-SMF (V-SMF)	Intermediate SMF. During mobility events such as handover or AMF change, if the service area of the SMF does not include the new UE location, then the AMF selects and inserts an I-SMF, which can serve the UE location and Single Network Slice Selection Assistance Information (S-NSSAI).
I-UPF	Intermediate UPF supports redundant transmission on *N3*/*N9* interfaces.
NSSAAF	Network Slice Specific Authentication and Authorization Function.
SCP	Service Communication Proxy is a decentralized solution and is composed of CP and data plane. It provides routing control, resiliency, and observability to the core network.
TNGF	Trusted Non-3GPP Gateway Function connects trusted non-3GPP access networks to the 5G core network. This extends the Release 15 N3IWF, which connects the untrusted non-3GPP access network to the 5G core network.
TWIF	Trusted WLAN Interworking Function enables Non-5G-Capable over WLAN (N5CW) devices to access 5GC via trusted WLAN access networks.
UCMF	UE radio Capability Management Function serves for storing dictionary entries related to PLMN-assigned or manufacturer-assigned UE radio capability IDs.
W-AGF	Wireline Access Gateway Function is an NF in the Wireline 5G Access Network (W-5GAN) that provides connectivity to the 5GC, 5G-Residential Gateway (5G-RG), and Fixed Network Gateway (FN-RG).

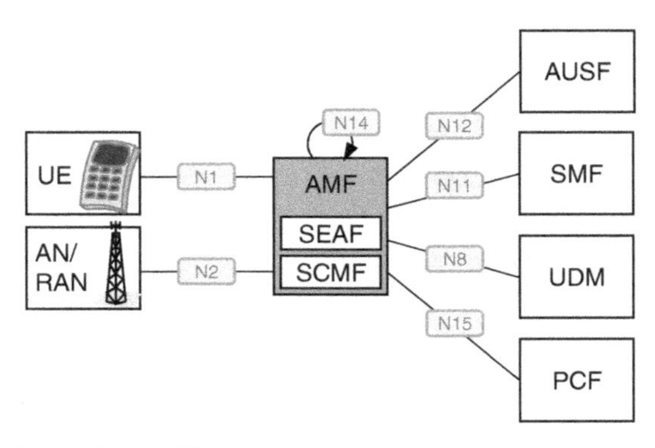

Figure 5.4 The interfaces of the AMF.

The AMF performs termination of the RAN CP *N2* interface and NAS *N1* interface. It also manages NAS ciphering and integrity protection, registration, connection, reliability, and mobility. It includes the LI interface. Furthermore, it provides transport for SM messages between UE and the SMF, and it functions as a transparent proxy for routing SM messages.

It accesses authentication and authorization and provides transport for SMS messages between UE and the SMSF. It embeds SEAF according to 3GPP TS 33.501. It also includes LCS management for regulatory services and provides transport for LCS messages for UE-LMF and RAN-LMF. It allocates EPS Bearer ID for EPS interworking, and manages UE mobility event notification.

In addition, the AMF may support other functions such as the *N2* interface with N3IWF/TNGF and NAS signaling with UE over N3IWF/TNGF. The AMF uses the *N14* interface for AMF reallocation and AMF-to-AMF information transfer. TS 23.501 provides further details of these additional features of the AMF.

5.5.4 AUSF

The 5G AUSF replaces LTE MME/AAA. It supports authentication for 3GPP access and untrusted non-3GPP access as per 3GPP TS 33.501. It has interfaces towards the AMF and UDM as depicted in Figure 5.5.

5.5.5 LMF

The LMF can determine the location for UE. It can obtain downlink location measurements or estimate the location from the UE and uplink location measurements from the NG-RAN. It can also obtain non-UE associated assistance data from the NG-RAN. Figure 5.6 depicts the interfaces of LMF. LMF functionality is defined in Clause 4.3.8 of TS 23.273.

5.5.6 N3IWF

N3IWF refers to the Non-3GPP Interworking Functions that take place when untrusted access is used. In this case, the N3IWF establishes an IPsec tunnel with the UE. The N3IWF

Figure 5.5 The interfaces of the AUSF.

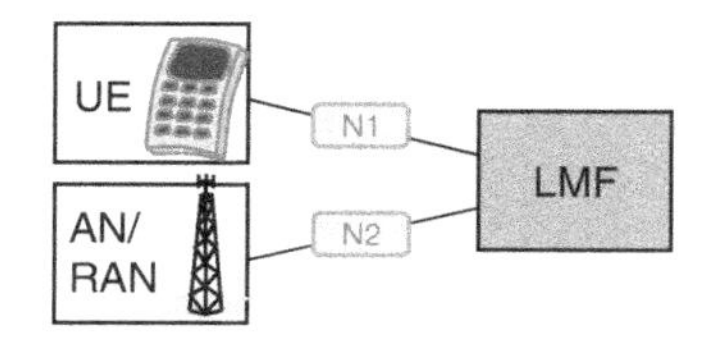

Figure 5.6 The interfaces of the LMF.

uses the IKEv2/IPsec protocols with the UE over *NWu* and relays *N2* information to authenticate and authorize access of the UE with the 5G core network. The N3IWF has termination points of *N2* and *N3* with 5GC. It also relays uplink and downlink CP NAS signaling of *N1* between the UE and AMF, handles *N2* signaling from the SMF, and is relayed by AMF related to PDU sessions and QoS. It establishes IPsec Security Association (IPsec SA) for PDU session traffic.

The N3IWF (Figure 5.7) relays uplink and downlink UP packets between the UE and UPF. This task includes the following: packet decapsulation and encapsulation for IPsec and *N3* tunneling; QoS enforcement for *N3* packet marking based on the information received from *N2*; *N3* UP packet marking in the uplink; and support in AMF selection. In addition, the N3IWF can act as a local mobility anchor within untrusted non-3GPP access networks using MOBIKE as described in IETF RFC 4555. Among other benefits, MOBIKE enables a remote access VPN user to move from one address to another without re-establishing all security associations with the VPN gateway [11].

5.5.7 NEF

The 5G NEF is an evolution of the SCEF and API layer of LTE. In 5G, it assists in storing and retrieving exposed capabilities and events of NFs via the *Nudr* interface with the UDR. This information is shared between NFs in the 5G network. The NEF can also expose this information securely to an entity residing outside the 3GPP network, such as third party and edge computing entities. As depicted in Figure 5.8, the NEF can access only the UDR residing within the same PLMN. For external exposure, the NEF can communicate with the respective entities via CAPIF and respective API. This procedure is detailed in 3GPP TS 23.222.

Via the NEF, 3GPP NFs can retrieve information needed for efficiently managing the connections. An example of such information is the capabilities and expected behavior of the UE. The NEF is also capable of translating information such as the AF service identifier and 5GC-related information, e.g., Data Network Name (DNN) and SNSSAI. One of the important tasks of the NEF is to mask sensitive information to external AFs as per the MNO policy rules.

The NEF stores the available information retrieved from NFs into the UDR in structured data format. This information can be exposed to other NFs via the NEF or be utilized for other tasks such as data analytics.

The NEF may also support a Packet Flow Descriptions Function (PFDF). In 5G, this could be the case, e.g., in the *Sx* Packet Flow Descriptions (PFDs) management procedure to provision PFDs for one or more application identifiers in the UP function (Sponsored Data Connectivity Improvements (SDCIs)) [12]. The PFDF is a repository that stores PFDs.

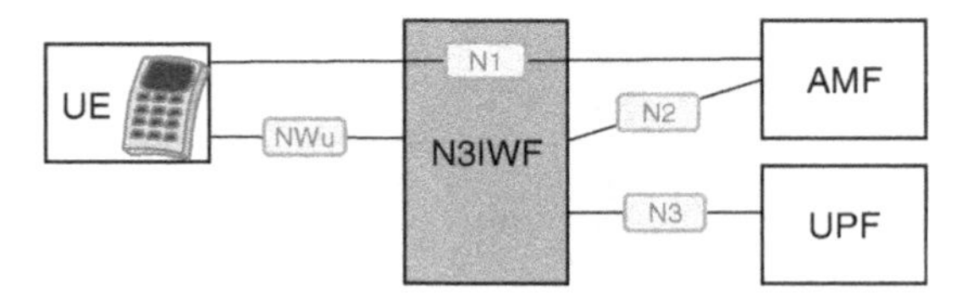

Figure 5.7 The interfaces of the N3IWF.

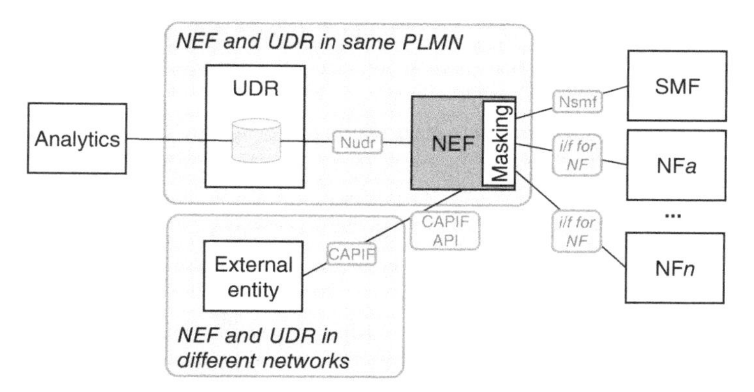

Figure 5.8 The interfaces of the NEF.

They can be added, updated, and removed by the Subcarrier Spacing/Access Stratum (SCS/AS) via the SCEF.

As an example, an IoT platform serving a business customer may want to add a new PFD allowing sensor traffic to pass to a new server. In this example, the PFD could be of a form consisting of PFD-ID, server-IP address, protocol, and port number. This, in turn, needs to be enforced and handled by the Policy and Charging Enforcement Function (PCEF) (which can be P-GW in the case of 4G). The PFDF thus enables the third party AS to provision, modify, and remove PFDs via the SCEF into the MNO network. The PFDF in return may store the PFDs or provision it to the PCEF via the JavaScript Object Notation (JSON) interface (*Gw*) [13].

In 5G, the PFDF in the NEF may store and retrieve PFDs in the UDR. The NEF provides these PFDs to the SMF on pull request. PFDs can also be delivered to the SMF triggered by the NEF push request. This procedure is detailed in 3GPP TS 23.503.

As for some of the additional functions detailed in Release 16, the NEF may support a 5G LAN Group Management Function as described in TS 23.502. It also can provide NWDAF analytics securely for an external party as per TS 23.288, as well as provide analytics data collection from a third party to the NWDAF. Furthermore, the NEF facilitates the management of Non-IP Data Delivery (NIDD) configuration and delivery of Mobile Originated (MO) and Mobile Terminated (MT) unstructured data so that it exposes the NIDD APIs as per TS 23.502.

The NEF resides in the HPLMN for external exposure of services of the UE. It is up to MNOs to agree if the NEF in the HPLMN has an interface with the VPLMN NFs.

CAPIF may be supported in the case of the external exposure of the NEF so that the respective NEF recognizes the CAPIF API provider domain functions as per TS 23.222.

5.5.8 NRF

The 5G NRF is part of the evolution of DNS utilized in LTE. In 5G, it supports the service discovery function for providing information of discovered NF instances or SCP to the requesting NF instance or SCP. The NRF thus maintains the NF profile and respective services of available NF instances, and in Release 16 it also maintains SCP profile of available SCP instances, and supports SCP discovery by SCP instances.

The NRF notifies registered, updated, and deregistered NF and SCP instances and the respective NF services to the subscribed NF service consumer or SCP, and it is aware of the status of NFs and SCP.

The UDR profiles are standardized, and in such scenario the NF profile contains the following: NF instance ID; NF type; PLMN ID; network slice identifiers such as S-NSSAI and Network Slice Instance (NSI) ID; Fully Qualified Domain Name (FQDN) or IP address of each NF; information on NF capacity; authorization information related to the NF; supported service names; endpoint address of each instance of supported services; and identification of stored data.

The applicability of this information as for other NF profiles is left up to implementation: notification endpoint for NF service notifications; routing ID of SUCI; a set of Globally Unique AMF Identifiers (GUAMIs) and Tracking Area Identities (TAIs) if the AMF is involved; UDM Group ID if UDM is involved; UDR Group ID if UDR is involved; and AUSF Group ID if AUSF is involved. The NRF is capable of mapping between the above-mentioned UDM, UDR, and AUSF Group ID and SUPIs to facilitate the discovery of UDM, UDR, and AUSF by SUPI.

For network slicing, multiple NRFs can be deployed at the PLMN level, shared-slice level, and slice-specific level.

Furthermore, multiple NRFs may be deployed in a multitude of networks: NRF in the visited PLMN is referred to as vNRF, and NRF in the home PLMN is known as hNRF, which is referenced by vNRF via the *N27* interface (Figure 5.9).

Release 16 NRF also supports Proxy Call Session Control Function (P-CSCF) discovery.

The NRF maintains a profile of the NF instances. The profile includes the following data elements:

- NF instance ID and NF type.
- PLMN ID.
- NSIs such as S-NSSAI and NSI ID.
- NF FQDN or IP address.
- NF capacity and priority information for AMF selection.
- NF set ID and NF service set ID of the NF service instance.
- NF-specific service authorization information.
- Names of supported services, if applicable.
- Endpoint addresses of instances of supported service.
- Identification of stored data and information for a UDR profile.

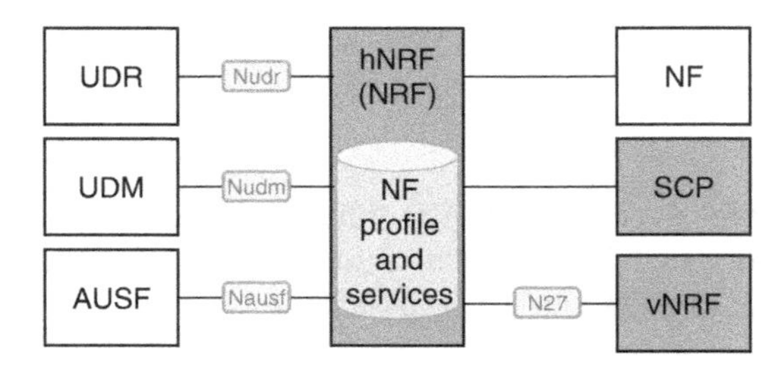

Figure 5.9 The interfaces of the NRF (which can be divided into home and visited NRF).

- Other service parameters such as a DNN list and notification endpoint for NF service.
- Location information for the NF instance such as geographical location and data center. Please note this information is operator specific.
- TAI list.
- NF load information.
- Routing indicator for UDM and the AUSF.
- GUAMI list for the AMF.
- SMF area identity list for the UPF.
- UDM Group ID, and ranges of SUPIs, Generic Public Subscription Identifiers (GPSIs), internal group identifiers, and external group identifiers for UDM.
- UDR Group ID, and ranges of SUPIs, GPSIs, and external group identifiers for the UDR.
- AUSF Group ID and ranges of SUPIs for the AUSF.
- PCF Group ID and ranges of SUPIs for the PCF.
- HSS Group ID, sets of IP Multimedia Private Identities (IMPIs) and IP Multimedia Public Identities (IMPUs) for the HSS.
- Supported Analytics IDs and NWDAF serving area information for the NWDAF.
- Event IDs and Application IDs supported by AFs, in the case of the NEF.
- Ranges of external identifiers, or ranges of external group identifiers, or the domain names served by the NEF, in the case of the NEF when it exposes AF information for analytics purposes.
- IP domain list, ranges of UE IPv4 addresses or IPv6 prefixes, in the case of the Binding Support Function (BSF).

5.5.9 NSSF

The 5G NSSF is a new function to support the 5G-specific network slicing concept. The NSSF selects a needed set of network slice instances serving the UE. It also determines the allowed and configured NSSAIs and if necessary maps them to the subscribed S-NSSAIs. Furthermore, it determines the set of AMFs serving the UE, or alternatively, the list of candidate AMFs. For the latter task, the NSSF may make a query to the NRF (Figure 5.10). Release 16 does not bring modifications to NSSF functionality.

5.5.10 NWDAF

NWDAF refers to network analytics logical function, which a 5G MNO manages. The task of the NWDAF is to provide slice-specific network data analytics to NFs that are subscribed to it. The data analytics may include, e.g., the metrics of the load of the network on slide instance level in such a way that the NWDAF does not need to be aware of the actual subscriber of the slice.

Figure 5.10 The interfaces of the NSSF.

Please note that in 3GPP Release 15, the only NFs defined to interface with the NWDAF via the standardized *Nnwdaf* are the PCF for policy decision making and NSSF for slice selection based on the load information (Figure 5.11). Furthermore, utilization of the NWDAF for non-slice-specific analytics is not supported in Release 15.

Release 16 defines the NWDAF support of data collection from NFs, AFs, and Operations Administration and Maintenance (OAM). It also defines the NWDAF service registration and metadata exposure to NFs and AFs, as well as the support of analytics information provisioning to NFs and AFs.

3GPP TS 23.288 details the NWDAF functionality.

5.5.11 PCF

The 5G PCF is an evolution of the LTE PCRF. In 5G, it supports the unified policy framework to govern network behavior (Figure 5.12). It also provides policy rules to CP functions to enforce them. Furthermore, it can access subscription information relevant for policy decisions in a UDR, which resides within the same PLMN. A description of the PCF is in 3GPP TS 23.503.

Release 16 does not bring modifications to the PCF.

5.5.12 SEPP

The 5G SEPP is a new element for securely interconnecting 5G networks, defined in Release 15. Figure 5.13 depicts the principle of the utilization of SEPP elements presented

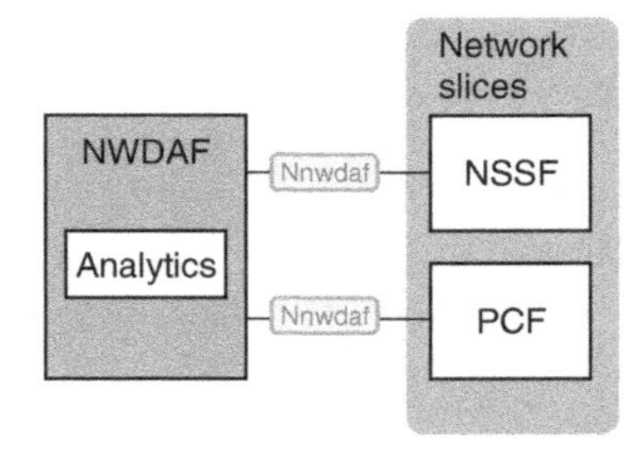

Figure 5.11 The interfaces of the NWDAF.

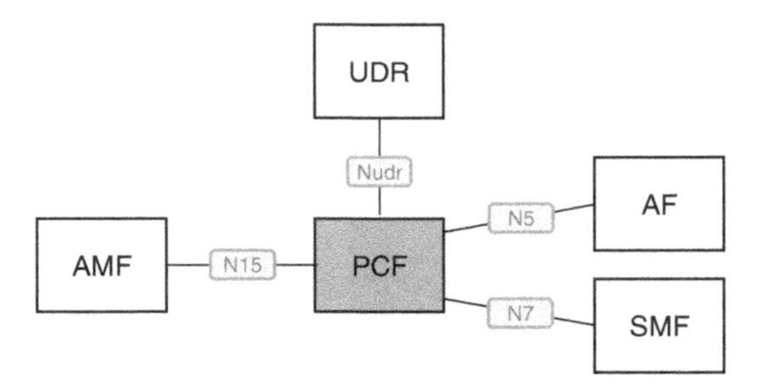

Figure 5.12 The interfaces of the PCF.

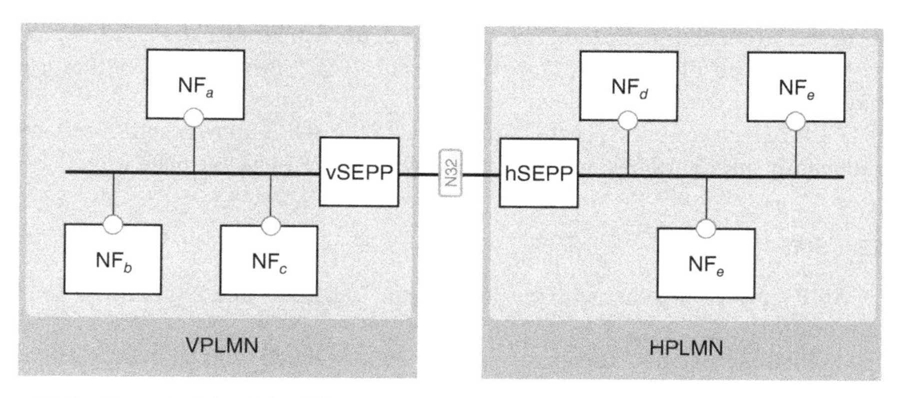

Figure 5.13 The principle of the SEPP for interconnecting 3GPP networks.

in an SBA model. The vSEPP refers to SEPP functionality in visited PLMN, while the hSEPP refers to the SEPP located at the home PLMN.

The 5G system is designed in such a way that the interoperator connection via the Internet Protocol Packet eXchange (IPX) is protected according to the SBA. IPX has emerged as a fundamental network backbone for LTE, and it continues being relevant in the 5G era, too. Details of IPX in practice can be found in GSMA reference [14].

The SEPP is, in practice, a non-transparent proxy. The role of the SEPP is as a service relay between service producers and service consumers, which is a comparable task to direct service interaction. CP messages between the SEPP elements may pass through IPX entities.

The SEPP elements are located at the edge of the 5G core network of the MNO. The tasks of the SEPP are to provision confidentiality and integrity protection of signaling between networks. The SEPP also takes care of the topology hiding of the firewalls according to the defined policies. It can also manage message filtering and policing on inter-PLMN CP interfaces [10].

The SEPP provides end-to-end application-layer security for signaling (e.g., security-sensitive IEs, i.e., Information Elements). The SEPP also takes care of transport-layer security, referring to hop-by-hop security of adjacent nodes of the IPX network.

3GPP TS 33.501 and TS 29.500 detail the functionality of the SEPP, its flows, and the *N32* reference point. Please note that further principles of SEPP deployment scenarios beyond Release 16 are ongoing in the responsible Fraud and Security Groups of the GSMA to interconnect the SEPP elements securely in the roaming scenario relying on connection to the IPX infrastructure. The GSMA released the respective latest 5G Roaming Guidelines of the GSMA version 2.0 on 28 May 2020 [15]. Please refer to the up-to-date version as there are further changes expected.

As summarized in Ref. [15], operators manually provision the functional SEPP elements with a protection policy based on bilateral agreements. Protection policies can be validated via *N32-c*, which is protected by Transport Layer Security (TLS). The SEPP ensures integrity and confidentiality protection for the elements that need to be protected and defines which parts the IPX provider, which is located between two SEPPs, it is allowed to modify. Some examples of SEPP tasks are message filtering and topology hiding. At the current stage, the SEPP relies on application-layer security defined in Protocol for N32 Interconnect Security (PRINS) on all Hypertext Transfer Protocol (HTTP) messages before they are sent externally over the roaming interface.

The IPX HTTP proxy, which is out of the scope of the 3GPP, allows the IPX service provider to modify information elements received by the SEPP in a controlled way.

5.5.13 SMF

The 5G SMF replaces, together with the 5G UDF, the LTE S-GW and P-GW. 3GPP TS 23.501 defines the SMF.

The SMF handles session management (Figure 5.14)-related tasks so it can perform session establishment, modification, and releasing. This includes the maintenance of a tunnel in the UPF–access network node interface. The SMF also performs UE IP address allocation and management, and optionally related authorization. The SMF may receive the UE IP address from a UPF or from an external DN. The SMF includes DHCPv4 and DHCPv6 server and client functions.

Furthermore, the SMF performs Allocation and Retention Priority (ARP) proxying according to IETF RFC 1027 and/or IPv6 neighbor solicitation proxying according to IETF RFC 4861 for Ethernet PDUs. It selects and controls UP function, makes traffic steering and routing at the UPF, and terminates interfaces with PCFs. It includes LI for SM events and the interface to the LI system. It also collects charging data, and controls charging data collection at the UPF. It terminates SM parts of NAS messages, has downlink data notification, and includes an initiator of access network-specific SM information (transported over *N2* via the AMF to the access network). It determines the Session and Service Continuity (SSC) mode of a session.

Furthermore, the SMF includes a roaming functionality such as handling of local enforcement for QoS Service Level Agreement (SLA) of VPLMN, collection and exposing of charging data (VPLMN), LI (VPLMN for SM events and interface to the LI system), and interaction with external DNs for signaling of PDU session authorization/authentication by external DNs.

Some of the Release 16 additions include C-IoT optimization, 5G Virtual Network (VN) group management on *N19* tunnels between PDU Session Anchor (PSA) UPFs, and *N6*-based forwarding or *N19*-based forwarding.

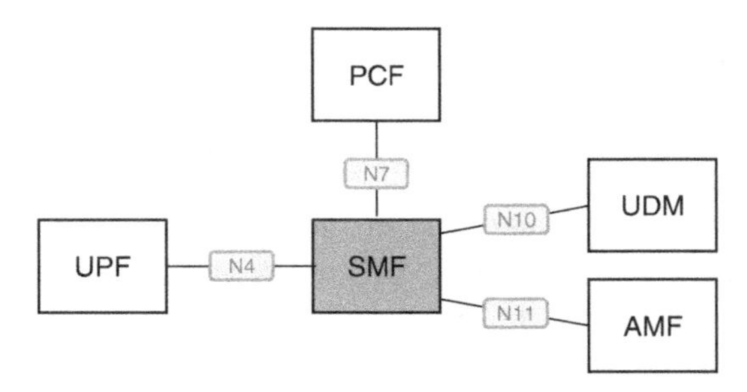

Figure 5.14 The interfaces of the SMF. Please note that the SMF also connects to the LI function upon its use.

5.5.14 SMSF

The SMSF supports SMS over NAS in the 5G infrastructure, managing SMS subscription data and its delivery. It includes Short Message Relay Protocol (SM-RP) and Short Message Control Protocol (SM-CP) with the UE as described in 3GPP TS 24.011 [16].

It includes the procedures necessary to support the SMS between the Mobile Station (MS) and the Mobile Switching Center (MSC), Serving GPRS Support Node (SGSN), MME, or SMSF, as described in 3GPP TS 23.040 [17]. It also relays the short message between the UE and Gateway Mobile Switching Center (SMS-GMSC), Interworking MSC (IWMSC), and SMS router. It collects Charging Data Records (CDRs) on SMS events and performs LI. It interacts with AMF and SMS-GMSC when the UE is unavailable for short message delivery by notifying SMS-GMSC, which in turn can notify UDM about the unavailability.

Figure 5.15 depicts the protocols for the *N1* mode used in 5G SMS delivery. In this mode, the Connection Management Sublayer (CM sub) provides services to the Short Message Relay Layer (SM RL). The SM RL provides services to the Short Message Transfer Layer (SM TL) at the MS side, while the Short Message Application Layer (SM AL) is the highest protocol layer. The short message user information elements are mapped to the Transaction Capabilities Application Part (TCAP) or Mobile Application Part (MAP) in the network side.

The MAP is an SS7 protocol providing an application layer for multiple mobile core network nodes to provide services to users. Among other tasks, the MAP can be used to deliver short messages within the network. The MAP relies on the TCAP, and can be transported using the old SS7 protocols, or nowadays over IP relying on the Transport Independent Signaling Connection Control Part (TI-SCCP), or via SIGTRAN, which is the default in current mobile networks, including 5G.

The protocol between the two SMC entities is denoted Short Message Control Protocol (SM CP), while there is an SM RP in use between two Short Message Relay (SMR) entities. The AMF transfers the SM CP messages between the SMSF and the UE.

As depicted in Figure 5.15, there is an SM RP between the SMR of the SMSF and the UE, whereas the SM CP is in use between the SMC entities of these same elements. Following the notation of Figure 5.15, MM sub refers to Mobility Management Sublayer, RR sub refers to Radio Resource Management Sublayer, and 5GMM-sub refers to 5G Mobility Management Sublayer.

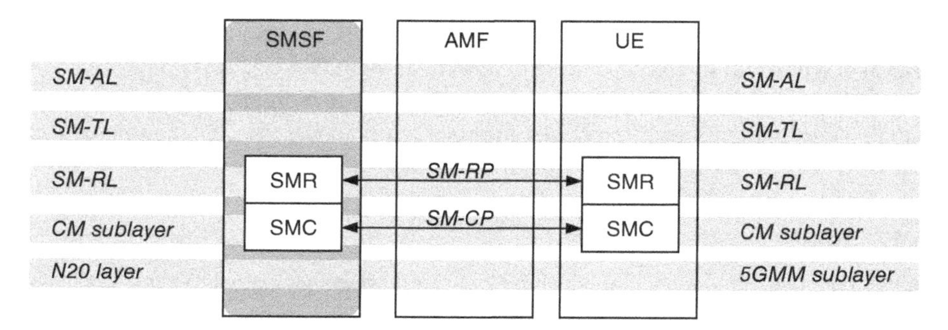

Figure 5.15 The protocols for the *N1* mode used in 5G SMS delivery.

5.5.15 UDM

As described in 3GPP TS 23.501, UDM generates the 3GPP AKA Authentication Credentials per user. UDM resides in the same HPLMN as the subscriber it serves and accesses the information of the UDR located in the same PLMN. It performs user identification, including storage and management of SUPI per each individual 5G subscriber. Furthermore, it is capable of de-concealing the SUCI, which is the privacy-protected subscription identifier, i.e., the secret version of the SUPI. UDM also manages access authorizations such as roaming restriction and UE serving NF registration management (e.g., storing serving AMF for UE and storing serving SMF for UE's PDU session). Furthermore, UDM supports service and session continuity keeping SMF or DNN assignment of sessions, and provides support to the delivery of MT short messages. It includes LI functionality, subscription, and SMS management. Release 16 also brings 5G LAN group management handling, and support of external parameter provisioning related to "expected UE behavior" or "network configuration" parameters.

UDM relies on subscription and authentication data, which may be stored in the UDR. This means that multiple UDM elements can provide service to the single user performing various transactions when the UDR is in the same PLMN as UDM. UDM can also interact with the HSS. The interaction between UDM and the HSS, when they are deployed as separate NFs, is defined in TS 23.632 and TS 29.563, or it is implementation specific.

5G UDM is an evolution of the HSS and UDR utilized in LTE. It is for 5G data storing and retrieving. As the 5G architecture is based on the separation of data plane and CP, e.g., network status information and other relevant data can be stored in a unified database. The 5G NFs have the right to access selected data of UDM via standard interfaces to locally store them in a dynamic way.

Distributed database synchronization allows a real-time backup procedure between data centers for storing, e.g., network status. The unified database simplifies the procedure for network information retrieval functions, the benefit being the reduction of signaling overhead upon data synchronization. Figure 5.16 clarifies the principle of UDM in connection with local data storage.

5.5.16 UDR

The 5G UDR is an evolution of the LTE SDS. As described in 3GPP TS 23.501, UDM can store and retrieve subscription data from and to the UDR. The UDR also takes care of storage and retrieval of policy data by the PCF, storage and retrieval of structured data for exposure, and application data by the NEF. Release 16 mentions that the UDR also handles the storage and retrieval of NF Group ID corresponding to subscriber identifier, e.g., IMPI, IMPU, or SUPI.

The UDR resides in the same PLMN as the NF service consumers. It stores and retrieves data via *Nudr*, which is an intra-PLMN interface. In practice, the UDR can be co-located with the UDSF. Figure 5.17 summarizes the interfaces of the UDR.

5.5.17 UDSF

The 5G UDSF is a function comparable with LTE SDSF. In 5G, the UDSF is left as an optional function for the storing and retrieval of information as unstructured data by any NF. Please note that in the context of 5G, the term "structured data" refers to data for which

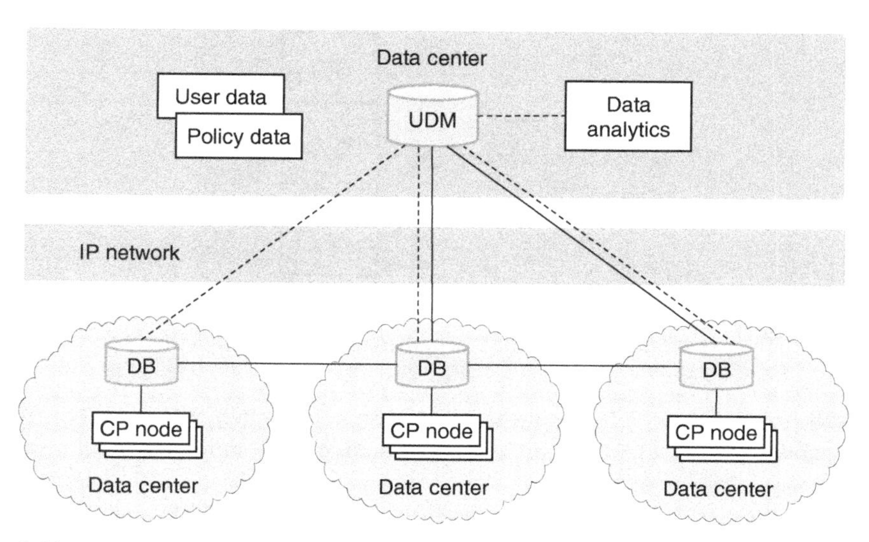

Figure 5.16 A conceptual example of UDM deployment.

the structure is defined in 3GPP specifications, whereas the structure of "unstructured data" is not defined in 3GPP specifications.

Figure 5.18 summarizes the interfaces of the UDSF.

5.5.18 UPF

The 5G UPF replaces, together with the SMF, the LTE S-GW and P-GW. The UPF acts as an anchor point for intra- and inter-RAT mobility as appropriate. It is also the external PDU session point of interconnect to DN and takes care of packet routing and forwarding. It performs packet inspection and takes care of the UP part of policy rule enforcement and LI. Furthermore, it performs traffic reporting and QoS handling for the UP, including data rate enforcement. The UPF verifies uplink traffic by mapping the SDF to QoS flow and makes transport-level packet marking. It buffers downlink packets and triggers downlink data notifications, and it also sends and forwards end markers to the source NG-RAN node.

3GPP TS 23.501 lists a variety of other tasks for the UPF such as high-latency communication, Access Traffic Steering, Switch and Splitting (ATSSS) steering functionality to steer

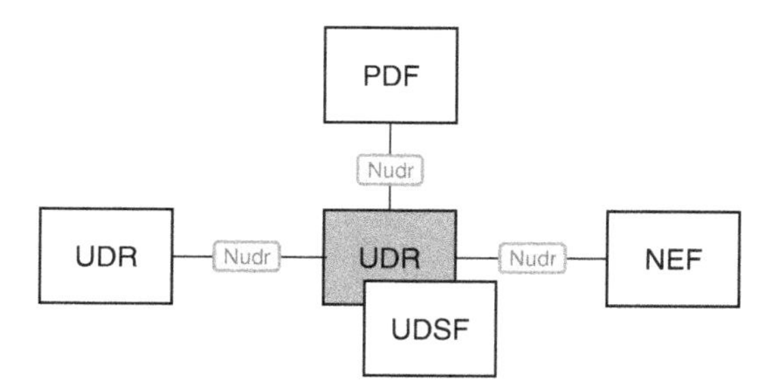

Figure 5.17 The interfaces of the UDR.

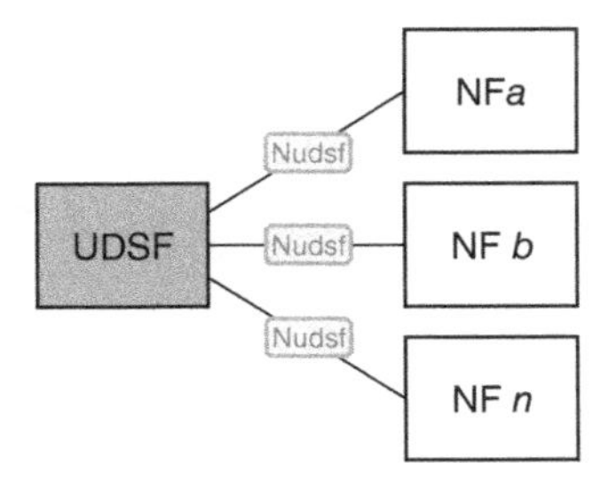

Figure 5.18 The interfaces of the UDSF. It is an optional functionality in 5G.

the Multi-Access (MA) PDU session traffic. Please note though that not all of the UPF functionalities that TS 23.501 summarizes are required to be supported in an instance of UPF of a network slice.

The UPF can perform ARP proxying according to IETF RFC 1027 and IPv6 neighbor solicitation proxying according to IETF RFC 4861. Figure 5.19 summarizes the interfaces of the UPF.

5.6 Additional NFs of Release 16

The following sections summarize the new NFs added in Release 16.

5.6.1 CAPIF

Related to the evolution of the NEF, Release 16 includes CAPIF. Release 16 allows the use of CAPIF and respective APIs in case the operator exposes the NEF to external entities. If the operator supports CAPIF, the NEF supports the CAPIF API provider domain functions. TS 23.222 defines CAPIF and the respective API provider domain functions.

A northbound API is an interface between an AS and the specified 3GPP functions of the operator as depicted in Figure 5.20 [18]. The AS can reside within or outside the operator's network.

The 3GPP SA6 working group develops the CAPIF further, considering common capabilities to unify the functioning of the northbound APIs as summarized in 3GPP TR 23.722 and depicted in Figure 5.21. The goal of the northbound APIs is to provide standardized integration of services with diverse service providers for interaction at the application layer.

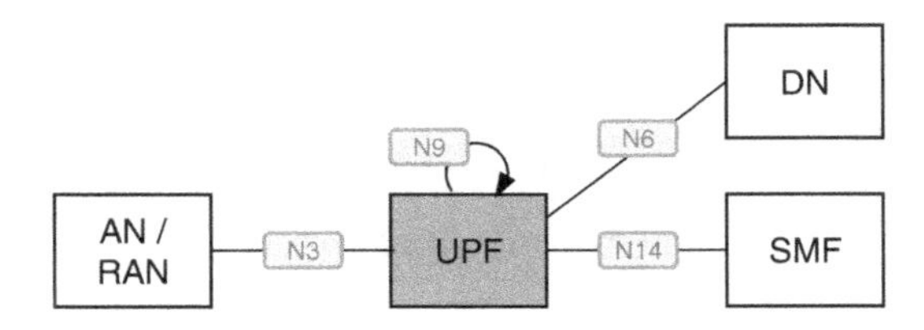

Figure 5.19 The interfaces of the UPF.

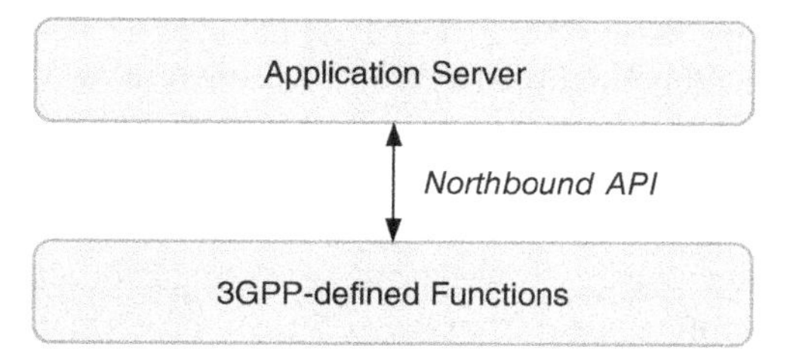

Figure 5.20 The 3GPP northbound interface.

The benefit of the northbound APIs is the opportunity for MNOs to offer a variety of services beyond the traditional teleservices by exposing them to third parties. Some of the motivations to develop further the northbound APIs are the merging of mobile communication networks and new solutions such as the further developed Machine Type Communication (MTC) and some of the sectors expanding the traditional stakeholder base, such as IoT, V2X, and CriC. Therefore, e.g., Release 15 defines a means to align with oneM2M release 2 exposing C-IoT and MTC capabilities of the operators via the northbound APIs.

5.6.2 GMLC

GMLC is related to the LMF defined in Release 15. Release 16 adds GMLC and its functionality in 3GPP TS 23.273 [19].

3GPP TS 22.071 presents a general description of LCS and service requirements. Furthermore, TS 23.271, TS 43.059, TS 25.305, and TS 36.305 present services for GSM/EDGE Radio Access Network (GERAN), UMTS Terrestrial Radio Access Network (UTRAN), and Evolved UMTS Terrestrial Radio Access Network (E-UTRAN).

UE positioning can be supported by RAT-dependent and -independent position methods, and either a 3GPP or non-3GPP access network can perform the UE positioning. An LCS

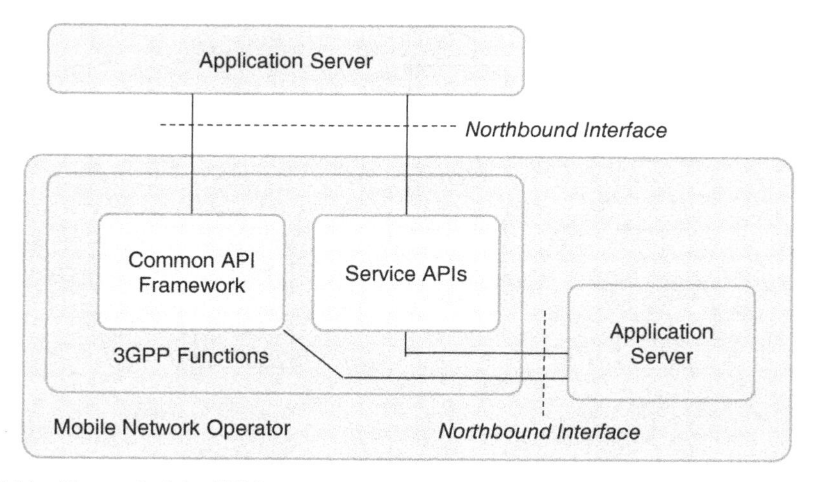

Figure 5.21 The goal of the 3GPP is to unify the northbound interface.

client or an AF in a home network or external to PLMN, or a CP NF within a PLMN, may request location information for target UE. A privacy verification and authorization procedure is applied to the location requests. More detailed principles depend on the local regulation.

Release 16 presents various architectural models for location-based services, including home network and roaming cases as well as interworking between 5GC and 4G EPC. For the latter case, the location requests from the LCS client or AF rely on a common interface in EPC and 5GC.

Figure 5.22 depicts an architectural reference model for 5GS LCS for a non-roaming UE in reference point representation, excluding the charging interface [19]. Please refer to TS 23.273 for other cases.

5.6.3 I-SMF and V-SMF

As interpreted from 3GPP TS 23.501, Release 16 adds I-SMF selection and V-SMF reselection [10].

The AMF adds and removes an I-SMF or a V-SMF for a PDU session based on the service area of the SMF as informed by the NRF. In mobility scenarios such as the handover procedure or AMF change, if the UE location is outwith the SMF service area, the AMF inserts an I-SMF that is capable of serving the UE and the respective S-NSSAI. Posteriorly, when the I-SMF is no longer required, the AMF removes it.

5.6.4 I-UPF

Release 16 adds an I-UPF, which is related to the support of redundant transmission on *N3/N9* interfaces as depicted in Figure 5.23.

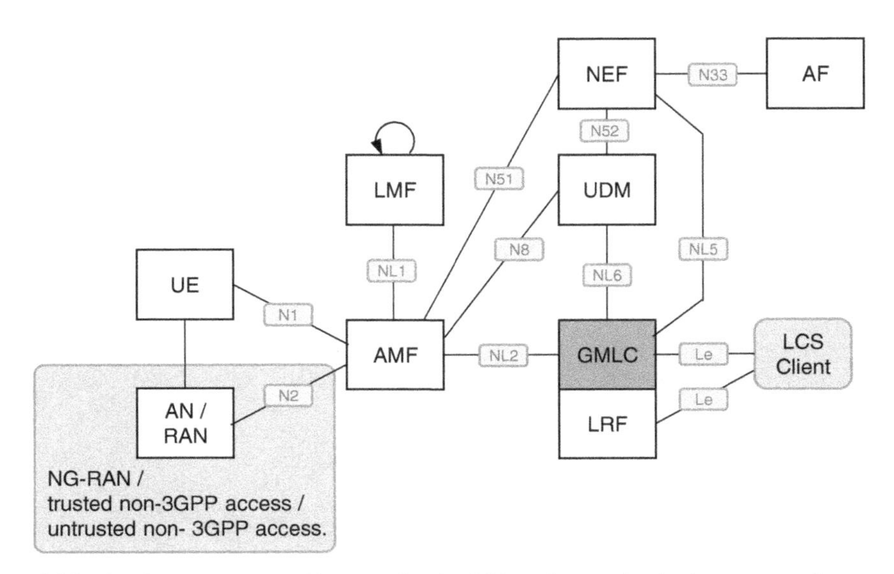

Figure 5.22 A reference point architecture for the 5G location service in the non-roaming case.

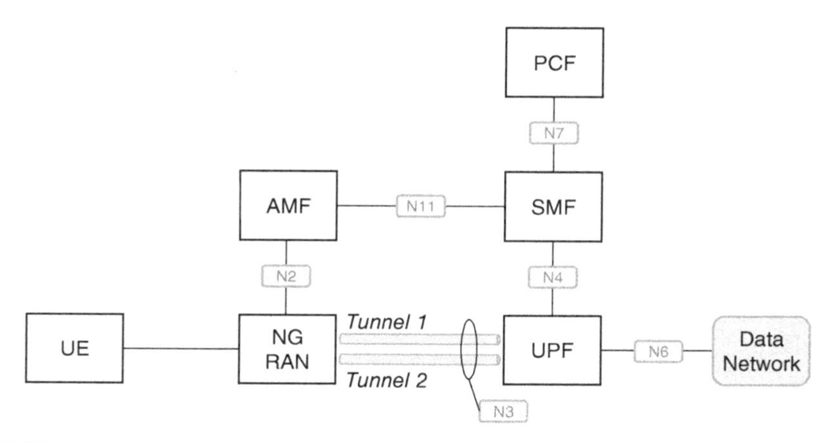

Figure 5.23 Redundant transmission with two *N3* tunnels between the PSA UPF and a single NG-RAN node.

If a single *N3* tunnel does not fulfill the reliability requirements of a URLLC use case, but the reliability of NG-RAN node, UPF, and CP NFs comply with them, there can be a redundant transmission between PSA UPF and NG-RAN based on two independent *N3* tunnels. These tunnels are associated with a single PDU session over different transport layer paths with the aim of enhancing the reliability of the connection.

Two I-UPF components can provide redundant transmission on two *N3* and *N9* tunnels between a single NG-RAN node and a PSA UPF. The NG-RAN node and PSA UPF are capable of performing a packet replication and elimination function as depicted in Figure 5.24.

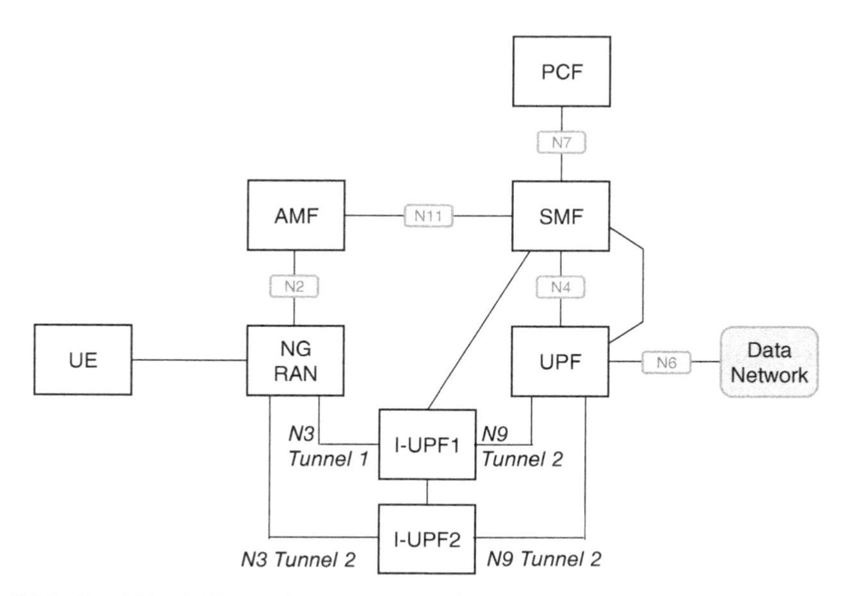

Figure 5.24 Two *N3* and *N9* tunnels between the NG-RAN and PSA UPF for redundant transmission.

5.6.5 NSSAAF

Interpreted from TS 23.501, Release 16 introduces the NSSAAF. It supports NSSAA as specified in TS 23.502 with an AAA-Server (AAA-S). If the AAA-S belongs to a third party, the NSSAAF may contact the AAA-S via an AAA-Proxy (AAA-P).

5.6.6 SCP

SCP is a decentralized solution composed of CP and data plane. This solution is deployed alongside 5G NFs for providing routing control, resiliency, and observability to the core network [20].

As interpreted in 3GPP TS 23.501, Release 15 defines the NF profile to the NRF, and Release 16 brings the new SCP profile to it. The SCP profile includes the following information:

- SCP ID.
- FQDN or IP address of the SCP.
- Indication that the profile is of an SCP.
- SCP capacity, load, and priority information.
- Location information for the SCP as per TS 29.510.
- Served locations as per TS 29.510.
- Network slice identifiers such as S-NSSAI or NSI ID.
- Remote PLMNs that are reachable through the SCP.
- Endpoint addresses accessible via the SCP.
- Interconnected SCP and NF IDs.
- NF sets served by the SCP.
- SCP domain the SCP belongs to.

5.6.7 TNGF

Release 16 defines a TNGF. As stated in 3GPP TS 23.501, the 5G core network supports UE connectivity via non-3GPP access networks, e.g., WLAN access networks. The 5G core network supports both untrusted non-3GPP access network and Trusted Non-3GPP Access Network (TNAN). The N3IWF connects the untrusted non-3GPP access network to the 5G core network, whereas the TNGF connects trusted non-3GPP access networks to the 5G core network. Both the N3IWF and the TNGF interface with the 5G core network CP and UP functions via the *N2* and *N3* interfaces, respectively.

3GPP TS 23.501 further states that a non-3GPP access network may advertise the PLMNs for which it supports trusted connectivity and the type of supported trusted connectivity such as 5G. The UE can discover the non-3GPP access networks that may provide trusted connectivity to one or more PLMNs.

The TNGF connected to a trusted non-3GPP access network includes the following functions:

- Implements a local Extensible Authentication Protocol (EAP) Re-authentication (ER) server, as per IETF RFC 6696, to facilitate mobility within the TNAN.
- Implements a local mobility anchor within the TNAN.
- Implements an AMF selection procedure.

- *N2* signaling with SMF, relayed by AMF, supports PDU sessions and QoS.
- Terminates EAP-5G signaling and acts as an authenticator of the UE that aims to register to 5GC via the TNAN.
- Terminates the *N2* and *N3* interfaces.
- Transparently relays NAS messages between the UE and the AMF via *NWt*.
- Transparently relays PDUs between the UE and UPF.

As stated in 3GPP TS 23.501, Section 4.2.8.1A, UE connected to a 5G Residential Gateway (5G-RG) or Fixed Network RG (FN-RG) can access the 5GC via the N3IWF or via the TNGF where the combination of 5G-RG/FN-RG, W-AGF, and UPF serving the 5G-RG or FN-RG acts respectively as an untrusted non-3GPP access network or as a TNAN. As an example of this scenario, UE can connect to 5G-RG via WLAN radio access and to 5GC via N3IWF as detailed in TS 23.316. As described in TS 23.501, an FN-RG and 5G-RG accessing the 5GS must be assigned a PEI.

Please note that Release 16 does not define the roaming architecture for the 5G Broadband Residential Gateway (5G-BRG), Fixed Network Broadband Residential Gateway (FN-BRG), 5G Cable Residential Gateway (5G-CRG), or Fixed Network Cable Residential Gateway (FN-CRG) with the W-5GAN. Release 16 supports the Home Routed (HR) roaming scenario for 5G-RG connected via NG-RAN, while the Local Breakout (LBO) scenario is not supported. The 5G Multi-Operator Core Network (5G MOCN) is supported for 5G-RG connected via NG-RAN.

Please refer to Chapter 6, Section 6.3.8, for an architectural overview of trusted and untrusted non-3GPP access, whereas Section 4.2.8 of 3GPP TS 23.501 presents all the other applicable cases for the support of non-3GPP access.

5.6.8 TWIF

3GPP TS 23.501, Section 4.2.8.5, presents TWIF and the 5GC architectural enhancements that enable N5CW devices to access 5GC via trusted WLAN access networks and their Trusted WLAN Access Point (TWAP).

A trusted WLAN access network is a type of TNAN that supports WLAN access technology such as IEEE 802.11. To facilitate the 5GC access support of N5CW devices, a trusted WLAN access network must support a TWIF.

When an N5CW device performs an EAP-based access authentication procedure to connect to a trusted WLAN access network, the N5CW device may register simultaneously to a 5GC. The TWIF performs 5GC registration in the trusted WLAN access network on behalf of the N5CW device. TS 33.501 specifies the EAP authentication procedure to authenticate the N5CW device during 5GC registration.

Figure 5.25 depicts the architecture of the trusted WLAN interworking and the main NFs required to support 5GC access from N5CW devices.

5.6.9 UCMF

As described in TS 23.501, the UCMF stores dictionary entries related to PLMN-assigned or Manufacturer-assigned UE Radio Capability IDs. An AMF may subscribe to the UCMF to obtain new values of UE Radio Capability ID. The UCMF assigns the IDs to cache them locally.

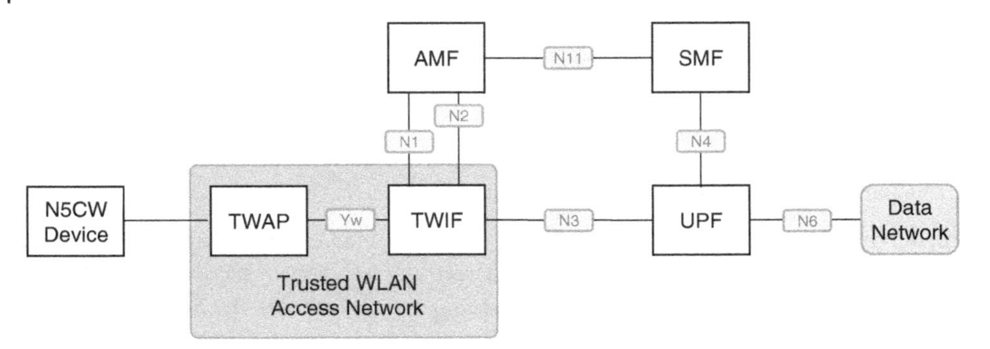

Figure 5.25 Non-roaming and LBO roaming architecture for supporting 5GC access from N5CW devices as interpreted from TS 23.501 [10].

AF provisions the Manufacturer-assigned UE Radio Capability ID entries in the UCMF. The AF interacts with the UCMF directly via network management or via NEF. 3GPP TS 23.502 defines the respective procedure.

The PLMN-assigned UE Radio Capability ID is associated with the Tracking Area Code (TAC) of the related UE model. In this scenario, the AMF requests the UCMF to assign a UE Radio Capability ID for a set of UE radio capabilities and indicates the related TAC. The UCMF also stores a Version ID value for the PLMN-assigned UE Radio Capability IDs.

The UCMF may be provisioned with a list of TACs and Manufacturer-assigned UE Radio Capability IDs, including a vendor ID. The UCMF stores the PLMN-assigned UE Radio Capability ID as long as it is associated with a TAC value.

5.6.10 W-AGF

3GPP TS 23.316 specifies the functionality of the W-AGF. It is an NF in the W-5GAN that provides connectivity to the 5GC to the 5G-RG and FN-RG.

The W-5GAN is connected to the 5G core network via a W-AGF. The W-AGF interfaces the 5G core network CP and UP functions via *N2* and *N3* interfaces, respectively. The case for UE connected to 5GC via the NG-RAN is defined in 3GPP TS 23.501, TS 23.502, TS 23.503, and TS 23.316.

When a 5G-RG connects via an NG-RAN and via a W-5GAN, multiple *N1* instances exist for the 5G-RG, served by a unique AMF. There is thus one *N1* instance over the NG-RAN and another one over the W-5GAN. 3GPP TS 23.316 specifies the 5G-RG connected to 5GC via the NG-RAN.

For a non-5G capable FN-RG that is connected via the W-5GAN to 5GC, the W-AGF provides the *N1* interface to the AMF on behalf of the FN-RG.

5.7 5GC Functionalities

5.7.1 Network Function Discovery

For the NFs to reach out to each other, the 3GPP 5G system includes procedures for NF discovery and NF service discovery. They enable the allowed core network entities, i.e., NF or SCP, to discover a set of NF instances and NF service instances for a specific NF service or NF type. 3GPP TS 23.502 (Sections 4.17.4, 4.17.5, 4.17.9, and 4.17.10) describes the NF service discovery and notification procedures.

The NRF serves NF and NF service discovery unless the requester's locally configured NF contains the needed NF and NF service information, e.g., in the same PLMN.

Prior to the discovery to function correctly, the NF profiles must be registered in the NRF upon the NF instance and its NF services becoming operative. The NF profile includes information on the NF instance such as its ID. 3GPP TS 23.502, Section 4.17.1, describes the NF service registration procedure, whereas TS 29.510 describes the respective validity period.

The requester NF or SCP may also receive notifications from the NRF about the updated NF profile of an NF, or newly registered NF instances. 3GPP TS 23.502, Clauses 4.17.7 and 4.17.8, describe the respective subscription and notification procedure.

The NF and NF service discovery can also take place across PLMNs. 3GPP TS 23.502, Section 4.17.5, describes the respective service discovery procedure, and TS 29.510 details the use of the target PLMN ID specific query to reach the NRF in the remote PLMN.

5.7.2 Network Slicing

5.7.2.1 Introduction

The 5G era has advanced and new business models. There are many more possibilities to cooperate between MNOs and third parties in such a way that some external organization has the right to utilize a dedicated part of the network. The respective use cases may relate to public safety and CriC to ensure people's safety. This type of dedicated resource utilization can be done via network slicing, which is one of the most essential concepts in the transition towards the 5G core network.

Network slicing provides the possibility to isolate network resources for certain services. The 5G system can execute resource provisioning so that a certain set of logical network nodes is assigned to the third party. Figure 5.26 depicts an example of such a principle by dedicating UPF, SMF, and AMF resources within a set of isolated network slices for each cooperating organization relying on those slices.

Also, the network slice concept allows differentiation of the level of provisioned security per service type, which is taken care of by different slices. This is a remarkable difference with previous mobile generations because before, the applied security would be the same for all users (devices) regardless of the service type. Although this has not been a limitation up to now, it is not optimal either as the level of the applied security in networks prior to 5G may be unnecessarily strong for certain cases such as simple IoT communications. Also, the uniformly applied security may be considered weak for some other service types such as CriC.

5G network slicing solves these problematics of ideal security level by distinguishing the v per use case. Furthermore, resource utilization is optimal as the CP of 5G can be managed by cloud, ensuring flexible and interrupt-persistent functioning via geo-redundant configuration. The UP nodes, in turn, can work on services within dedicated slices. The slice can be instantiated, updated, and deleted according to the NFV concept, which makes it possible to differentiate the QoS in a fluent way. Figure 5.27 presents an example of such implementation.

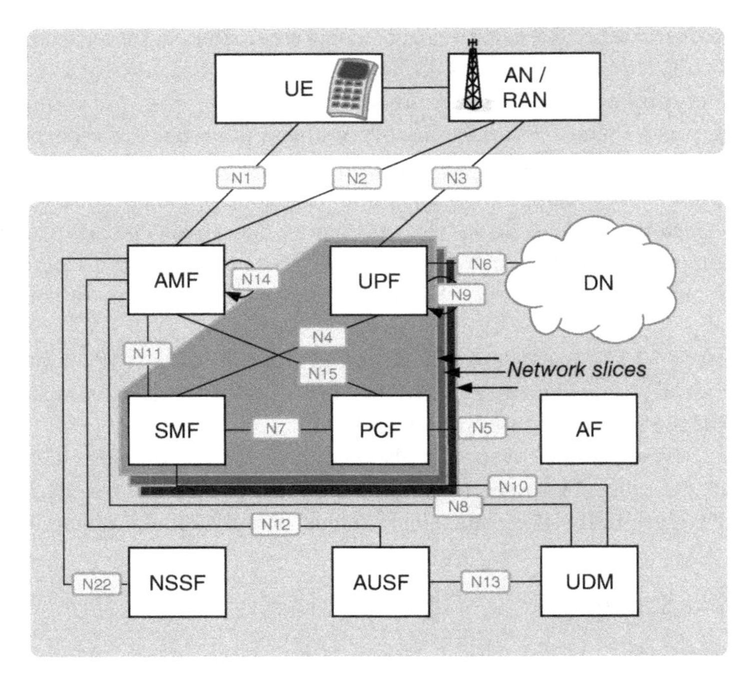

Figure 5.26 The principle of network slicing in core network deployment.

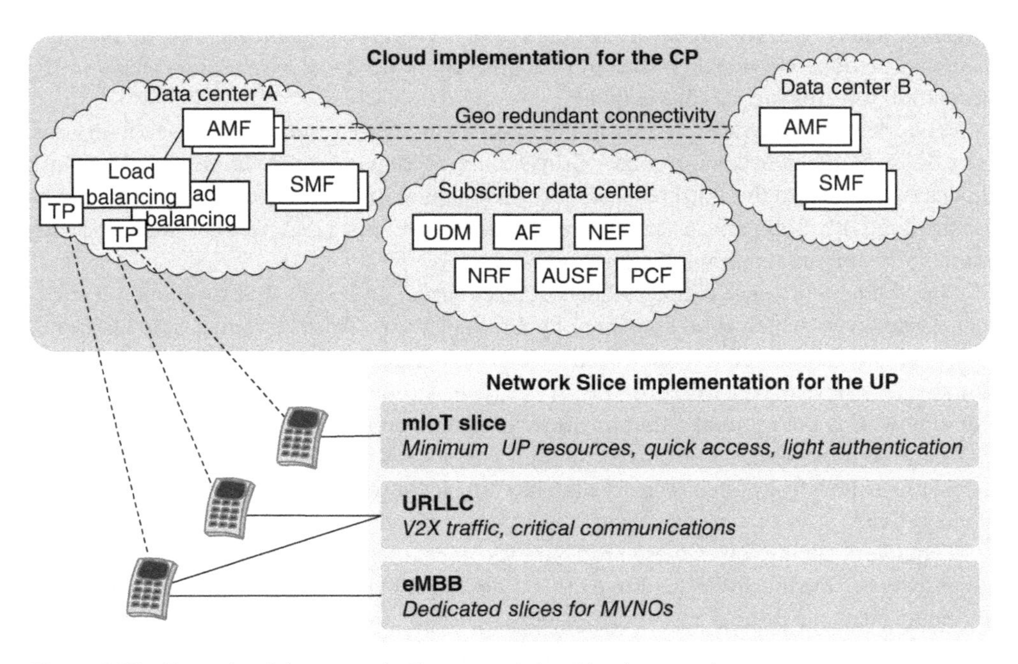

Figure 5.27 Example of the network slice set and cloud implementation.

In 5G, one of the ways to meet the demanding performance requirements is to rely on network slices that run in an end-to-end chain, including radio and core networks as well as application logic [21]. To make this happen, service and resource orchestration is needed across the main functional domains. The problematics of such orchestration arise from the fact that each of these domains has its own control system:

- Transport network can rely on an SDN for traffic segmentation and policy selection;
- RAN can rely on a radio network controller or base station for the management of packet scheduling;
- Cloud core network can rely on an NFV orchestrator.

In normal operations, the service assurance of 5G network slices needs to happen automatically without human intervention except for special cases. The above-mentioned domains thus need service layer assistance in interdomain coordination, as can be seen in Figure 5.28.

Each 5G network slice equals in practice a separate, self-contained network for specific use cases. In 5G, each slice can have a separate SLA, which is determined per each customer with the service provider. This is the reason why service management and service assurance need to be divided into each network slice separately.

It is also worth mentioning that the lifetime of some of the slices may be quite short, which can happen, e.g., when they are used for some special event for a limited time only. Thus, the SLA is accordingly short term, and considers the initial and termination phase of the slice setup. The challenge of this case is the fact that the same resource layer is utilized as a basis for multiple slices, and the assurance of adequate resources per slice may not be possible to achieve in some of these special cases due to unforeseen peak capacity demand.

The solution for this issue is the deployment of an adequate service assurance tool that can correlate service experience with the respective resource utilization. In an optimal

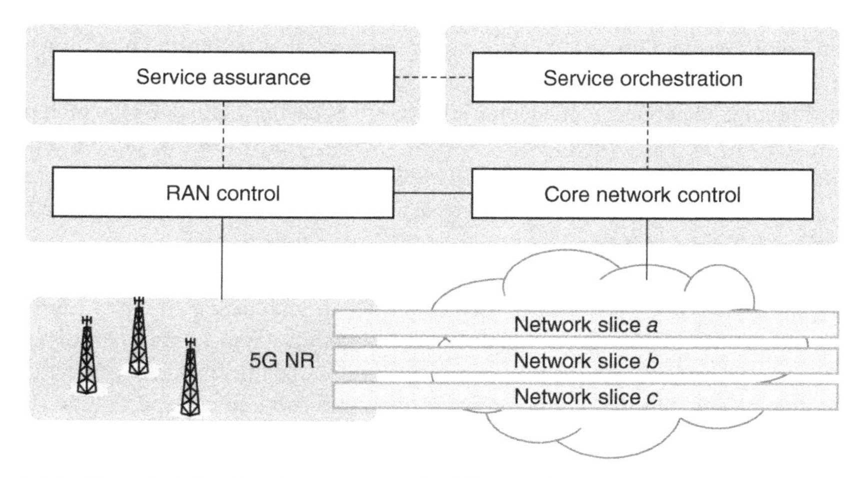

Figure 5.28 The principle of service assurance for 5G network slicing.

setup, this ensures that a single slice does not consume more resources than needed to fulfill the respective SLA, considering the performance and capacity. The strategy for such balancing depends on the use case, so the service priority for, e.g., a set of non-critical mIoT sensors is lower than for applications requiring ultra-low latency. This type of balancing is known as contextual service assurance between the Network Management and Orchestration (NMO) plane and the slice layer [21].

5.7.2.2 Network Slicing in Practice

Ref. [22] has identified examples of network slice utilization in practice. One of them is an automotive slice for connected vehicles. This use case requires a network that can deliver simultaneously high bit-rate data for the use of in-car entertainment, as well as data of the URLLC category for autonomous driving of the vehicle, accompanied by telemetry data from the vehicle's sensors and device-to-device communication. In this type of environment, it is of utmost importance to ensure service continuity in all situations, including internal roaming. For more information on commercial solutions, please see more details from, e.g., Refs. [23, 24].

The network slice is a set of NFs, which, in turn, represents a functional block. Each function can be provided by an independent vendor providing a wider grade of independency between infra vendors. Network slicing offers important benefits such as dynamic, customizable end-to-end operations to fulfill the special needs of highly versatile users – including traditional subscribers and new verticals alike – such as law enforcement entities, drones, the VR/AR industry, and autonomous vehicles.

Being highly dynamic, network slicing requires advanced integrated and automated techniques to handle the flexibility and efficiency. The development of supporting solutions includes areas such as AI, which eases the service orchestration throughout the life cycle of the slices, from concept testing, commercial operations and maintenance, up to the end of the product.

For network slicing to benefit best in the ecosystem, including the roaming environment of MNOs, it is important to consider common principles that ensure fluent user experiences on a global scale. The GSMA is one of the entities investigating and planning together such strategies with members. One concrete example of the work of the GSMA Future Networks team is the creation of a Generic Network Slice Template (GST), which presents attributes for key services relying on network slicing. The respective document, GSMA Permanent Reference Document (PRD) NG.116, "Generic Network Slice Template," is an evolving document of which version 3.0 was available in 2020. There will be more network slice attributes and respective value ranges identified in future versions.

The document provides a description of the GST, and presents attributes that characterize the type of network slice. These attributes refer to technical parameters resulting in certain performance for the slices. The document also shows typical examples of respective Network Slice Types (NESTs) and proposes a recommended minimum set of attributes and their feasible values that the mobile network operators can deploy especially in the interoperable environment to ensure fluent user experiences.

This released document represents the initial results of the work so far. Nevertheless, 5G will open new business and technical opportunities for many verticals that might not be present in this document yet, so the work continues while the team is seeking indications from the practical field, and aims to ensure that respective needs are taken into account in network slice deployment.

Network slicing allows operators to use their physical mobile network partitioned into multiple virtual networks. 5G is a logical base to house network slicing thanks to network virtualization and sufficient performance to manage such a service. Different customer segments, or verticals, benefit from this solution because they are able to receive focused support for the use of different types of services. This is one of the key differentiators of 5G compared to legacy systems. By deploying the network slicing feature, the operator may configure each slice to better comply with the requirements of different user segments. Operators may use network slicing for supporting an as-a-service model for customers. This enhances operational efficiency and speeds up time-to-market for new services. Network slicing technology is possible thanks to SDN, NFV, and advanced 5G network orchestration. In fact, these features are essential to apply and manage the network slicing.

The operator can form, manage, and terminate network slices when needed in a highly dynamic fashion, and each slice can be optimized for a specific use case. This makes a big difference compared to previous systems in which the mobile network is adjusted in a more static manner and offers certain uniform performance to be shared among the users in that area. Instead, 5G network slices can be tuned to offer different user experiences within a selected area. Figure 5.29 depicts the principle of network slicing, each using only the needed NFs per each slice.

Furthermore, 5G UE is capable of joining one or multiple network slices at the same time, supporting different parallel services. As an example, smart city sensors might not require the highest data speeds but can benefit from the possibility of sharing low bit-rate resources among a large number of other devices communicating simultaneously, so the sensors can subscribe to an mMTC network slice offering such desired performance. Meanwhile, some other users within the very same area, such as CriC applications, may want the highest reliability, so operators can offer an adjusted URLLC network slice for them. Figure 5.30 depicts this principle.

The benefit of this approach is that differently adjusted network slices optimize the network resources while providing a good user experience.

Each 5G network slice equals a separate, self-contained virtual network for specific use cases. Each slice can have separate SLA determined individually per each customer with the service provider. This is the reason why service management and service assurance need to be divided into each network slice separately.

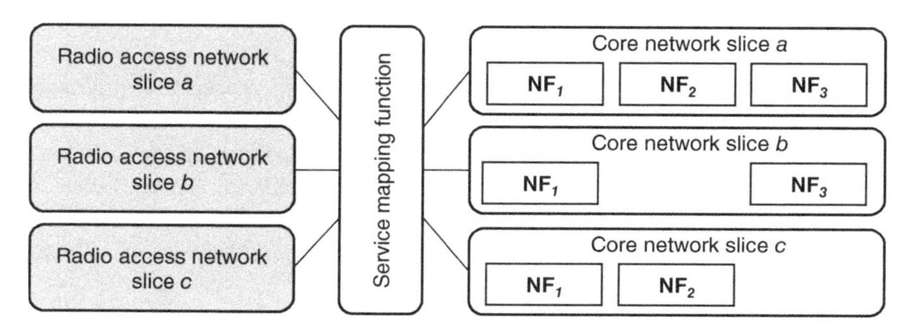

Figure 5.29 The principle of network slicing in 5G.

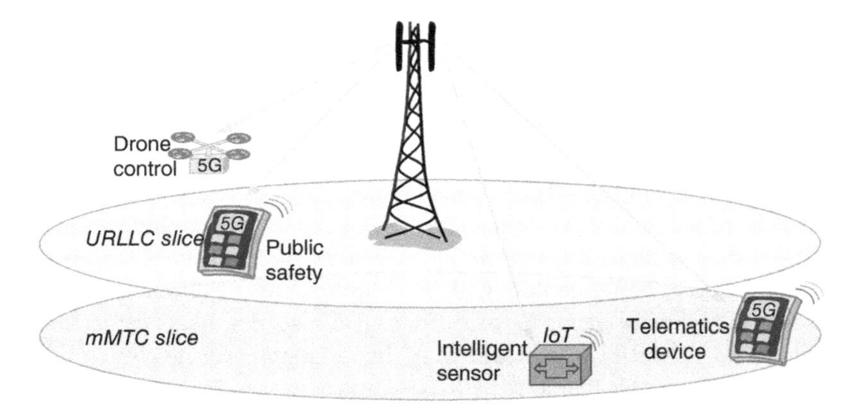

Figure 5.30 Network slice forms virtual networks optimized for different users and respective use cases.

The lifetime of some of the slices may be quite short in practice, such as in some special, temporal events. The challenge is that the same resource layer is used as a basis for multiple slices, and it might not be possible to comply with the original assurance of adequate resources per slice due to unforeseen peak capacity demand during the event.

5.7.2.3 Procedures

As per the Release 16 aspects, 3GPP 38.300, Section 16.3, outlines the general principles and requirements related to network slicing in the NG-RAN for the scenarios of NR connected to 5GC, and E-UTRA connected to 5GC. As stated in 3GPP TS 38.300 and TS 33.305, a network slice always consists of a RAN part and a core network part. The separate PDU sessions handle traffic for different network slices. The network can schedule and configure on L1/L2 level a variety of network slices. The S-NSSAI identifies slices as stated in 3GPP TS 23.501. The NSSAI contains a list of a maximum of eight S-NSSAIs.

The S-NSSAI contains a mandatory Slice/Service Type (SST) field, which identifies the slice type, and an optional Slice Differentiator (SD) field, which differentiates among slices with the same SST field. The UE signals the NSSAI to the network slice selection. In practice, the network can support hundreds of slices, whereas the UE can support a maximum of eight slices simultaneously. This applies also to a Bandwidth reduced Low complexity UE (BL UE) and a Narrow-Band IoT (NB-IoT) UE. Network slicing provides the MNOs with a way to distinguish customers between different tenant types so that each slice complies with varying service requirements, based on the agreed SLA.

3GPP TS 23.502 and TS 38.300 detail the network slicing procedures and aspects in NG-RAN. These aspects include RAN awareness of slices, selection of the RAN part of the network slice, resource management between slices, QoS support, and RAN selection of the core network entity. TS 38.300 also defines the resource isolation between slices, access control, network slice availability for partial coverage, support for UE associating with multiple network slices simultaneously, granularity of slice awareness, and validation of the UE rights to access a network slice.

3GPP TS 38.300 defines procedures for the core network–RAN interaction and describes aspects related to internal RAN. In Release 16 of the 5G network, NG-RAN selects the AMF based on a Temp ID or NSSAI that UE provides over Radio Resource Control (RRC). TS 38.300 also describes the mechanisms used in the RRC protocol. Some of the important aspects include resource isolation and management and signaling for the AMF and network slice selection. In the latter scenario, the RAN selects the AMF based on a Temp ID or NSSAI provided by the UE.

More information on network slicing can be found at 3GPP TS 28.530 (aspects, management, and orchestration; concepts, use cases, and requirements in Release 16), and GSMA NG.116, which presents the GST.

5.7.2.4 Release 15 Foundation

The 5G system provides the possibility for the MNO to deploy a network slicing concept. The network slice is a logical network that is deployed to serve a defined business purpose or customer, and it contains all the required network resources that are configured as a set to serve that specific purpose.

The network slice is an enabler for the services, and can be created, modified, and deleted via the respective network management functions. The network slice is thus a provider's managed, logical network, and it can be deployed technically in any network type, including mobile and fixed networks.

A network slice instance is selected by authorized UE, allowing UE to associate with multiple slices simultaneously. The network instance is security isolated from other slice instances. It provides access to common network functions of the core network and access network enabling network slicing. The network slice instance also supports roaming scenarios.

Network slicing provides the MNO with the possibility to deploy multiple, independent PLMNs. Each of these PLMNs can be customized by instantiating only the features, capabilities, and services that are needed for the subset of the served UE.

The requirements for network slice selection are defined in 3GPP TR 23.799, which shows examples of the procedures. There are three types of network slicing scenarios identified, which are referred to as group A, B, and C, to support more than one network slice per device (Figure 5.31):

- Group A refers to situations where the device is consuming services from multiple network slices and different core network instances that are logically separated. Each network slice serves the UE independently for subscription and mobility management. The drawback of this scenario is the increased signaling in the core and radio interface, although isolation is probably easiest to achieve.
- Group B is a combination of shared NFs among slices, while other functions are in individual slices.
- Group C refers to the scenario where the CP is common among slices, and individual network slices handle the UP.

5.7.2.5 Release 16 Enhancements

Related to the further enhancement of network slicing, 3GPP TR 21.916 summarizes its Release 16 aspects. Release 16 network slicing addresses accordingly the major limitations that Release 15 has in 5GC. As a result, there is:

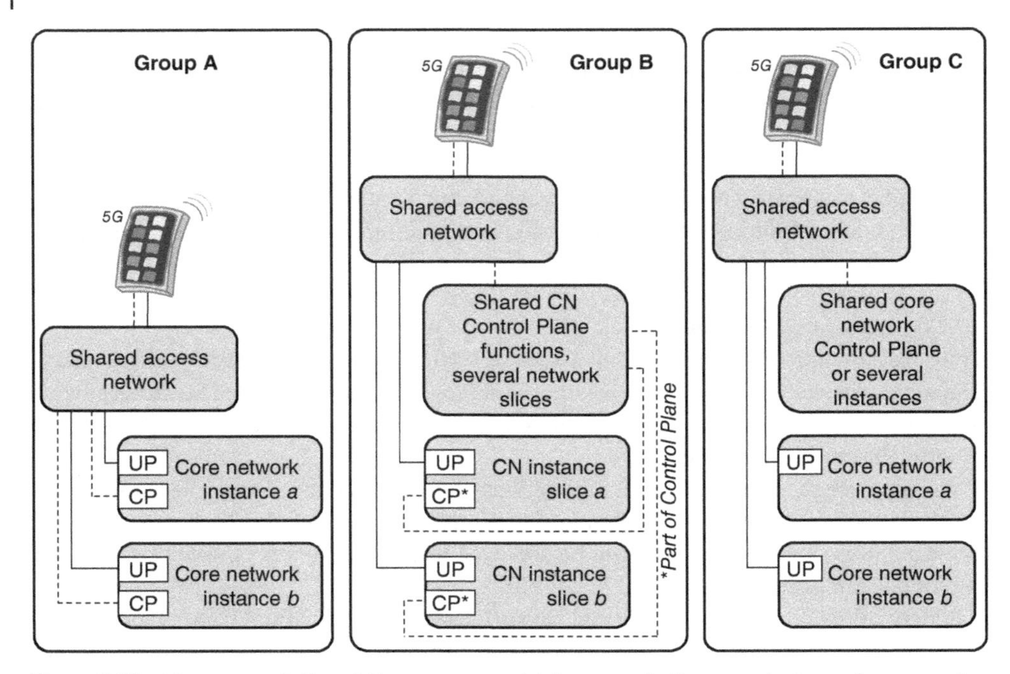

Figure 5.31 The groups A, B, and C to support multiple network slices per device, as interpreted from 3GPP TR 23.799.

- Enhancement of interworking between EPC and 5GC when the UE moves from EPC to 5GC, and when the target AMF may not be able to serve all the PDU sessions that the UE intends to move to the 5GC.
- Support for NSSAA. There is a need to enable support for separate authentication and authorization per network slice. The AMF thus performs the role of the EAP authenticator and communicates with the AAA-S via the AUSF. Consequently, the AUSF undertakes AAA protocol interworking with the AAA protocol supported by the AAA-S.

These enhancements are reflected in Release 16 in 3GPP TS 23.501 (5G system architecture, stage 2), and TS 23.502 (5G procedures, stage 2).

5.8 Transport Network

Dimensioning of the transport network is one of the essential tasks of 5G operators. The fast development of mobile communication systems has resulted in largely virtualized network architecture models and respective cloud infrastructure, which can be used in core and radio systems. The transport network infrastructure develops accordingly to ensure adequate end-to-end performance.

5G-XHaul was originally a European-wide project for converged Fronthaul (FH) and Backhaul (BH) networks of 5G. It describes a logical transport architecture integrating multiple wireless and optical technologies under a common SDN CP [25]. 5G-XHaul refers

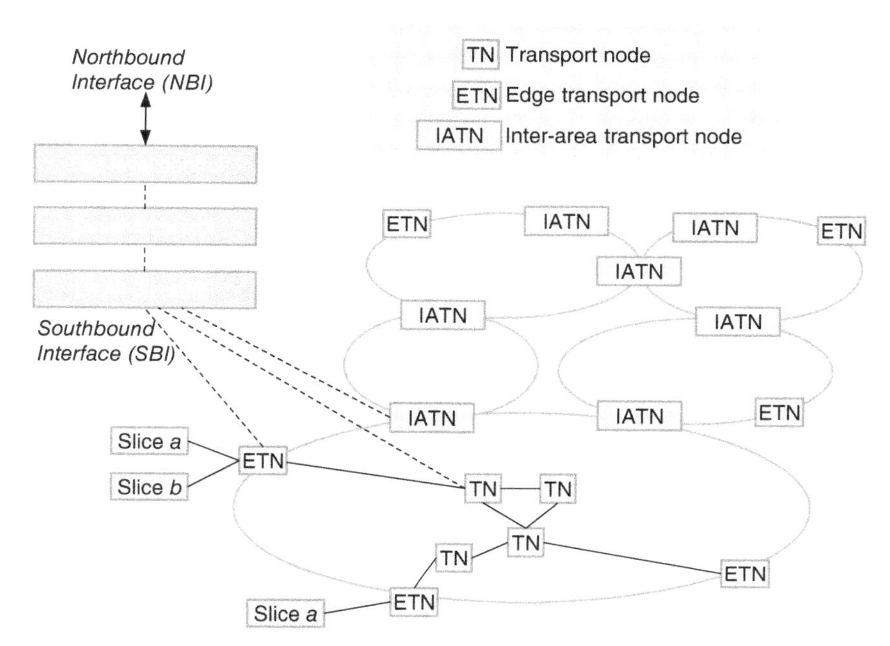

Figure 5.32 Principle of the 5G-XHaul CP.

to a unified transport supporting split concept of flexible RAN. The European Union is planning to use XHaul as a basis for the 5G transport infrastructure.

There is a unified SDN XHaul CP designed for wireless and optical networks. Figure 5.32 depicts the principle of XHaul control.

The data plane of XHaul unifies the transport of wireless networks such as Point-to-Multipoint (P2MP) mmWave radio, sub-6 GHz mobile networks, optical networks, Time Shared Optical Network (TSON), and Wavelength Division Multiplexing (WDM)-based Passive Optical Network (PON) referring to optical fiber network architecture. Figure 5.33 depicts the UP of XHaul.

These networks tolerate much greater data loads than the older systems were able to, and can provide wider bandwidth compared to the copper-based legacy networks.

XHaul serves as an example of some of the practical options and respective performance and functional requirements for 5G transport network data transfer. More up-to-date information on the architecture of XHaul is available at the web page of 5G PPP [26].

XHaul thus refers to a new RAN transport infrastructure that supports FH, mid-haul, and BH. Its physical layer is based on fiber optics, metro SDN, and edge cloud solutions for serving 5G signaling and data traffic.

As stated in Ref. [27], there are various requirements for Open XHaul in the 5G environment, including the following:

- **Bandwidth**: 5G requires interfaces that are capable of supporting 100 Gb/s data speeds for RANs based on cost-effective fiber optics.

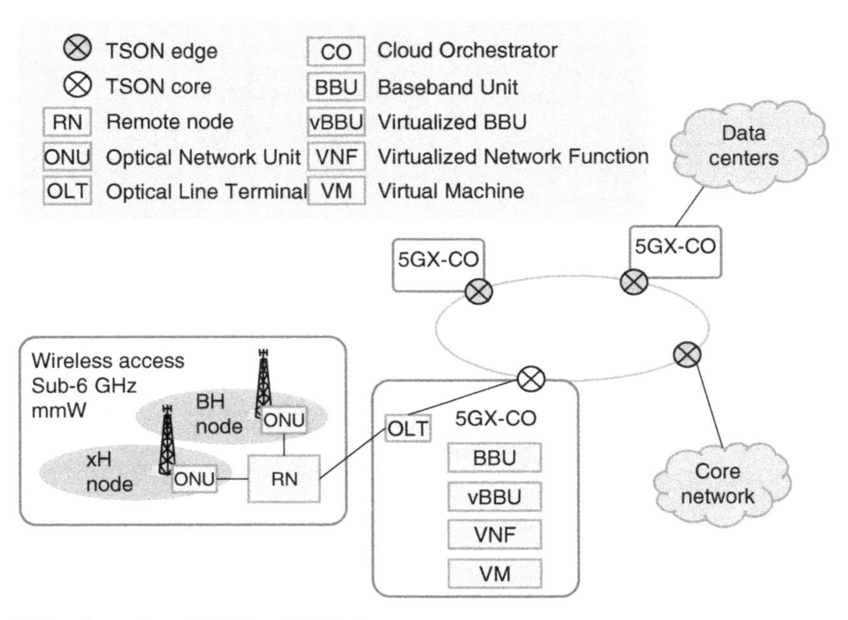

Figure 5.33 Example of 5G-XHaul UP deployment.

- **Interoperability**: 5G FH needs to support a combination of legacy protocols such as Common Public Radio Interface (CPRI), and new ones such as evolved CPRI (eCPRI) and Time-Sensitive Networking (TSN).
- **Latency**: The round-trip time for many 5G applications needs to be in the sub-10 ms range, whereas the most demanding applications require latency values of down to 1 ms. FH schemes may require even faster responses as stated in IEEE 1914.3.
- **Synchronization**: 5G brings strengthened requirements for synchronization based on Time Division Multiplexing (TDD).

The role of the transport network is to interconnect the radio access and core network. To comply with the strict performance requirements for 5G networks, optimal deployment can be done by relying on increased intelligence, flexibility, and automation in the transport network [28].

As the RAN and core network develop, the transport network also needs to follow the evolution to avoid creating a bottleneck similar to the QoS level and performance. This requires a higher grade of flexibility, simple service configurations, as well as new operations model support and cross-domain orchestration. The dynamic and virtualized RAN requires that transport connectivity is also more dynamic than in previous generations, which requires inevitably more automatization to manage the transport.

As an example of commercial productization, Ericsson has developed an SDN accompanied by an intelligent application referred to as Transport Intelligent Function (TIF), in an effort to seek optimal 5G transport network design. In this solution, the 5G transport network is based on a self-contained infrastructure underlay with an SDN-controlled overlay for a variety of RAN and user services. The distributed CP in the underlay maintains the basic infrastructure and handles redundancy and quick restoration in case of network failures. The service and characteristics- aware overlay is handled by the SDN controller with the TIF application, and this creates a dynamically controlled and orchestrated transport network that requires minimum manual interaction [28]. The respective automation framework of the solution is based on the Open Network Automation Platform (ONAP).

5.9 IMS for 4G and 5G Voice Service

5.9.1 IMS Architecture

Voice over LTE (VoLTE) relies on the IMS. The IMS as such is not LTE specific, but LTE can benefit from its services. The IMS works as a base for many other types of services such as Voice over Wi-Fi (VoWiFi) and Video over LTE (ViLTE).

Figure 5.34 outlines the key components of a typical IMS core network. Table 5.3 summarizes the IMS core components of Figure 5.34.

IMS communications are based on the SIP. The IETF has developed this for initiating, modifying, and terminating an interactive user session that involves multimedia elements such as video, voice, instant messaging, online games, and VR (RFC 3261). It is the signaling solution within the IMS, so it also serves the VoLTE ecosystem. SIP can establish sessions for features such as audio and videoconferencing, interactive gaming, and call forwarding over IP networks. It enables service providers to integrate basic IP telephony services with the Web, e-mail, and chat services.

The main objective of SIP is communication between multimedia devices. SIP makes communication possible thanks to the supported Real-time Transport Protocol/Real-time Control Protocol (RTP/RTCP) and SDP protocol. VoLTE uses the RTP, which is a network protocol for delivering audio and video over IP networks.

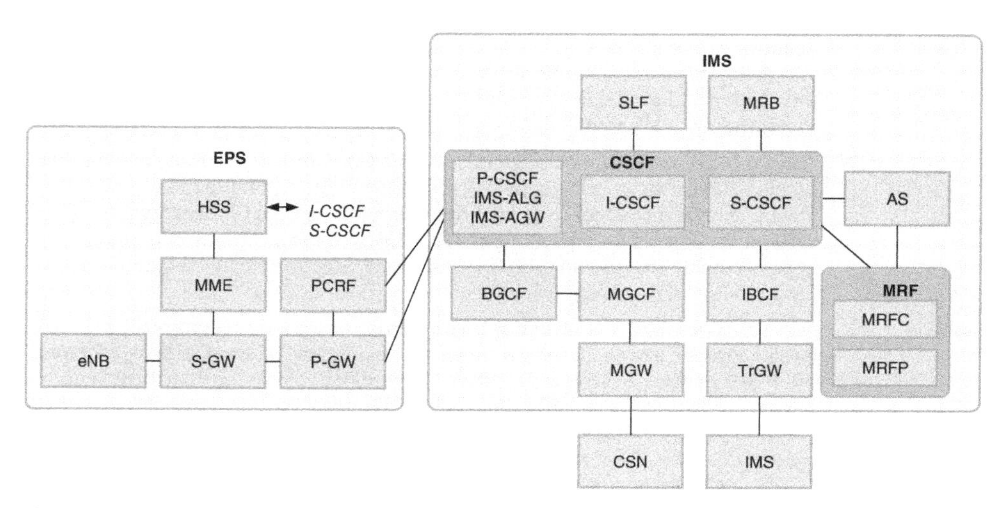

Figure 5.34 The conceptual presentation of the main IMS components that interconnect to the 4G system.

Table 5.3 The IMS core elements. Please refer to Figures 5.34 and 5.35 for the respective IMS scenarios.

Element	Description
P-CSCF	*Proxy Call Session Control Function* is the initial point of contact for the session signaling of the IMS-enabled VoLTE UE. The P-CSCF functions as a SIP proxy. It forwards SIP messages between the UE and the IMS core network, and maintains security associations between itself and the VoLTE UE. The P-CSCF may be implemented in an access session border controller, which may also incorporate the IMS Application Level Gateway (ALG)/IMS Access Gateway (AGW).
I-CSCF	*Interrogating Call Session Control Function* connects with the mobile network for all the traffic within that specific network. On IMS registration, it interrogates the HSS to determine a suitable S-CSCF for routing the registration requests. For terminating calls, it interrogates the HSS to determine on which S-CSCF the user is registered.
S-CSCF	*Serving Call Session Control Function* provides session setup, termination, control, and routing functions. It generates records for billing purposes, and invokes ASs. It acts as a SIP registrar for the VoLTE UE that the HSS and I-CSCF assign to it. It queries the HSS for the applicable subscriber profiles and handles calls involving these endpoints.
AS	*Application Server* provides IMS applications.
MRF	*Media Resource Function* provides media plane processing such as transcoding, multiparty conferencing, and network announcements/tones, under the control of IMS ASs. It also serves for basic media processing functions to the CSCF. The MRF has two components: the *Media Resource Function Controller* (MRFC) and the *Media Resource Function Processor* (MRCP). The MRFC receives SIP requests and information from the AS and S-CSCF. It manages media resources and controls the MRFP.
IBCF/TrGW	*Interconnection Border Control Function/Transition Gateway* is responsible for the control/media plane at the network interconnect point to other PLMNs. The IBCF/TrGW may be implemented as a part of the interconnect session border controller.
IMS-ALG/ IMS-AGW	*IMS Application Level Gateway/IMS Access Gateway* may be a standalone function or co-located with the P-CSCF. The IMS-ALG/IMS-AGW is responsible for the control/media plane at the access point to the IMS network. It provides functions for gate control and local Network Address Translation (NAT), IP realm indication and availability, remote NAT traversal support, policing, QoS packet marking, and IMS media plane security.
MGCF/MGW	*Media Gateway Control Function/IMS Media Gateway* is responsible for control/media plane interworking at the network interconnect point to the Circuit-Switched (CS) networks. This includes interworking with CS networks based on Bearer Independent Caller Control/ISDN User Protocol/Session Initiation Protocol with Encapsulated ISUP (BICC/ISUP/SIP-I) and may include transcoding of the media plane.
BGCF	*Breakout Gateway Control Function* determines the next hop for routing of SIP messages. This happens based on the information it receives from the Session Description Protocol (SIP/SDP), or from internal or Electronic Number Mapping System/Dynamic Name Server (ENUM/DNS) lookup routing configuration data. For CS terminations, the BGCF determines the network in which the CS domain breakout will take place and selects the appropriate MGCF. For terminations with interconnecting peer IMS networks, the BGCF selects the appropriate IBCF. The BGCF may also provide directives to the MGCF/IBCF on which interconnect or next network to select.

Table 5.3 (Continued)

Element	Description
SLF	The *Subscriber Location Function* of IMS is typically a database and provides information about the HSS and respective user profiles. If the home network has more than one HSS, the I-CSCF and S-CSCF need to find the correct HSS via the SLF.
MRB	The *Media Resource Broker* manages the collection of published MRF information and supplying of MRF information to consuming entities such as the AS.

Another important IMS protocol is the SDP. While SIP establishes, modifies, and terminates sessions, SDP's sole function is to inform the media within those sessions. SDP thus describes the media of a session, but it does not negotiate it.

5.9.2 Voice Service

VoLTE is a GSMA profile for the delivery of services of CS networks – mainly voice and SMS – over the packet-switched connectivity of the LTE network. VoLTE leverages the core network IMS. The VoLTE profile is defined in the GSMA PRD based on 3GPP specifications.

Conformity to the VoLTE profile provides operators with a fluent interworking between their LTE networks and VoLTE-capable devices. It also ensures the expected experience for users of the Multimedia Telephony Service (MMtel) and SMS.

The VoLTE service relies on LTE and IMS infrastructure, and the respective SIP signaling. Figure 5.35 summarizes the elements in a typical VoLTE setup.

Table 5.4 summarizes the additional elements compared to the base IMS for supporting VoLTE. There is no need to modify the 4G radio network due to VoLTE. The 4G core network

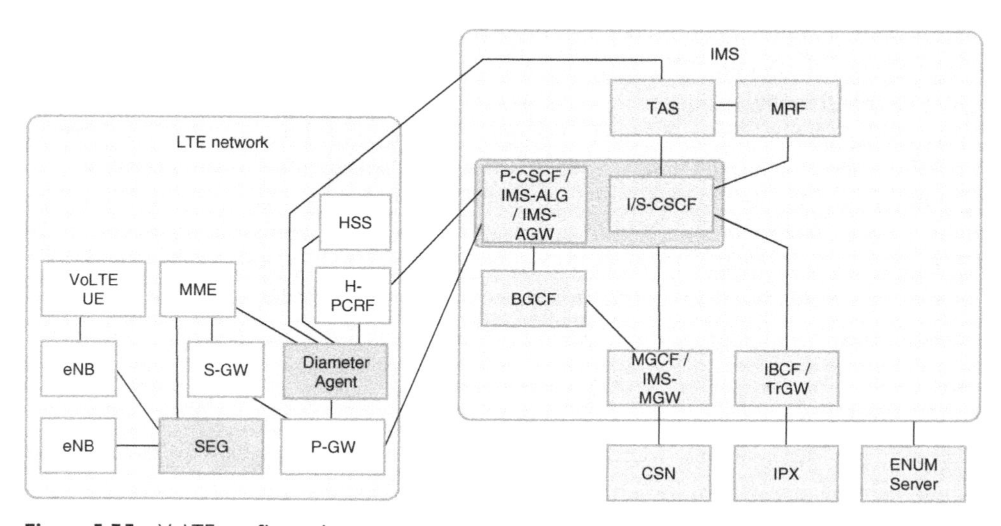

Figure 5.35 VoLTE configuration.

Table 5.4 The additional elements that can be deployed to support VoLTE in IMS.

Element	Description
IPX	*IP Packet Exchange* transit network provides an interconnect capability between Private Mobile Networks (PMNs), as defined in GSMA PRD IR.34.
Diameter agent	Defined by IETF RFC 3588 and utilized by GSMA PRD IR.88, the diameter agent controls diameter signaling. It enables seamless communication and control of information between network elements within LTE or IMS networks and across network borders. The diameter agent reduces the mesh of diameter connections that would negatively impact network performance, capacity, and management.
SEG	*Security Gateway* can be used to originate and terminate secure associations between the eNodeB and the EPC network. IPsec tunnels are established with pre-shared security keys that can take a number of different formats. IPsec tunnels enforce traffic encryption for increased protection based on the parameters exchanged between the parties during the tunnel setup. This enables secure communications between the eNodeB and EPC across the *S1-MME*, *S1-U*, and *X2* interfaces.
TAS	*Telephony Application Server* provides support for a minimum set of mandatory MMTel services as defined by the 3GPP. These services include supplementary service functionality, and they are profiled in GSMA PRD IR.92 and GSMA PRD IR.94.
ENUM	*Electronic Number Mapping System* was developed by the IETF. It translates the E.164 numbers to Uniform Resource Identifiers (SIP URIs) using the DNS to enable message routing of IMS sessions, as defined in GSMA PRD IR.67 and NG.105.

works as a base for VoLTE, too, but VoLTE can be deployed by introducing an additional secure gateway and diameter agent as depicted in Figure 5.35. These elements provide an additional security layer for VoLTE communications between the LTE and IMS networks.

The VoLTE UE needs to support VoLTE communications. This happens via an embedded VoLTE client, which is a module on a device that provides the VoLTE (or VoWiFi) service. This client verifies with the network's entitlement configuration server if it is entitled or not to offer the VoLTE service to end-users.

VoLTE functionality resides in the TAS located in the IMS core network. The VoLTE UE refers to the mobile device capable of communicating with the IMS core and serving VoLTE. As depicted in Figure 5.35, VoLTE is based on the LTE and IMS architectures.

Entitlement refers to the applicability, availability, and status of a service, needed by the client before offering that service to end-users. Entitlement configuration refers to the information returned to the client by the network, providing entitlement information on a service. The entitlement configuration server is a network element that provides entitlement configuration for different services to clients [29].

5.9.3 Roaming

VoLTE roaming extends the benefits of VoLTE, including higher quality and better spectrum efficiency, to the roaming subscribers. It provides customers with fluent user experience regardless of the location, wherever VoLTE and respective roaming with the home operator has been set up.

The GSMA-defined VoLTE roaming architecture includes some of the following:

- The device attaches to IMS and the local P-CSCF via the LBO visited routed or LBO home routed option;
- The device attaches to the P-CSCF of the IMS of the home network directly via the S8 Home Routed (S8HR) option.

The S8HR VoLTE roaming deployment model provides fast time-to-market implementation, and may provide cost benefits depending on the regulatory requirements. The rationale of this statement is the lower number of entities that need to be enhanced, lower grade of implementation, and reduced testing to support VoLTE roaming.

Figure 5.36 presents the S8HR VoLTE roaming architecture. This model provides the LTE device with the means to attach to the home IMS network's P-CSCF based on the functional procedures for resolving the Access Point Name (APN) to the P-GW in the HPLMN as described in GSMA IR.65. This model requires the use of QoS-level roaming. This means that the service-specific QoS level[1] need to be supported on the home-routed PDN connection when roaming takes place.

In the VoLTE scenario, the charging model needs to be considered by the involved parties as there are options for both QoS Class Identifier (QCI)-based and APN-based charging. In practice, to cope with these two different models, operators deploying the S8HR option should ensure that their billing system and clearing house support Transferred Account Procedure (TAP) with QCI and with APN. If a roaming partner does not support both options, it is of utmost importance to agree mutually whether the QCI or APN is applied.

5.9.4 Key Definitions

GSMA PRD IR.92 defines the User-Network Interface (UNI) profile for IMS voice and SMS. It describes the minimum mandatory set of 3GPP features that the UE and network are

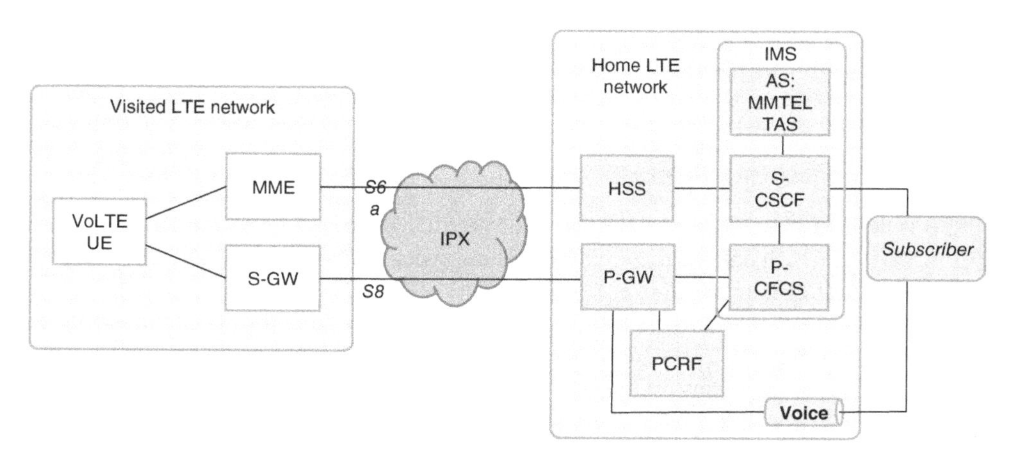

Figure 5.36 The S8HR roaming architecture.

[1] Some examples of the service-specific QoS are the QCI 1 for voice media and QCI 5 for IMS signaling.

required to implement to guarantee an interoperable, high-quality IMS-based voice telephony service over LTE:

- GSMA PRD IR.94 defines a minimum mandatory set of 3GPP features. It is produced on top of GSMA PRD IR.92 to ensure that the UE and network are required to guarantee an interoperable, high-quality IMS-based conversational video service over LTE or HSPA radio access.
- GSMA PRD IR.65 provides guidelines for Network-Network Interface (NNI) VoLTE IMS roaming and interworking.
- GSMA PRD IR.88 gives guidelines for VoLTE roaming.
- In addition, the "VoWiFi and VoLTE Entitlement Configuration" document of GSMA (TS.43) describes the needed VoLTE configuration [29].

5.9.5 VoLTE Infrastructure Options

An MNO wanting to offer VoLTE may deploy and operate its own IMS core, or rely on a third party offering the IMS functions. Technically, both solutions are equally valid and transparent for end-users. Business-wise, personal deployment requires investment and management, which has an impact on the CAPEX and OPEX models, while the business model for outsourced infrastructure can be based on, e.g., the volume of the VoLTE traffic.

5.9.6 Fallback Mechanisms

Although the LTE specifications do not support CS connectivity any more, traditional CS voice calls can still be formed via the previous 2G and 3G systems via the following means:

- CS Fallback (CSFB) mechanism;
- Single Radio Voice Call Continuity (SRVCC).

Figure 5.37 shows an example of the combined architecture of 2G, 3G, and 4G as a basis for CSFB scenarios.

5.9.7 Circuit-Switched Fallback

CSFB is defined in 3GPP TS 23.272, Circuit-Switched Fallback in Evolved Packet System (EPS), stage 2, which defines the architecture and specification for CSFB and for SMS over SGs for EPS or CSFB and SMS over S102.

CSFB and SMS over SGs in the EPS function is realized by using the SGs interface mechanism between the MSC Server and the MME. SGs interface functionality is based on the 3GPP-specified mechanisms for the *Gs* interface. Figure 5.38 depicts the principle of CSFB and SMS over SGs as interpreted from 3GPP TS 23.060.

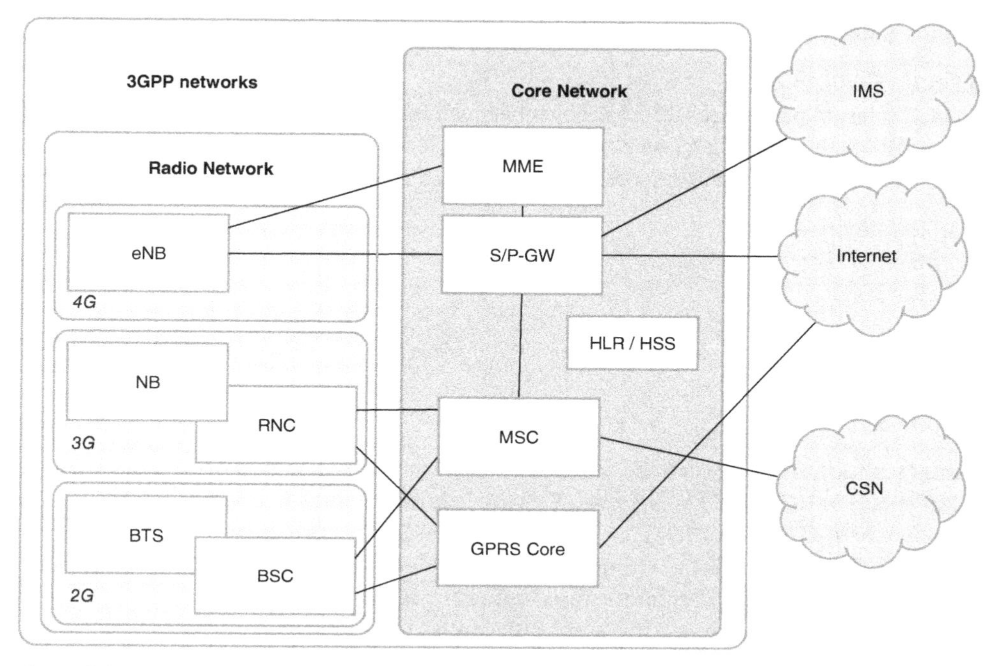

Figure 5.37 The connectivity between 4G, 3G, and 2G.

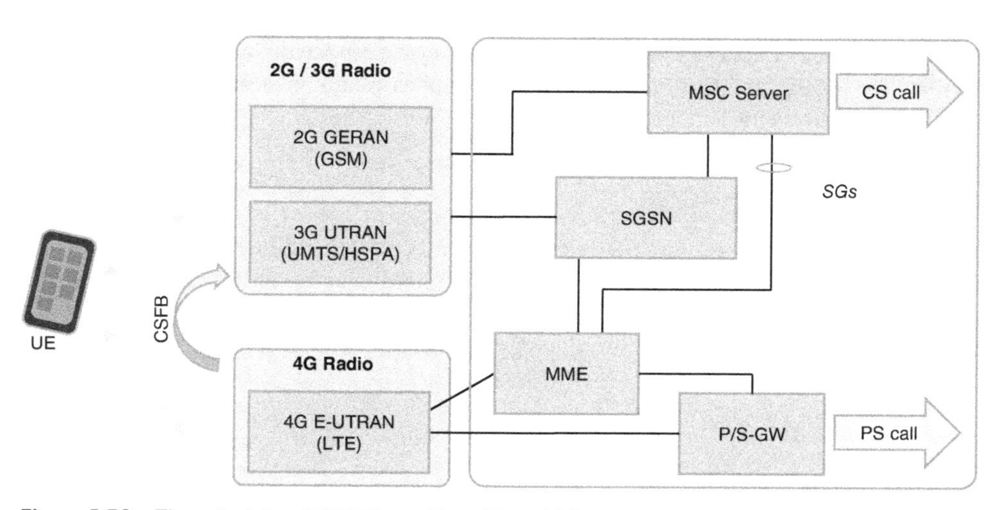

Figure 5.38 The principle of CSFB from 4G to 2G and 3G.

5.9.8 Single Radio Voice Call Continuity

5.9.8.1 SRVCC to 2G/3G

SRVCC refers to procedures for a handover between the IMS-based VoIP call in LTE and CS voice call of a legacy system. SRVCC is defined in 3GPP TS 23.216, as of Release 8. It presents various scenarios and respective architectures for SRVCC:

- LTE – American 2G[2];
- LTE – 2G/3G[3];
- LTE – 3G[4];
- 3G HSPA – 3G UMTS/2G[5];
- 2G/3G – LTE/3G HSPA[6];
- 2G to 3G HSPA[7];
- 5G to 3G[8]

Let us outline a roaming scenario as an example. Figure 5.39 presents a SRVCC procedure taking place in LTE to 2G/3G systems. This procedure is defined in 3GPP TS 23.216, which presents the E-UTRAN to 3GPP UTRAN/GERAN SRVCC architecture.

For SRVCC to work, TS 23.216 introduces support for SRVCC in the MSC Server element, enhancing the E-UTRAN architecture as defined in 3GPP TS 23.401.

SRVCC handover presented in Figure 5.39 applies to a home network as well as to communications while roaming.

5.9.8.2 SRVCC from 5G to 3G

Figure 5.40 presents the SRVCC procedure for the handover from 5G to 3G. The procedure is defined in 3GPP TS 23.216.

The NG-RAN to 3GPP UTRAN 5G-SRVCC architecture introduces an additional function in the MSC Server and MME SRVCC enhancing the service defined in the 5GC architecture TS 23.501 for 5G-SRVCC. 5G-SRVCC applies to home networks and roaming networks as long as the service is deployed.

5.9.9 Interworking in 4G/5G

Interworking between 4G and 5G is especially important in the early phase of the 5G deployment relying on LTE as an anchoring system. The principle of VoLTE as defined and deployed in 4G networks continues to be valid also in 5G networks with only minor modifications in the setup. As for the terminology, the service is referred to as Voice over New Radio (VoNR) in 5G networks.

---

[2]E-UTRAN to 3GPP2 1xCS SRVCC.
[3]E-UTRAN to 3GPP UTRAN/GERAN SRVCC.
[4]3GPP E-UTRAN to 3GPP UTRAN vSRVCC.
[5]UTRAN [HSPA] to 3GPP UTRAN/GERAN SRVCC.
[6]3GPP UTRAN/GERAN to 3GPP E-UTRAN or UTRAN (HSPA) SRVCC.
[7]3GPP GERAN to 3GPP UTRAN (HSPA) SRVCC.
[8]NG-RAN to 3GPP UTRAN 5G-SRVCC architecture.

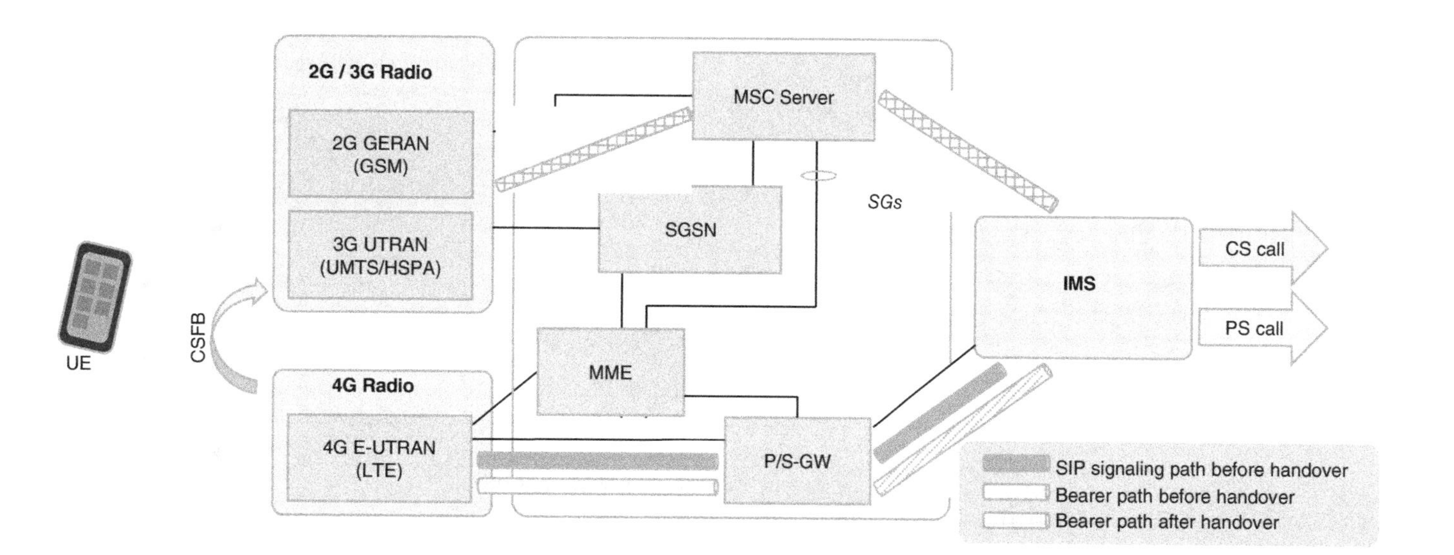

Figure 5.39 The principle of SRVCC from 4G to 2G/3G. The MSC Server in this scenario is enhanced for supporting SRVCC handover.

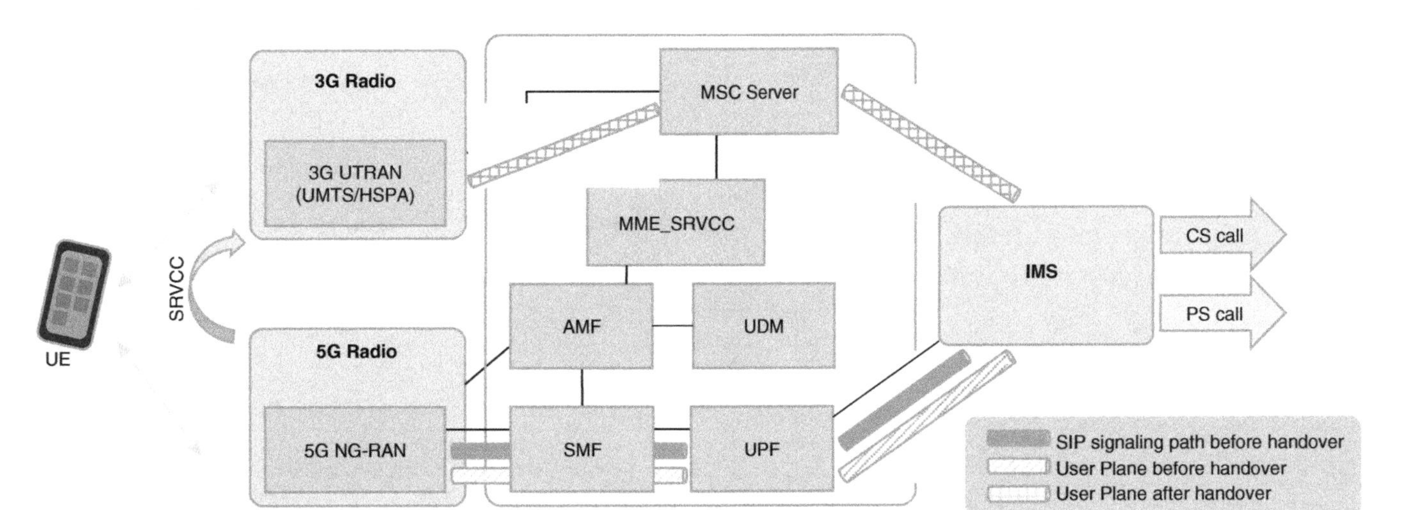

Figure 5.40 The principle of SRVCC from 5G to 3G.

In the SA deployment of 5G, legacy 4G and new 5G core network interworking and respective mobility are thus essential. The 4G network elements supporting such interworking include HSS, PCRF, and P-GW, while the respective NFs of 5G include UDM, PCF, SMF, and UPF. Figure 5.41 depicts an example of non-roaming architecture for 4G and 5G interworking. Of the elements presented in this example, 4G has introduced Control and User Plane Separation (CUPS) of P-GW and S-GW.

5G is purely a packet-switched system, like 4G LTE. Both lack native CS connectivity. Instead, VoLTE is used for voice calls in 4G. The native voice call service in 5G is referred to as VoNR. Figure 5.41 depicts the principle of VoLTE and VoNR interconnection.

The initial 5G deployment relies on the 4G core network, while 5G base stations are added gradually in the field. This refers to the Non-Standalone(NSA) deployment scenario. The respective 4G core services like IMS and VoLTE remain as such or with small modifications. In fact, in this phase, IMS has no means to recognize the presence of new 5G radio.

In SA deployment referring to Option 2 of the 3GPP, the native 5G VoNR starts to be relevant. Also, the fallback from VoNR to VoLTE is possible based on the 3GPP Release 15 definitions. From this point on, Release 15 no longer supports the CSFB fallback mechanism from 5G to 2G or 3G, which was still the case in LTE. There is, though, an option to perform SRVCC from 5G to 3G as described in Figure 5.40.

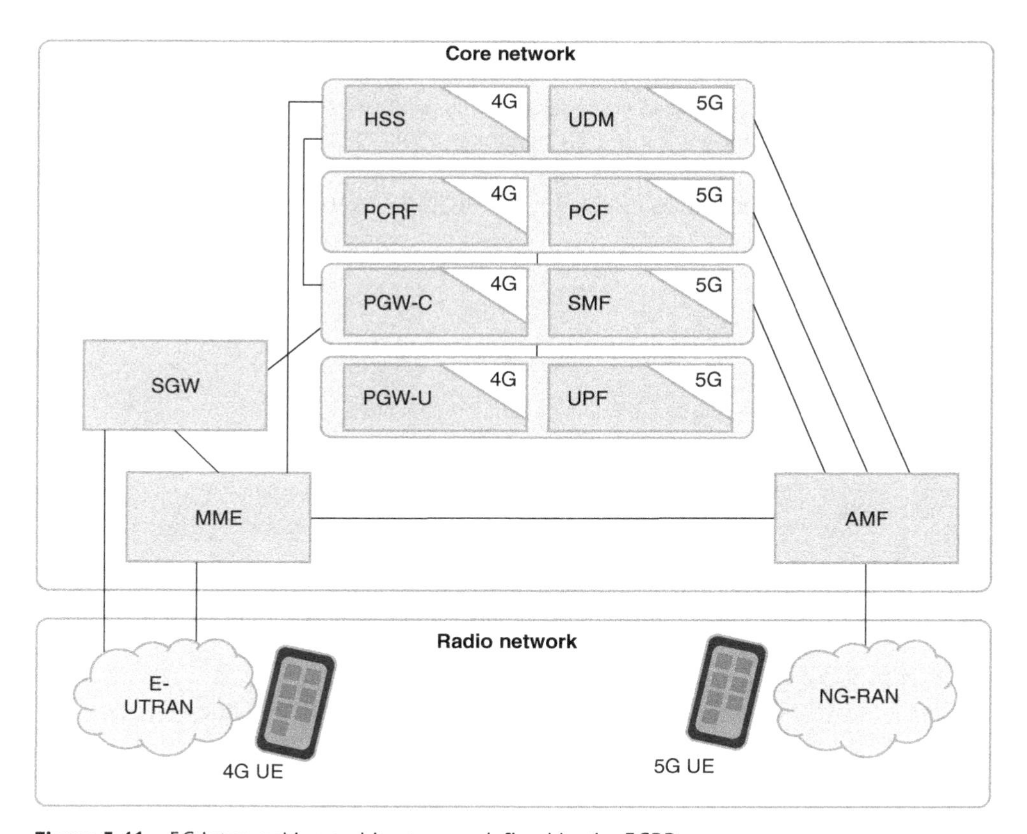

Figure 5.41 5G interworking architecture as defined by the 3GPP.

As the deployment of 5GC with full 5G radio connected is gradual, 5GC still relies largely on the 4G EPC and respective VoLTE service. This combination provides seamless voice service based on the defined fallback mechanism. In this scenario, it is essential that both 4G and 5G radio, core, and IMS support the voice service and respective capabilities (Figure 5.42).

The VoLTE and VoNR deployment phases can be divided into the following:

- **5G NSA**: 5G services need to include VoLTE.
- **5G SA, first phase**: It is important to activate Evolved Packet System Fallback (EPS FB) for voice service establishment through VoLTE.
- **5G SA, second phase**: Along with the increasing number of UE supporting SA mode, the default voice service would be based on VoNR. As soon as the user moves away from the 5G radio coverage during the call, packet-switched handover can make the user experience seamless for the NR-to-LTE session switch and provide VoLTE to take over. This procedure is similar to SRVCC of 4G.

According to the GSMA, when evolving to 5G based on 5GC and if NR coverage is continuous, then IMS voice and video call on NR via 5GC with Option 2 is the most applicable, as the cost of NR is much lower than LTE and the cell capacity of NR is much higher than LTE.

5.9.10 Requirements for IMS Voice

For IMS voice support in NG-RAN, the following assumptions apply:

- Network ability to support IMS voice sessions, i.e., ability to support QoS flows with 5G QoS Identifier (5QI) for voice and IMS signaling (3GPP TS 23.501), or through EPC system fallback;
- UE capability to support "IMS voice over PS" (3GPP TS 24.501).

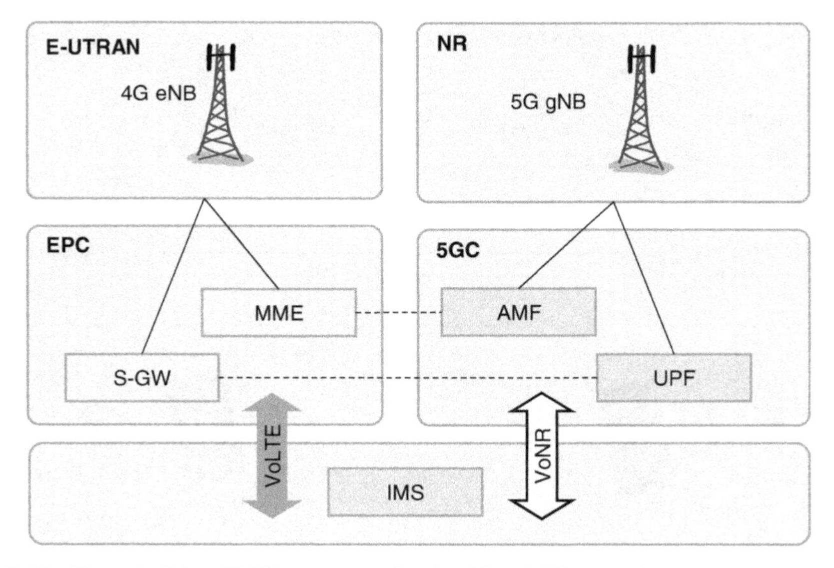

Figure 5.42 The principle of IMS interconnection for 4G and 5G networks.

The NAS layer handles capabilities indication checking. To maintain the voice service in NG-RAN, the UE provides additional capabilities over RRC (3GPP TS 38.331) to determine the NR voice support options.

Release 16 enhances the support for MMTel IMS voice and video. One of the items is RAN-assisted codec adaptation, which provides a means for gNodeB to send codec adaptation indication with recommended bit rate to assist the UE to select or adapt to a codec rate for MMTel voice or MMTel video.

References

1 Parssons, G., "5G, SDN and MBH," Ericsson, 2014.
2 3GPP, "TR 22.891; Feasibility Study on New Services and Market Technology Enablers, Release 14," 3GPP, 2016.
3 Cranford, N., "The Role of NFV and SDN in 5G," RCR Wireless News, 4 December 2017. [Online]. Available: https://www.rcrwireless.com/20171204/fundamentals/the-role-of-nfv-and-sdn-in-5g-tag27-tag99. [Accessed 31 July 2018].
4 SDX Central, "How 5G SDN Will Bolster Networks," 2019. [Online]. Available: https://www.sdxcentral.com/5g/definitions/5g-sdn.
5 ETSI, "Network Functions Virtualisation (NFV)," ETSI, 30 August 2020. [Online]. Available: https://www.etsi.org/technologies/nfv. [Accessed 30 August 2020].
6 Miller, R., "With 5G on the Horizon, Telcos Embrace Open Compute," Data Center Frontier, 21 June 2018. [Online]. Available: https://datacenterfrontier.com/with-5g-on-the-horizon-telcos-embrace-open-compute. [Accessed 21 June 2018].
7 Raynovich, R.S., "How IoT & 5G Will Change the Data Center," Futuriom, 27 February 2018. [Online]. Available: www.futuriom.com/articles/news/how-5g-iot-will-change-the-data-center/2018/02. [Accessed 21 June 2018].
8 Carugi, M., "Key Features and Requirements of 5G/IMT-2020 Networks," ITU, 14–15 February 2018. [Online]. Available: https://www.itu.int/en/ITU-D/Regional-Presence/ArabStates/Documents/events/2018/RDF/Workshop%20Presentations/Session1/5G-%20IMT2020-presentation-Marco-Carugi-final-reduced.pdf. [Accessed 21 June 2018].
9 ITU-T, "Cloud Computing – Functional Architecture of Network as a Service (Recommendation Y.3515)," ITU-T, July 2017.
10 3GPP, "TS 23.501 V16.5.1 (2020–08); System Architecture for the 5G System; Release 16," 3GPP, August 2020.
11 Eronen, P., "IKEv2 Mobility and Multihoming Protocol (MOBIKE)," IETF, June 2006. [Online]. Available: https://www.ietf.org/rfc/rfc4555.txt. [Accessed 23 June 2018].
12 Schmitt, P. (Huawei), Landais, B. (Nokia), and Yong Yang, F. (Ericsson), "Control and User Plane Separation of EPC Nodes (CUPS)," 3GPP, 3 July 2017. [Online]. Available: http://www.3gpp.org/cups. [Accessed 23 June 2018].
13 Rabie, K., "Core Network Evolution (RCAF, PFDF, & TSSF) – 3GPP REST Interfaces," LinkedIn, 23 July 2017. [Online]. Available: https://www.linkedin.com/pulse/core-network-evolution-rcaf-pfdf-tssf-3gpp-rest-interfaces-rabie. [Accessed 23 June 2018].
14 GSMA, "IR.34 – Guidelines for IPX Provider Networks; Version 9.1," GSMA, 2013.
15 GSMA, "5G Roaming Guidelines Version 2.0," GSMA, 28 May 2020.

16 3GPP, "TS 24.011; Point-to-Point (PP) Short Message Service (SMS), Release 15," 3GPP, 2018.
17 3GPP, "TS 23.040; Technical Realization of the Short Message Service (SMS), Release 15, V15.1.0," 3GPP, 2018.
18 3GPP, "3GPP Initiates Common API Framework Stud," 3GPP, 9 May 2017. [Online]. Available: https://www.3gpp.org/news-events/1854-common_ap. [Accessed 14 August 2020].
19 3GPP, "TS 23.273 V16.4.0, 5G System (5GS) Location Services (LCS), Stage 2, Release 16," 3GPP, July 2020.
20 Oracle, "Oracle Communications Service Communication Proxy (SCP) Cloud Native User's Guide; Service Communication Proxy System Architecture," Oracle, 2019. [Online]. Available: https://docs.oracle.com/communications/F21353_01/docs.10/SCP/GUID-6C3A599D-AB9F-430D-B7E8-66AD51B69006.htm#:~:text=The%20Service%20Communication%20Proxy%20is,observability%20to%20the%20core%20network. [Accessed 14 August 2020].
21 Brown, G., "Contextual Service Assurance in 5G: New Requirements, New Opportunities," Heavy Reading, September 2017.
22 GSMA, "An Introduction to Network Slicing," GSM Association, 2017.
23 Nokia, "Cloud Mobile Gateway," Nokia, 21 June 2018. [Online]. Available: https://networks.nokia.com/products/mobile-gateway. [Accessed 21 June 2018].
24 Ericsson, "Ericsson Virtual Evolved Packet Core," Ericsson, 21 June 2018. [Online]. Available: https://www.ericsson.com/ourportfolio/core-network/virtual-evolved-packet-core?nav=fgb_101_256. [Accessed 21 June 2018].
25 5G PPP, "5G-XHaul," 5G PPP, 31 December 2016. [Online]. Available: https://5g-ppp.eu/new-deliverables-available-for-5g-xhaul. [Accessed 6 July 2019].
26 "5G-XHaul," New Deliverables Available for 5G-XHaul. [Online]. Available: https://5g-ppp.eu/new-deliverables-available-for-5g-xhaul. [Accessed 7 July 2019].
27 Brown, G., "New Transport Network Architectures Fro 5G RAN," 2019. [Online]. Available: https://www.fujitsu.com/us/Images/New-Transport-Network-Architectures-for-5G-RAN.pdf.
28 Ericsson, "Intelligent Transport in 5G (Ericsson Technology Review)," Ericsson, September 2017.
29 GSMA, "VoWiFi and VoLTE Entitlement Configuration, Version 2.0," GSMA, 3 October 2018.

6

Release 16 Features and Use Cases

6.1 Introduction to Release 16 Use Cases

6.1.1 5G Pillars

This chapter extends the features chapter of the previously published *5G Explained* of Release 15 that outlined expected services and application types for the 5G era. The previous book also detailed the initial building blocks of the 5G system that are Evolved Multimedia Broadband (eMBB) communications, Massive IoT (mIoT) communications that can also be called massive Machine Type Communications (mMTC), Critical Communications (CriC), Network Operations (NEO), and Vehicle-to-Vehicle (V2V) communications.

This chapter details further the descriptions and extends the summaries to cover the key features of Release 16, walking the reader through the use cases relevant to Phase 2 of 5G. Of various references, the key source of information forming the foundation of this chapter is the 3rd Generation Partnership Project (3GPP) use case study [1]. It contains more than 70 use cases that form the above-mentioned five categories of eMBB, mIoT/mMTC, CriC, NEO, and V2X. Recommendations on four of the first of these categories are also in Technical Report (TR) 22.861, TR 22.862, TR 22.863, and TR 22.864. In addition, these references present respective security considerations and in the consolidated document, TR 22.864.

Release 15 of the 3GPP standardization focused on eMBB use cases, while it only considered the very basic level of the other pillars, and left evolved V2X (eV2X) for Release 16. Regardless of the feasibility of the publishing of this book, the role of vehicle communications is under constant debate as the car industry aims to have a long-term solution as a basis for advanced functionality satisfying strict requirements, including self-driving vehicle use cases [2]. One of the remarkable decisions made happened in 2019 when the European Union indicated that IEEE 802.11p would be the basis for the future of V2X in Europe, leaving Cellular-based V2X (C-V2X) based on 5G and previous generations “in the air.” Nevertheless, this decision does not necessarily mean that 5G would be out of the question for serving complex automotive ecosystems, including autonomous vehicle communications [3].

5G Second Phase Explained: The 3GPP Release 16 Enhancements, First Edition. Jyrki T.J. Penttinen.

Release 16 represents the second phase of the 5G system as defined by the 3GPP with further evolved and new functions. In fact, Release 15 is not capable of providing performance and functionality to comply with the ITU IMT-2020 requirements for the 5G-connected society, so Release 15 works as an intermediate phase towards the full version of 5G.

Release 16 provides a platform for many new use cases. The initial list of the expected use cases is included in TR 22.891 (Release 14). This TR presents 74 use cases that represent five categories:

- **Enhanced Mobile Broadband**: Examples of this category include mobile broadband, Ultra High Definition (UHD)/hologram, high mobility, and virtual presence.
- **Critical Communications**: This category includes interactive games, eSports, industrial control, drones, robotics, vehicles, and emergency scenarios.
- **Massive Machine Type Communications**: This category includes communications for subways, stadium services, eHealth, wearables, and inventory control.
- **Network Operation**: This category includes scenarios for network slicing, routing, migration and interworking, and energy saving.
- **Enhanced Vehicle-to-Everything**: This category includes cases for autonomous driving, safety aspects, as well as other less critical aspects associated with vehicles.

The related recommendations are in 3GPP TR 22.861, TR 22.862, TR 22.863, and TR 22.864. Furthermore, 3GPP TR 22.891, Section 6.3.2, presents high-level requirements for vertical groups (eMBB, Mobile IoT (MIoT), CriC, and eV2X).

6.1.2 Technical Reports as a Foundation

5G will form an important communications platform throughout the decade of 2020–2030. While 3GPP Release 15 functions as a foundation for the first phase of 5G, Release 16 adds elemental definitions and thus enhances the specifications for tackling a variety of industry verticals.

At the beginning of the 5G specification work, the 3GPP identified a set of use cases from different points of view to define 5G [4]. As an example, 3GPP SA1 designed four TRs that outline the new Services and Markets Technology Enablers (SMARTER) for 5G. The 3GPP SA#72 meeting approved these reports.

The identified, potential 5G requirements resulted in 3GPP TR 22.891, which includes over 70 categorized use cases. The foundation of the four TRs is the mIoT, CriC, eMBB, and NEO. The following TRs describe the respective principles:

- 3GPP TR 22.861, SMARTER: massive Internet of Things. This category outlines use cases where a huge number of devices are present. Some examples of the related equipment include sensors and wearable devices. These use cases are important to verticals such as smart homes, smart cities, and eHealth.
- TR 22.862, SMARTER: Critical Communications. The key point of this TR is to outline the needed improvements for CriC in terms of latency, reliability, and availability. These, in turn, enable industrial control applications and tactile Internet, among other highly critical applications. The practical ways to achieve the respective requirements may include architectural optimization, improved radio interface, core, and radio resources.

- TR 22.863, SMARTER: enhanced Mobile Broadband. This category includes use cases that rely on high data rates, device density, and mobility. The devices of this category typically use and produce variable data speeds, and can have a mixture of mobile and fixed mobile environments. A typical scenario for eMBB is a dense city center with a set of small cells.
- TR 22.864, SMARTER: Network Operation. This use case category outlines requirements for the functional system. Some examples of the needs are the flexibility of the functions and capabilities, fluent interworking, optimization, and security.

3GPP SA1 has consolidated the above-mentioned TRs into a single Technical Specification (TS), 3GPP TS 22.261.

6.1.3 Use Cases Identified by Industry

Many of the current and completely new verticals can benefit from the 5G-specific use cases. The specification community as well as industry have discussed the needs of these use cases. As an example, Ref. [5] summarizes some of the estimated top use cases that assume support for a considerable diversity of devices and services, e.g., mIoT equipment. The source mentions, e.g., the interaction between humans and IoT devices, critical control of remote devices, media that are available within wide areas, applications for smart vehicles, transport and respective infrastructure, and broadband experience that is available "everywhere, anytime." As another practical example, the Global System for Mobile Communications (GSMA) working groups such as Global and North Americas Network Slicing setups have interpreted and assessed different worldwide and regional vertical needs to design the network slicing attribute template accordingly in a form of the NG.116 documentation [6].

It is important to note, though, that the most advanced solutions and applications will not be available anytime soon at the very beginning of 5G services even in the Release 16 era. As an example, fixed wireless access may be one of the most concrete and typical 5G use cases as it provides wireless Internet access at home and office relying on 5G infrastructure instead of the typically utilized fixed lines. Another early use case is eMBB. The ultimate goal, as per the ITU IMT-2020 requirement set, is to offer up to 20 Gb/s peak throughput in static environments, and 1 Gb/s throughput in high mobility cases, although in non-congested networks and in the best cases.

In the latter phase, as soon as the coverage area of 5G as well as the system's functions evolve, some of the typical use cases can be assumed to include mMTC. This case has been generating wide discussions, being one of the most anticipated 5G use cases. It is able to connect fluently a vast number of sensors and other IoT devices in very dense areas. ITU IMT-2020 defines a requirement for the respective use case demanding up to 1 million simultaneously connected devices within an area of 1 km^2. The benefit of 5G networks becomes especially relevant in the Industrial IoT (I-IoT) environment. mMTC enables highly efficient communications infrastructure for, e.g., smart cities with many subverticals such as smart utilities and transportation, advanced agriculture, and security solutions. Another mid-term use case that became a reality by Release 16 was Ultra-Reliable Low Latency Communications (URLLC). This includes use cases such as remote control of critical infrastructure, drones, industrial automation, and the control and management of autonomous vehicles.

The longer-term, future 5G use cases include a vast number of all kinds of applications of which the following sections summarize some of the most important ones.

6.1.4 Market Needs for Release 16

The communication needs of consumers, businesses, verticals, and other users have increased steadily. The LTE networks can deliver an impressive capacity and their service areas have developed considerably, so it is justified to question the need for yet another generation.

Nevertheless, as has been the case with all previous generations, the increased performance of the new networks enables novelty applications and services that gradually set the new level of expectations by their users. 5G is capable of offering more fluent user experiences for many use cases such as remote education and virtual training.

Not only is technological performance important per se, but 5G is expected to contribute to the overall development of society across nations. 5G is expected to generate new jobs and provide a variety of direct and indirect business opportunities to many new stakeholders within the telecommunication industry and beyond.

Yet another aspect is the role of 5G in increasing people's security and well-being. The new networks are formed by many types of cells of which the dense small cell areas are in a special position. As one of their benefits, they can provide enhanced location accuracy information in special cases such as a medical emergency, expediting first responders to find patients thanks to the increased accuracy of horizontal and vertical cellular-aided positioning solutions enabled by the customer.

Continuing with the summary of the benefits of 5G, 5G small cells can help in lifeline communications by offering network infrastructure redundancy if, say, natural disaster or other issues damage it. This may be of utmost importance as the probability for disasters damaging the telecommunications infrastructure is high and all the additional access points, transmission lines, and core network resources are thus beneficial on such occasions.

6.2 Use Cases for 5G Release 16

The following sections present the key use cases of the 5G system, with the emphasis on Release 16.

6.2.1 Network Slicing

5G opens a new era along with the first phase of technical specifications as presented in 3GPP Release 15, whereas Release 16, aka the second phase of 5G, will enhance the system further. 5G represents a major facelift compared to any previous mobile communications generation, and as a result, 5G performance will outperform the capabilities of all the older systems [7].

Some of the key items making this modernization happen include wider and higher Radio Frequency (RF) bands, more advanced Quality of Service (QoS) management, edge computing, Virtualized Network Functions (VNFs), and software-based networking. Remarkably, the philosophy of the core network will change completely as 5G will rely on service-based architecture, which takes advantage of the abstraction of the interfaces.

5G will also optimize the experienced service relying on network slicing. A network slice is a kind of "network within a network." Network slicing is thus a technology that allows an operator to form a number of logical networks on top of a common physical infrastructure. Furthermore, 5G User Equipment (UE) is capable of activating one or multiple network slices at the same time for different parallel services.

The operator can form, manage, and terminate network slices when needed in a highly dynamic fashion, and each slice can be optimized for specific use cases. This is monumental compared to any previous systems in which the mobile network is adjusted in a more static manner and offers certain uniform performance to be shared among users in that area. Instead, 5G network slices within the selected area can be tuned to offer different user experiences.

As an example, a smart city sensor might not require the highest data rates but benefits from the possibility of sharing low bit-rate resources among a huge number of other, simultaneously communicating devices, so that sensors can subscribe to the network slice offering such desired performance. Meanwhile, other users within the very same area, such as CriC applications, may want the highest reliability, so an operator can offer respectively adjusted network slices for them. The benefit of this approach is that differently adjusted network slices optimize the utilization of the network resources, while the user experience is optimal for a variety of users and use cases.

6.2.2 Network Functions Virtualization

Network Functions Virtualization (NFV) refers to a principle of separating Network Functions (NFs) from the hardware they run on by using virtual hardware abstraction (Figure 6.1).

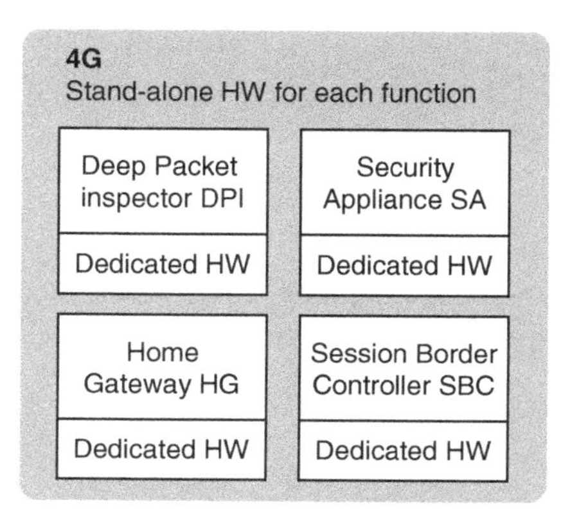

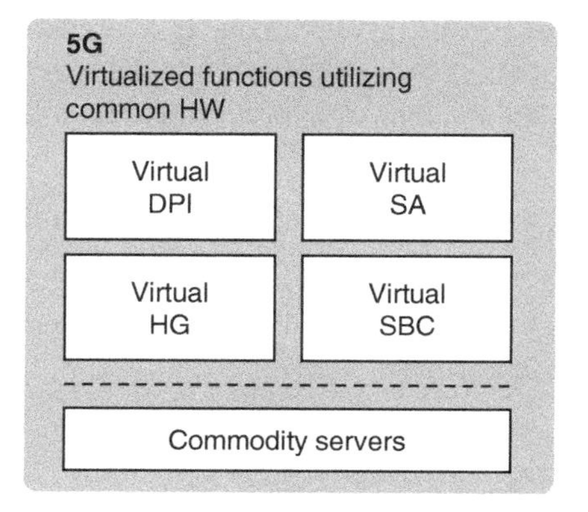

Figure 6.1 The principle of the NFV concept.

Current network technology is based on NFs that are implemented as a combination of vendor-specific software and hardware. These NFs are statically chained or connected to achieve overall functionality. This static combination of NFs is called the *network function forwarding graph* or NF set. In 5G, as defined by the 3GPP, these forwarding graphs will be composed dynamically, which makes it possible to address the current demands.

The NFV concept provides the possibility to decouple software from hardware, which allows a smooth evolution of both software and hardware independently. Furthermore, the flexible and automated NF software instantiation deployment leverages available cloud and network technologies. As an additional benefit of NFV, the dynamic operation and management of NF sets enhances the scaling of the network performance.

The service-based architecture facilitates a highly modular and reusable approach for 5G network deployments. It allows Mobile Network Operators (MNOs) to deploy 5G gradually. The service-based architecture contains the core network functions by interconnecting them with the rest of the system, as described in Chapter 5. NFV also enables distributed cloud for optimizing further the deployment of services in a flexible way. Distributed cloud refers to the possibility of treating a variety of data centers as a single, virtual data center. This, in turn, provides a means for stakeholders to serve a variety of verticals and use cases.

6.2.3 SDN

Software Defined Networking (SDN) is a network architecture model that overcomes hardware limitations.

SDN thus provides MNOs with the possibility of offering 5G services via a centralized control plane in a hardware-agnostic manner. It enhances data flows via lower bandwidth and lower latency, and also network redundancy can be managed more efficiently.

SDN results in more flexible networks. The SDN network architecture supports the 5G ecosystem requirements and can be used to design, build, and manage 5G networks. As the control and user planes are separated, the control plane is directly programmable, while the underlying infrastructure is abstracted for applications and network services to create various network hierarchies.

While control is distributed in the traditional networks, SDN detaches the control plane from the network hardware, enabling packet data flow control through a controller. As a consequence, network control is programmable, facilitating the management of the 5G network as well as modifying and adding services [8].

SDN together with NFV form an enabler for a variety of user types and use cases.

6.2.4 Use Cases for Cloud-Based Functions

The Open Radio Access Network (O-RAN) is described in Chapter 4. The O-RAN Alliance has listed O-RAN use cases and their respective specifications [9]. These use cases include the following:

- Low-cost RAN white-box hardware;
- Traffic steering;
- Quality of Experience (QoE) optimization;

- Massive Multiple In, Multiple Out (MIMO) optimization;
- QoS-based resource optimization;
- RAN sharing;
- RAN slice Service Level Assurance (SLA);
- Context-based dynamic handover management for V2X;
- Flight path-based dynamic Unmanned Aerial Vehicle (UAV) resource allocation;
- Radio resource allocation for UAV applications.

The following sections summarize their principle as interpreted from Ref. [9].

6.2.4.1 Low-Cost RAN White-Box Hardware

This use case is about white-box hardware design, and it has an impact on the open fronthaul interface, O-RAN Distributed Unit (O-DU), and O-RAN Radio Unit (O-RU).

The white-box hardware approach may help reduce the cost of 5G deployment. One of the goals of O-RAN is thus to specify a reference design of a high-performing white-box base station that is capable of increasing spectral and energy efficiency. This aims to benefit the ecosystem in terms of the scale effect of hardware selection, and optimize the R&D challenges and costs. The expected impact is the increased interest of small and medium enterprises to join the ecosystem with increased innovation.

The white-box hardware design focuses on (1) O-DU (including Uplink/Downlink (UL/DL) baseband processing, synchronization, signal processing, Operations Management (OM) function, interface with O-RAN Central Unit (O-CU), and the fronthaul gateway); (2) O-RU (including DL baseband/RF signal conversion, UL RF/baseband signal conversion, and interface with the fronthaul gateway); and (3) fronthaul gateway when needed (including DL broadcasting, UL combining, power supply for O-RU, cascade with other fronthaul gateway, and synchronization).

6.2.4.2 Traffic Steering

This use case is about Artificial Intelligence (AI)-enabled RAN and OpenRAN interfaces (*O1*, *A1*, *E2*), and has an impact on *O1*, *A1*, *E2*, non-Real-Time Radio Access Network Intelligent Controller (non-RT RIC), near-RT RIC, and O-CU.

Network traffic steering is an evolved version of mobile load balancing, and helps operators enhance traffic distribution. Along with increasing network traffic and the respective need for capacity, networks become more complex, and this has set new challenges on traffic steering to maintain or increase network efficiency and to keep enhancing user experiences.

The respective traffic optimization requires typically manual work, and in addition the current Radio Resource Management (RRM) features tend to be radio cell centric. Some of the respective challenges are accompanied by changes in the radio coverage of adjacent cells, signal strength, and interference levels, which require manual intervention to adjust the radio network plan accordingly. Alternatively, the operator may simply assume average, defaulted cell performance values to keep offering users adequate service levels even with changing situations. In other words, the operator may select a strategy based on average cell-centric rather than UE-centric performance assumption.

O-RAN architecture improves the flexibility of the network via RAN automation, and lowers the need for manual intervention for traffic steering, which, in turn, has a positive impact on Operating Expenditure (OPEX), human errors, response, and processing time to fix issues. RAN intelligence is also capable of predicting network conditions and UE performance levels, improving user experience thanks to the inclusion of AI and Machine Learning (ML) capabilities in non-RT RIC and near-RT RIC control traffic steering. ML consumes data originated via the *O1* interface from the O-CUs and O-DUs.

6.2.4.3 QoE Optimization

This use case is about AI-enabled RAN and OpenRAN interfaces (*O1*, *A1*, *E2*), and has an impact on *O1*, *A1*, *E2*, non-RT RIC, near-RT RIC, and O-CU.

Demanding 5G native applications like cloud Virtual Reality (VR) are highly interactive and traffic intensive, and they thus consume much more bandwidth and are more sensitive to latency than other typical applications. Semi-static QoS cannot optimally serve these types of applications because users may generate highly dynamic and fluctuating traffic volumes affecting the performance of the radio interface. Therefore, there is no way to guarantee the service of QoE in traditional network architecture models in a cost-effective way. Instead, network optimization that is based on vertical application-specific QoE prediction, respective estimated priority need assessment, and allocation in a form of a RT network optimization would better suit VR applications.

Software-defined RIC and the open interfaces of O-RAN, combined with AI models, are capable of optimizing the QoE of VR and other highly demanding new vertical services. O-RAN is capable of processing complex data via ML algorithms, while the *O1* interface delivers the needed data for training in the non-RT RIC.

6.2.4.4 Massive MIMO Optimization

This use case is about AI-enabled RAN and OpenRAN interfaces (*O1*, *A1*, *E2*), and has an impact on *O1*, *A1*, *E2*, non-RT RIC, near-RT RIC, and O-CU.

Dynamic beamforming and massive MIMO are some of the key technologies for 5G to increase offered capacity, while reducing interference levels.

Massive MIMO optimization improves cell-centric network QoS. The O-RAN architecture can manage the respective complexity of the configuration parameters and their values accompanied by analytics and ML functions. The non-RT RIC and near-RT RIC are able to monitor traffic, coverage, and interference levels over the areas of interest. *O1*, *A1*, and *E2* will support the exchange of data, policy, and configuration information between the architectural elements, thus facilitating massive MIMO optimization in the O-RAN networks.

6.2.4.5 QoS-Based Resource Optimization

This use case is about AI-enabled RAN and OpenRAN interfaces (*O1*, *A1*, *E2*), and has an impact on *A1*, *O1*, *E2*, non-RT RIC, near-RT RIC, and O-CU.

For the high bandwidth and low latency of 5G services and applications, operators need to carry out in-depth planning and configuration to provide adequate RF coverage and a sufficient level of offered capacity. The challenge arises because the demand for capacity and coverage of the radio network is not static, and the default configuration may not optimally serve users in all situations. In those cases, the non-RT RIC and near-RT RIC can rely on the resource policies based on the QoS to seek the optimal level of radio resource allocation for user segments based on their requirements and priorities. As an example, assessing the overall environment, non-RT RIC analytics has a concrete means to decide the prioritization of users consuming similar types of services in a given time. In this way, the non-RT RIC uses the *A1* QoS policies for reallocating RAN resources to users consuming the same service, while the near-RT RIC receives the respective policies via the *E2* of the O-CU and O-DU.

6.2.4.6 RAN Sharing

This use case is about the virtual RAN network, and has an impact on *O1*, *O2*, *E2*, non-RT RIC, near-RT RIC, O-DU, and O-CU.

RAN sharing is aimed at reducing network deployment costs, while increasing network capacity and coverage. In addition, the open nature of the O-RAN architecture can accelerate the development of RAN sharing solutions. As an example, a home operator can make available its RAN infrastructure and computing resources to host virtual RAN functions (in a form of VNF) of a host operator in such a way that each VNF refers to a logic implementation of O-DU and O-CU functionalities. The O-RAN model provides the host operator with the possibility to monitor, configure, and control VNF resources in this infrastructure that another operator owns via the open interfaces *E2*, *O1*, and *O2*. O-RAN thus enables operators to configure the shared network resources independently from the configuration of other sharing operators.

6.2.4.7 RAN Slice SLA

This use case is about AI-enabled RAN and OpenRAN interfaces (*O1*, *A1*, *E2*), and has an impact on *O1*, *O2*, *A1*, *E2*, non-RT RIC, near-RT RIC, O-DU, and O-CU.

Network slicing is a key feature of 5G networks to comply with service level agreements with different business customers, i.e., verticals in terms of data rates, traffic density, latency, and reliability. The open interfaces of O-RAN enable the respective SLA mechanisms relying on AI and ML of the architecture. As an example, the non-RT RIC and near-RT RIC can adjust RAN behavior to assure dynamic radio network slice SLA because the non-RT RIC is able to monitor long-term trends related to RAN slice performance metrics via *E2* nodes while it feeds the AI and ML models at the near-RT RIC.

6.2.4.8 Context-Based Dynamic Handover Management for V2X

This use case is about AI-enabled RAN and OpenRAN interfaces (*O1*, *A1*, *E2*), and has an impact on *O1*, *A1*, *E2*, non-RT RIC, near-RT RIC, and O-CU.

V2X is a key service that car manufacturers can benefit from while it increases road safety and reduces emissions. V2X delivers RT information on road conditions, traffic status, and drivers' actions that V2X UE and V2X Application Servers (ASs) can communicate with. The flow includes Cooperative Awareness Messages (CAMs) from UE to V2X ASs (detailed in ETSI EN 302 637-2), radio cell ID, connection ID, and radio measurements such as received power levels and quality.

In practical environments, high-speed vehicles may generate rapid back-and-forth handovers between radio cells and other non-optimal behaviors, while ideal V2X communications should be continuous and reliable to work properly. The O-RAN helps mitigate these non-ideal conditions, e.g., by collecting historical traffic and radio-related data, evaluating AI and ML assessment to predict non-ideal handover occurrences, and facilitating RT monitoring of traffic and radio conditions.

6.2.4.9 Flight Path-Based Dynamic UAV Resource Allocation

This use case is about AI-enabled RAN and OpenRAN interfaces (*O1*, *A1*, *E2*), and has an impact on *O1*, *A1*, *E2*, non-RT RIC, near-RT RIC, and O-CU.

The UAV is becoming increasingly popular and many verticals such as agriculture, power line companies, and law enforcement rely on it as part of their activities. The performance of 5G is beneficial for the low-altitude UAV and can replace the single point-to-point communication link between the UAV and the ground control station. One of the challenges is radio network planning of traditional systems, which is based on the optimal coverage goals at the ground level, i.e., in the two-dimensional domain. This is in contradiction with the needs of, e.g., drones, which may not obtain an optimal radio signal in the three-dimensional environment while flying above the base station masts and thus outside of practical, directive antenna patterns. On the other hand, the transmission of drones may cause widespread interference for ground base stations.

O-RAN architecture facilitates multidimensional data acquiring. As an example, a non-RT RIC can retrieve aerial vehicles' measurement metrics from the network as part of the respective UE measurement report, combining information on the flight path, climate and weather forecast, and limited areas via Unmanned Traffic Management (UTM). Based on this multisourced data, the O-RAN can train AI and ML models and facilitate the near-RT RIC to allocate radio resources based on very specific demands for UAV coverage.

6.2.4.10 Radio Resource Allocation for UAV Applications

This use case is about AI-enabled RAN and OpenRAN interfaces (*O1*, *A1*, *E2*), and has an impact on the *O1*, *A1*, *E2*, non-RT RIC, near-RT RIC, and O-CU.

O-RAN architecture can adjust radio resources for supporting UAV control vehicles' communication requirements, e.g., for video and sensors. UAV communications take place on the 5.8 GHz band, and applications include a wide range of solutions such as remote control of the UAV for area inspection, surveillance, mapping, and video broadcast. For illegal UAV clearance in special areas, control stations and anti-UAV weapons may form a joint solution together with a UAV control vehicle based on 5G connectivity. In this scenario, the UAV control vehicle deploys the non-RT RIC, near-RT RIC, O-CU, O-DU, and the ASs serve the actual data such as video content. Furthermore, the near-RT RIC for the O-CU and O-DU provides radio resource management.

6.2.5 Quality of Service

The integrated QoS mechanism of 5G is an enhanced version from previous generations. 5G customers benefit from this evolution as the services can be used in a more flexible and efficient manner. For fluent user experiences, it is essential that customers have seamless Internet Protocol (IP) session continuity when the device is moving within the service area. In the LTE system, this refers to maintaining the IP address of the PDN Gateway (P-GW) and Packet Data Unit (PDU) session while on the move. It should be noted that there are also applications that do not require actual IP session continuity for the smooth experience.

5G QoS is based on further developed QoS flows for Guaranteed Bit Rates (GBRs) and on flows not requiring GBRs. This approach provides QoS differentiation for PDU sessions at the signaling plane level.

6.2.6 Session Continuity

5G provides a set of different types of session continuity modes coping well with UE and service types. 5G QoS is designed to differentiate data services, and to adapt the performance requirements of a variety of applications. This is important because – regardless of the considerably increased capacity of 5G – the radio resources are limited as soon as users demand data from the network simultaneously.

5G QoS supports different requirements of other access technologies, too. Wi-Fi access points are an example of these. 5G can cooperate with them thanks to the integrated, native 5G security functions for such scenarios [10].

Operators can deploy application functions in a flexible manner. 5G deploys QoS via a set of Session and Service Continuity (SSC) modes. The SSC modes provide applications with the possibility of influencing the selection of adequate data service characteristics.

As depicted in Figure 6.2, there are three modes in 5G:

- **SSC 1** is a familiar mode from LTE networks. It ensures that the IP anchor remains stable supporting applications and maintaining the link for UE in location updates.
- **SSC 2** is a new, "break-before-make" mode introduced in Release 15. It means that a network may break connectivity and release the data session and possibly the IP address before creating a new connection.
- **SSC 3** is another Release 15 model, which can be characterized as "make-before-break." It means that the network makes sure that the 5G device does not lose connectivity. In this mode, before breaking the previous connection, the network creates a new connection, thus allowing service continuity. The IP address will change in this mode.

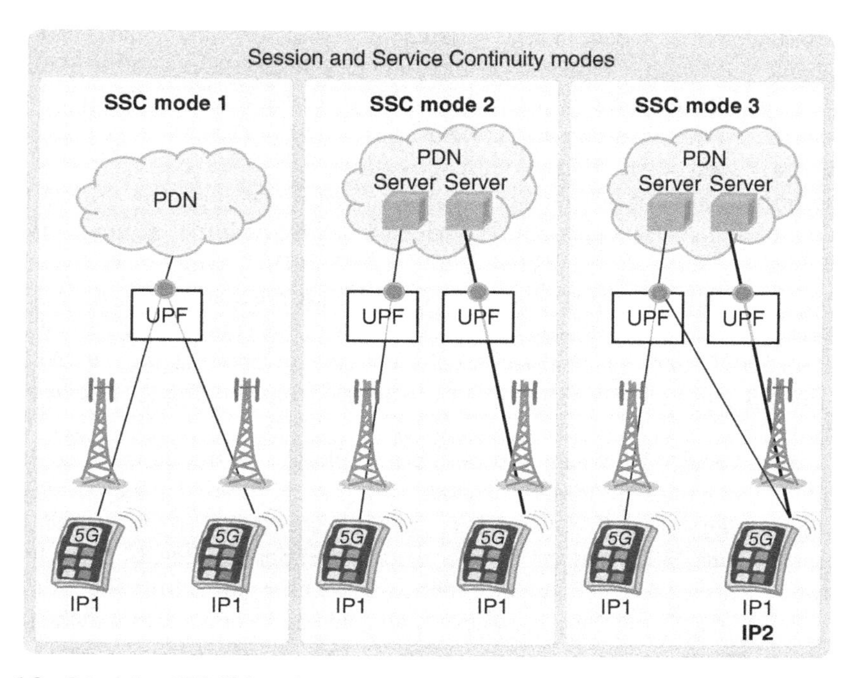

Figure 6.2 Principle of 5G SSC modes.

Both SSC 2 and SSC 3 facilitate the relocation of the IP anchor in the 5G environment.

Once the SSC mode is associated with a PDU session of 5G it will not change during the rest of the lifetime of the PDU session. The 5G architecture allows applications to influence the selection of SSC modes as needed for required data service.

5G also has QoS solutions called Uplink Classifier (UL-CL) and branching point. The 5G User Plane Function (UPF) supports UL-CL and it is designed to divert traffic to local data networks based on traffic matching filters of UE traffic.

The branching point, in turn, refers to UPF's generalized logical data plane function for the UE PDU session. Both UL-CL and branching point methods allow the injection of traffic selectively to and from application functions on the user plane links.

The QoS maps with the user's QoE although their definitions differ. While QoS refers to the technical performance of the system and connectivity, QoE is a more customer-focused measure and provides an understanding of the human interpretation of the quality of the service. It indicates the customer's "happiness" – or annoyance – based on subjective experiences related to a service such as voice call or Internet browsing.

QoE thus indicates the quality level of the whole service experience. The relation of QoS and QoE can be seen from the fact that a poor QoS also lowers the perceived QoE.

More details on the QoS of 5G can be found at 3GPP TS 38.300 and 3GPP TS 23.501, while Section 4.4.3 of *5G Explained* describes the Release 15 aspects of the QoS.

6.2.7 IMS Voice Calls in 5G

The IP Multimedia Subsystem (IMS) and voice service in 5G is summarized in Chapter 5. The native IP voice call service of 5G is referred to as Voice over New Radio (VoNR). Initial 5G deployment is typically based on Option 3. It gives MNOs the possibility of relying on their legacy 4G core networks by gradually attaching NR elements of 5G. The respective 4G platforms and services like IMS and Voice over LTE (VoLTE) remain largely valid for 5G users, either as such or with only minor modifications.

In the initial phase of 5G, IMS is not able to distinguish between the 4G Evolved NodeB (eNB) and 5G Next Generation NodeB (gNB). In the initial deployment, the 5G voice service still relies on the 4G Evolved Packet Core (EPC) and VoLTE to facilitate a seamless voice call experience thanks to the fallback mechanisms. Both 4G and 5G radio, core, and IMS networks can support the voice service and respective capabilities. In the final Standalone (SA) deployment phase, based on Option 2, the native 5G VoNR serves the NR voice call users with call continuity from the VoNR back to previous systems.

VoLTE and VoNR deployment can be done in phases. In the network on 5G Non-Standalone (NSA), the VoLTE service can be used as such. In the first stage of 5G SA, for voice calls through VoLTE, Evolved Packet System Fallback (EPS FB) can be used, whereas on the second stage, 5G SA, as the number of devices supporting the SA mode increases, VoNR can be selected as a default voice service. If the user moves from the 5G to the 4G radio coverage area during the call, a packet-switched handover takes care of the seamless voice service continuity from 5G NR to 4G VoLTE. This procedure is comparable with the Single Radio Voice Call Continuity (SRVCC) of 4G.

According to the GSMA, as soon as the 5G radio and core networks are deployed widely enough facilitating a continuous service, IMS voice and video calls via the 5G infrastructure are superior for capacity as compared to the same services on LTE [11].

3GPP TS 23.501, Section 4.4.3, and TS 23.502, Section 4.13.6, define the 5G IMS call, and TS 23.228, titled IP Multimedia Subsystem, Stage 2 of Release 16, defines the actual IMS support for 5G Core (5GC). For more details, please refer to Chapter 5.

6.2.8 SMS in 5G

3GPP TS 23.501, Chapter 4.4.2, describes the Short Message Service (SMS) over Non-Access Stratum (NAS), and the procedures for SMS over NAS are in 3GPP TS 23.502, Section 4.13.3.

The Mobile Originated (MO) and Mobile Terminated (MT) SMS over NAS can rely on either 3GPP or non-3GPP networks. If the security context is already activated prior to the transfer of the short message, the NAS transport message needs to be ciphered and integrity protected. It is based on the UE and Access and Mobility Management Function (AMF) NAS security context as described in ETSI 133 501, V.15.1.0, Chapter 6.4.

6.2.9 Dual Connectivity

3GPP TS 23.502 details the support of Dual Connectivity (DC) in 5G.

6.2.10 Network Exposure

5G network exposure functionality is defined in 3GPP TS 23.502, Chapter 4.15.

6.2.11 Policy

5G policy scenarios are defined in 3GPP TS 23.502, Chapter 4.16.

6.2.12 Network Function Service Framework

The NF service framework is defined in 3GPP TS 23.501, Chapter 4.17.

6.3 5G Use Cases

6.3.1 Overview

Release 16 key features and work items are summarized in TR 21.916. Release 16 was planned to be frozen in March 2020, and completed by June 2020. This TR is thus one of the useful sources of information for understanding the overall Release 16 features and guidelines for those wanting to seek more details about their main purpose and state, and the feasibility of the technical foundation to tackle various use cases. More detailed descriptions can be found separately via the 3GPP portal in a form of studies, work item descriptions, and feature descriptions, while TR 21.916 presents the initial state of each feature. Interpreting the above-mentioned sources, the following sections summarize key features and probable use cases of 5G as per Release 16.

6.3.2 Use Cases of 3GPP TR 22.891

Figures 6.3 and 6.4 summarize potential use cases as identified from 3GPP TR 22.891.

6.3.3 The 3GPP Use Cases of SMARTER

The following sections collect service requirements from the main categories of SMARTER use cases related to verticals. These serve as examples to distinguish between the verticals and their grand variety of needs.

6.3.3.1 Evolved Mobile Broadband

The main differentiator of the eMBB use case is the highest data speed. For a slowly moving user device, the peak rate can be up to 20 Gb/s in DL and 10 Gb/s in UL. This value is applicable to a non-congested, ideal network. In practice, the data speed may oftentimes be 300 Mb/s in DL and 50–100 Mb/s in UL.

The latency of eMBB is typically very low, while the device's speed is high. An example of the practical environment is low-latency connectivity between aerial objects such as drones and their ground control station.

The traffic density of eMBB may be very high, such as in the range of Tb/s/km^2, while the connection density for the UE could be typically in the range of hundreds to thousands per km^2 of which dozens of devices communicate simultaneously.

The mobility of eMBB can be anything between the extreme values allowed by the 3GPP, i.e., 0–500 km/h.

eMBB does not necessarily require high reliability, communications efficiency, or position accuracy.

For more details on eMBB, please refer to 3GPP TR 22.863.

6.3.3.2 Critical Communications

CriC is characterized mainly by low latency and ultra-high reliability.

The latency of CriC is practically close to RT, down to 1 ms end-to-end value in extreme cases. As an example, a smart grid use case may result in 8 ms or less, while round trip latency is up to 150 ms, low latency (~1 ms). Communications between devices (Digital Down Converter (DDC)) results UE-UE latency of 1–10 ms with 0.5 ms one-way delay, while round trip latency can be less than 150 ms.

CriC is characterized by ultra-high reliability and high availability. The mission critical service type may have a packet loss rate down to $1 \cdot 10^{-4}$ with a delivery time of 8 ms.

The traffic density of CriC is typically the high-density distribution type. An example of such an environment could be represented by 10 000 sensors within 10 km^2.

CriC benefits from precise positioning with 10 cm raster values in densely populated areas, while it is not critical for data rate, communications efficiency, connection density, or mobility.

For more details on CriC, please refer to 3GPP TR 22.862.

Ultra-reliable communications	Virtual presence	IoT Device Initialization	Temporary Service for Users of Other Operators in Emergency Case
Network Slicing	Connectivity for drones	Subscription security credentials update	Improvement of network capabilities for vehicular case
Lifeline communications and natural disaster	Industrial Control	Access from less trusted networks	Connected vehicles
Migration of Services from earlier generations	Tactile Internet	Bio-connectivity	Mobility on demand
Mobile broadband for indoor	Localized real-time control	Wearable Device Communication	Context Awareness to support network elasticity
Mobile broadband for hotspots	Coexistence with legacy systems	Best Connection per Traffic Type	In-network and device caching
On-demand Networking	Extreme real-time communications and the tactile internet	Multi Access network integration	Routing path optimization when server changes
Flexible application traffic routing	Remote Control	Multiple RAT connectivity and RAT selection	ICN Based Content Retrieval
Flexibility and scalability	Light weight device configuration	Higher User Mobility	Wireless Briefcase
Mobile broadband services with seamless wide-area coverage	Wide area sensor monitoring and event driven alarms	Connectivity Everywhere	Devices with variable data

Figure 6.3 Use cases as identified from 3GPP TR 22.891 [1].

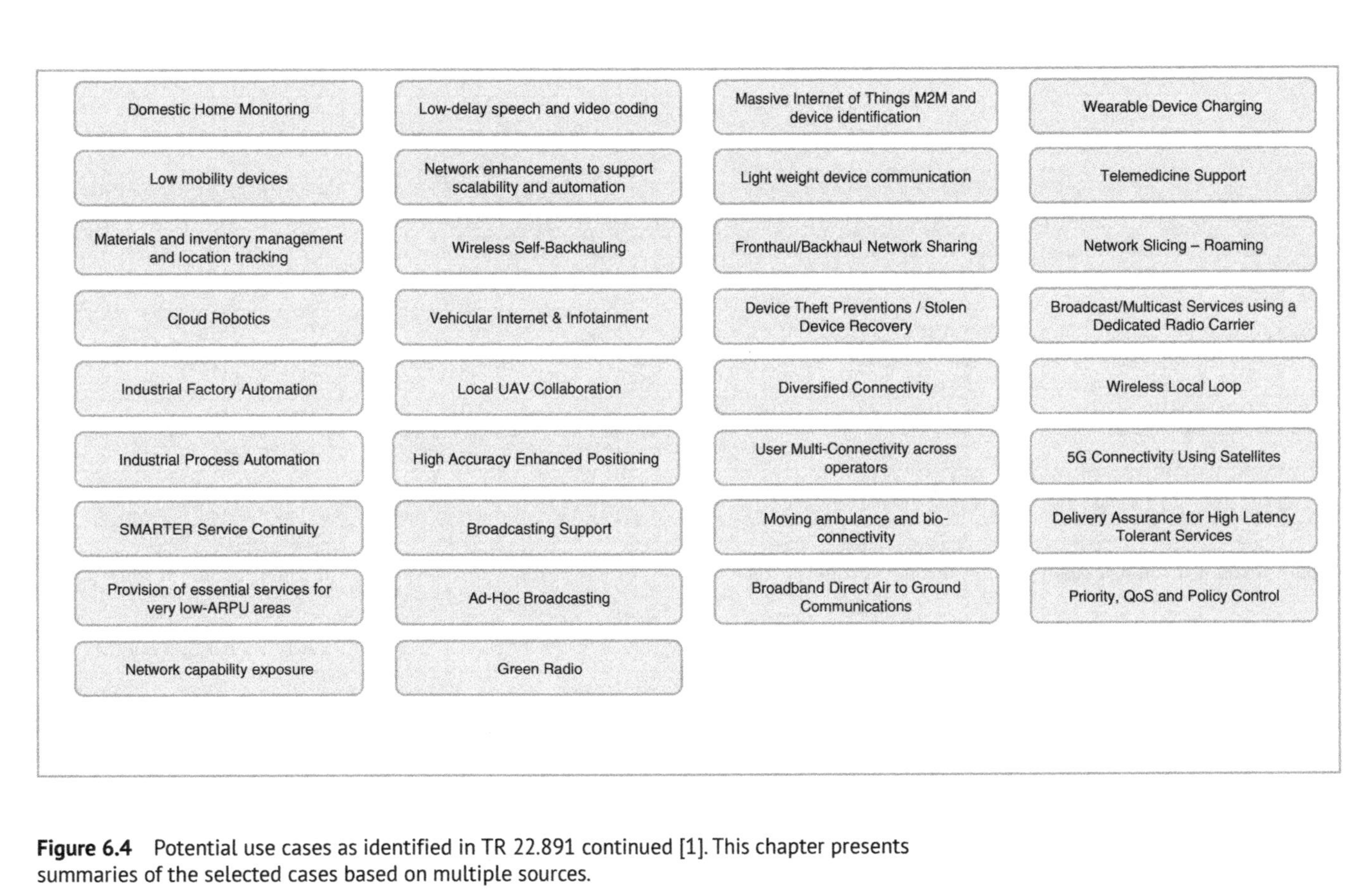

Figure 6.4 Potential use cases as identified in TR 22.891 continued [1]. This chapter presents summaries of the selected cases based on multiple sources.

6.3.3.3 Mobile IoT/Cellular IoT

mIoT, also known as mMTC, refers to the ability of 5G to deliver contents among a vastly increased number of devices communicating simultaneously such as intelligent sensors. MIoT refers to the combination of Cellular IoT (C-IoT) modes such as LTE-M and NB-IoT. It is typically not critical for data rate, latency, reliability, or mobility.

The communications efficiency of MIoT includes aspects of enhanced coverage as well as efficient resource and signaling to support low-power devices such as smart meters. These devices may not have too strict communication requirements and have typically only limited capabilities.

Traffic density of MIoT can be characterized by high-density massive connections. As per ITU IMT-2020 requirements, this type may have up to 1 million simultaneously connected devices within a 1 km^2 area.

Connection density of group MIoT can be characterized as low mobility in typical MTC cases, inventory being one of the exceptions.

MIoT is of high positioning accuracy. This applies to outdoor and indoor scenarios, the value being, e.g., 0.5 m.

5G has thus been designed taking into account the forthcoming volume of mIoT. In fact, the ITU-R IMT-2020 requirement set includes a value for the number of simultaneously communicating mIoT devices, which is 1 million devices per km^2.

The generic term for the IoT communication system is Low Power Wide Area Network (LPWAN). The name indicates that the respective terminals are highly optimized to support long battery lifetime. Oftentimes, the communication of such devices is based on low bit rate and occasional transmissions.

Communications for IoT can be referred to as C-IoT when a mobile communication network is involved. There are also alternative technologies developed such as LoRa, SigFox, and RPMA. The factor that distinguishes C-IoT is the presence of International Mobile Equipment Identity (IMEI), which is not found in alternative devices for their own system.

The evolution of C-IoT, as defined by the 3GPP, has been ongoing since LTE Release 8. There are a few variants in the markets based on LTE as summarized in Table 6.1. There is also a mode defined for GSM, Extended Coverage GSM IoT (EC-GSM-IoT), which is defined in Release 8.

- **LTE-M** (or Cat M1) is an LTE MTC LPWA standard as per 3GPP Release 13. LTE-M device complexity is low, and it supports massive connection density, low device power consumption, low latency, and extended coverage allowing reuse of the LTE installed base.

Table 6.1 C-IoT categories for LTE.

Metrics	**Cat 1** **LTE Rel. 8**	**Cat 0** **LTE Rel. 12**	**Cat M1** **LTE Rel. 13**	**NB-IoT** **LTE Rel. 13**
DL peak data Mb/s	10	1	1	0.2
UL peak data Mb/s	5	1	1	0.2
Complexity	High	Medium	Low/Medium	Very low

- **Narrowband IoT** (NB-IoT) is a 3GPP Release 13 solution that provides improved indoor coverage, support for a massive number of low throughput devices, low delay sensitivity, low cost, and low device power consumption.

As a continuum to the Release 13 definitions, LTE Release 14 developed NB-IoT further via new features like enhanced positioning accuracy and peak data rates. It also defines a low device power class, enhanced non-anchor carrier operation, multicast mode, and radio coverage enhancements [12].

As stated in Ref. [13], NB-IoT and LTE-M support both the LTE and 5G LPWA requirements. The 3GPP has agreed that LPWA use cases will continue to be addressed by evolving NB-IoT and LTE-M as part of the 5G specifications. NB-IoT and LTE-M defined prior to the 5G era are thus considered to be part of the 5G family; in fact, one of the 5G NR deployment scenarios is to place LTE-M or NB-IoT directly into a 5G NR frequency band.

More details on NB-IoT radio transmission and reception prior to integrated 5G modes can be found in 3GPP TR 36.802. *5G Explained*, Section 3.3, details the role of the 3GPP in LPWA and IoT. For more details on mIoT, please refer to 3GPP TR 22.861.

Release 16 brings support for I-IoT. The specifications facilitate various new use cases such as factory automation and transport industry via the latency and reliability enhancements. In addition, Release 16 supports Time-Sensitive Networking (TSN) and the respective requirement for highly precise time synchronization.

6.3.3.4 Evolved V2X

The data speed of eV2X represents medium rate, the practical value being in a range of tens of Mb/s per device. Some key requirements mentioned in SMARTER are:

- The latency of eV2X is low, a typical value being 1 ms for end-to-end cases.
- The reliability of eV2X needs to be the highest possible, such as 99.999%.
- The traffic density of eV2X use cases can be assumed to be typically medium.
- eV2X uses medium connection density. A typical scenario could be 10 000 or more when multiple lanes and multiple levels and types of roads are involved, according to the SMARTER documentation.
- eV2X needs to support high mobility, up to the maximum 5G-supported value of 500 km/h.
- eV2X needs to support high positioning accuracy, a typical value being 0.1 m.
- eV2X is not critical for communications efficiency.

6.3.4 Enhancement of Ultra-Reliable Low Latency Communications

URLLC is the foundation and one of the most important benefits of 5G. 3GPP TR 23.725 V16.2.0 details a study on enhancement of URLLC support in the 5G core network, and 3GPP TR 38.824 V16.0.0 presents a study on physical layer enhancements for the NR ultra-reliable and low latency case. Also, the latest versions of 3GPP TR 23.725 and 3GPP TR 38.824 present URLLC.

URLLC serves many verticals that have not been able to take full advantage of the performance of the previous mobile communication generations. Ref. [14] presents the results from the 3GPP study item titled "Study on physical layer enhancements for NR Ultra Reliable Low Latency Communication (URLLC)." It considers the Release 15 NR URLLC baseline for prioritized URLLC use cases, and presents respective enhancements as one part of the Release 16 URLLC descriptions.

Depending on the sources investigated in Ref. [14], the setup and, considering the above-mentioned baseline, the percentage of UE satisfying the latency and reliability requirements by Release 15 NR can vary for the investigated URLLC use cases, which are electrical power distribution, transport industry, Release 15-enabled use case, and factory automation. These studies have been the basis for further enhancements of URLLC to satisfy the more demanding requirements in Release 16.

6.3.4.1 Electrical Power Distribution

The baseline performance achievable with Release 15 for electrical power distribution considers latency values of 6 ms for differential protection or 3 ms for grid fault of power distribution and outage management, and reliability of 99.999% for differential protection or 99.9999% for power distribution grid fault and outage management.

6.3.4.2 Transport Industry

The baseline performance achievable with Release 15 NR for the transport industry considers the latency value of 3 ms for remote driving and 7 ms for Intelligent Transportation System (ITS), and reliability of 99.999%.

6.3.4.3 Release 15-Enabled Use Case

The Release 15-enabled use case with urban macro considers data packet sizes of 32 bytes and 200 bytes, and indoor hotspot. The baseline performance achievable with Release 15 NR for this use case assumes a latency of 1 ms and reliability of 99.999%.

6.3.4.4 Factory Automation

The baseline performance achievable with Release 15 NR for factory automation considers the latency value of 1 ms for the air interface and reliability of 99.9999%.

6.3.4.5 Enhancements in Release 16

Release 16 URLLC includes enhancements in Physical Downlink Control Channel (PDCCH) reliability and blocking by applying Downlink Control Information (DCI) format compression, increased PDCCH monitoring capability, and PDCCH repetition. Other enhancements include:

- **Uplink Control Information (UCI) enhancements**: Hybrid Automatic Repeat Request (HARQ) feedback with more than one Physical Uplink Control Channel (PUCCH) for HARQ Acknowledgment (ACK) transmission within a slot, and evolved reporting procedure and feedback for HARQ-ACK. Furthermore, enhanced Channel-State Information (CSI) feedback includes Demodulation Reference Signal (DMRS)-based CSI measurement, Aperiodic CSI (A-CSI) on PUCCH, and enhanced CSI reporting mode.
- **Physical/Primary Uplink Shared Channel (PUSCH) enhancements**: (1) mini-slot level repetition; (2) multisegment transmission; (3) one or more actual PUSCH

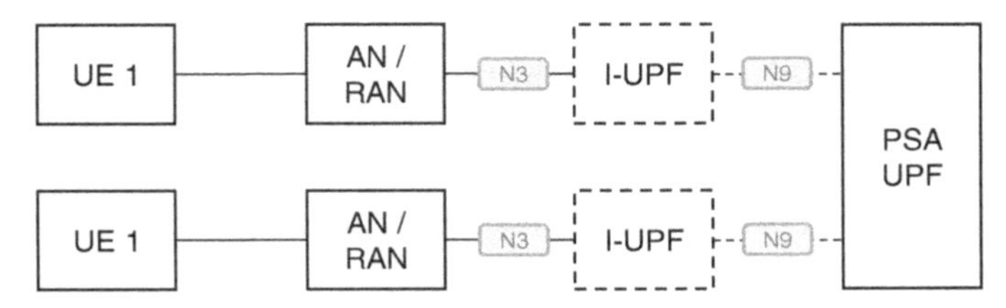

Figure 6.5 Local switch-based user plane architecture in a non-roaming scenario. In this scenario, the PDU Session Anchor (PSA) UPF serves as a local switch.

repetitions in one slot, or two or more actual PUSCH repetitions across the slot boundary in consecutive available slots is supported using one UL grant for dynamic PUSCH, and one configured grant configuration for configured grant PUSCH; (4) one or more actual PUSCH repetitions in one slot, or two or more actual PUSCH repetitions across the slot boundary in consecutive available slots is suppvorted using one UL grant for dynamic PUSCH, and one configured grant configuration for configured grant PUSCH; and (5) one or more PUSCH repetitions in one slot, or two or more PUSCH repetitions across the slot boundary in consecutive available slots, supported using one UL grant for dynamic PUSCH, and one configured grant configuration for configured grant PUSCH.

- **Enhancements to scheduling/HARQ/CSI processing timeline**: (1) enhancements to scheduling/HARQ processing timeline; and (2) out-of-order HARQ-ACK and PUSCH scheduling.

6.3.5 5GS Enhanced Support of Vertical and LAN Services

As stated in 3GPP 23.501, 5G LAN-type services, the basic user plane's architectural 5G models for roaming and non-roaming scenarios also apply to 5G LAN-type services. The additional options are a non-roaming user plane architecture to support 5G LAN-type service using local switch and *N19* tunnel, as depicted in Figures 6.5 and 6.6.

The 3GPP defines the *N19* reference point between two UPFs for direct traffic routing between PDU sessions without the use of the *N6* reference point.

6.3.6 Advanced V2X Support

The importance of V2V technologies will increase rapidly in the near future [15, 16]. The designed architecture and functionality will be enhanced as the system will be

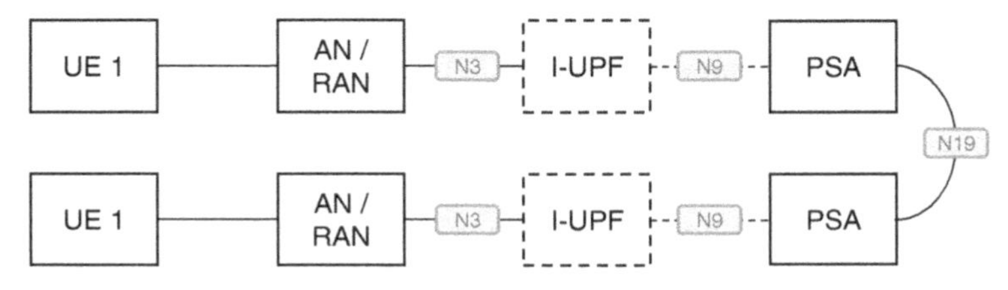

Figure 6.6 *N19*-based user plane architecture in a non-roaming scenario.

implemented in new environments. Likewise, the importance of the respective security increases for communications links internally within the vehicle, between different vehicles, as well as with external entities. The technology is going to be increasingly popular along with the new CriC technologies such as self-driving cars. As the 3GPP has been active in developing the foundation for IoT via cellular networks, 5G could also play a major role in vehicle communications.

3GPP TS 23.287 presents architecture enhancements to the 5G system to facilitate vehicular communications for V2X services. The following reference points detail the *PC5* reference point (NR, LTE), *Uu* reference point (NR, E-UTRA), and interworking between EPS V2X and 5GS V2X as defined in TS 23.287, TS 22.185, and TS 22.186, respectively.

Some of the most relevant other specifications for V2X are:

- TS 23.285: V2X over LTE PC5;
- TS 23.287: NR-based PC5;
- TS 23.501: Latency reduction for V2X (unicast, edge);
- TS 23.501 and TS 23.503: QoS profiles;
- TS 23.501: Standardized slice/service type for V2X services;
- TS 33.536: Security of advanced V2X;
- TS 24.587, TS 24.588: Stage 3 new specifications for V2X;
- TS 23.287: Architecture enhancements for 5G system to support Vehicle-to-Everything services;
- 3GPP TS 22.185: Service requirements for V2X services; Stage 1;
- 3GPP TS 22.186: Enhancement of 3GPP support for V2X scenarios; Stage 1;
- 3GPP TS 23.285: Architecture enhancements for V2X services;
- 3GPP TS 23.501: System architecture for the 5G system; Stage 2;
- 3GPP TS 23.503: Policy and charging control framework for the 5G system; Stage 2;
- 3GPP TS 33.536: Security aspects of 3GPP support for advanced Vehicle-to-Everything (V2X) services;
- 3GPP TS 24.587: Vehicle-to-Everything (V2X) services in 5G System (5GS); Stage 3;
- 3GPP TS 24.588: Vehicle-to-Everything (V2X) services in 5G System (5GS); user equipment (UE) policies; Stage 3;
- 3GPP CP-192078: WID: CT aspects of architecture enhancements for 3GPP support of advanced V2X services.

V2V refers to the direct links between vehicles. Vehicle-to-Infrastructure (V2I) involves other components such as roadside elements. Vehicle-to-Pedestrian (V2P) means communication between vehicles and nearby persons, while V2N refers to Vehicle-to-Network communications. V2X consists of a combination of V2N, V2V, V2I, and V2P. One of the important use cases of 5G is related to V2V, or more generally, Cellular-based Vehicle-to-Everything (C-V2X).

The strict telecommunication requirements needed for automotive environments, including self-driving vehicles, map with the full version of 5G performance and capabilities. In 5G, sidelink is the term used for V2X.

V2X evolves further. It will include enhanced concepts such as Advanced Driver Assistance System (ADAS). ADAS refers to a system that provides a means for vehicles to cooperate, coordinate, and share sensed information. V2X will grow into Connected and

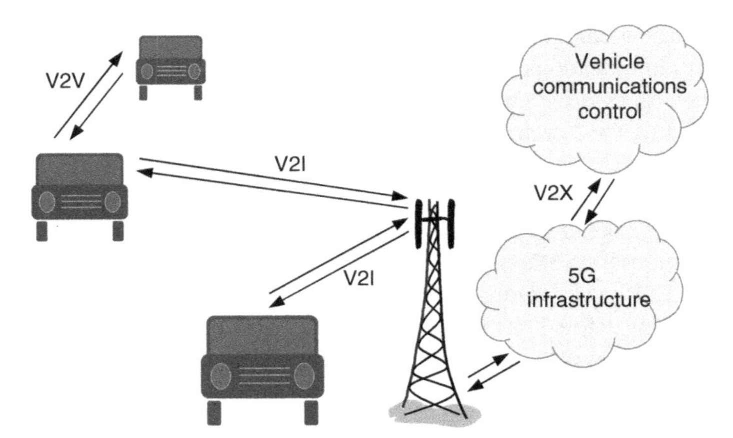

Figure 6.7 5G serves as a platform for vehicle communications.

Automated Driving (CAD). As soon as 5G networks and corresponding reliable, low-latency, and mission critical services are available for V2X applications, ADAS and CAD can be enhanced further.

V2V communication market size was over US$15 billion in 2015 and is likely to witness gains at an estimated annual growth rate of more than 5% up to 2023 [17]. Connected vehicles may generate a significant amount of data, which brings new market potential for the data owner. It can also be assumed that the developing ecosystem provides new business opportunities to involved and third parties.

The 5G system can serve as a service for V2V communications between cars and V2X communications within the ecosystem (Figure 6.7).

Combining the URLLC-based communications of 5G with a vehicle's internal Augmented Reality (AR) functions, the new automotive ecosystem may turn out to be highly interesting, including 5G-based AR-projected windshields, autonomously driving cars, and advanced in-car entertainment. 5G could also offer a platform for further developed concepts such as see-through cars. As an example, the 5G Automotive Association (5GAA) has investigated the benefits of 5G and considers it to serve as a feasible platform for advanced vehicle communications [18].

The NR of 5G can enable V2X communications via the 5G infrastructure, including Multimedia Broadcast Multicast Service (MBMS) support. 5G's V2X is not planned to replace the V2X services offered by LTE, but they have been designed to complement each other by offering interconnectivity [19].

6.3.7 Satellite Access in 5G

As stated in Ref. [20], 5G offers improvements over 4G in terms of performance, and also introduces new enabling technologies and services. One of these is potential satellite components.

3GPP TS 22.261, Release 14, presents a requirement to support satellite access in 5G. Consequently, along with 5G Release 17, the 3GPP is planning to facilitate convergence of

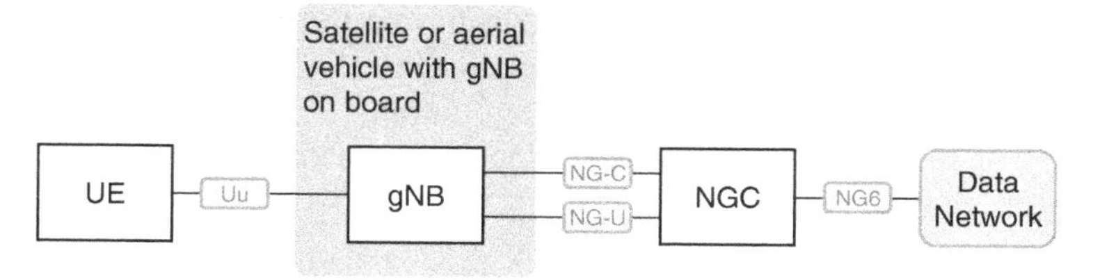

Figure 6.8 Example of the NTN as per 3GPP TR 38.811 [21].

the terrestrial networks and satellite component. The goal of these Non-Terrestrial Networks (NTN) is to help 5G to also serve regions lacking terrestrial infrastructure. Another benefit of the NTN is the assurance of the service continuity for either static or moving Machine-to-Machine (M2M) and IoT devices. It also extends both connectivity for mission critical communications on unpredicted areas and radio coverage for airplanes and trains.

In fact, NTN can be deployed to both satellites and high-altitude platforms such as balloons and Unmanned Aerial Systems (UAS). The NTN brings important additional benefits for industries such as transportation, public safety, medicine, energy, agriculture, finance, and automotive.

NTN use cases of 5G are not able to comply with the strictest latency requirements due to the physical distance between the devices, even in low Earth orbit cases. The 5G NTN is thus most suitable for non-critical applications where latency is concerned. The NTN also refers to other types of non-traditional network access methodologies for airborne or spaceborne vehicle communications. An airborne vehicle is part of High Altitude Platforms (HAP) serving UAS. 3GPP TR38.811 presents a related study on NR to support non-terrestrial networks (Figure 6.8) [21].

Furthermore, 3GPP TR 22.822 details the satellite access of 5G in Release 16 [22]. It defines three categories of use cases for satellite access in 5G: service continuity, service ubiquity, and service scalability. The document also presents practical use cases for satellite access such as roaming between terrestrial and satellite networks, broadcast and multicast with a satellite overlay, IoT with a satellite network, temporary use of a satellite component, optimal routing or steering over a satellite, and indirect connection through a 5G satellite access network. Please refer to TR 8.22 for more use cases and details.

6.3.8 Wireless and Wireline Convergence Enhancement

To continue with the inclusion of WLAN as an additional access into 5GC via the Non-3GPP Interworking Function (N3IWF), 3GPP SA2 has studied Wireless and Wireline Convergence (WWC). Document S2-172164 of the 3GPP SA2 meeting 120 states that the objective of this study is to enhance the common 5G core network defined in Phase 1 to natively support non-3GPP access networks, specifically fixed access networks [23].

3GPP TR 21.916 V0.5.0 (2020-07) presents the 3GPP Release 16 description as per the frozen stage, 31 July [24]. One of the items of the document is coexistence with non-3GPP systems, based on a multitude of proposals and discussion papers. The key papers are "Wireless and Wireline Convergence Enhancement," "Study on the Wireless and Wireline

Convergence for the 5G system architecture," "Study on the security of the Wireless and Wireline Convergence for the 5G system architecture," and "Wireless and Wireline Convergence for the 5G system architecture" by Huawei.

The following sections summarize the key aspects of the impact on overall architecture, radio network, and charging.

6.3.8.1 Support of Wireline Access

As the 3GPP SP-200253 document states, the aim of the 3GPP is to enhance the 5G core network to support connectivity via a residential gateway's wireline access network and via 3GPP RAN access [25]. In addition, the goal is also to support Non-3GPP (N3) access in a trusted network mode.

The respective outcome includes a new access network node W-AGF (Wireline Access Gateway Function). At the same time, the specifications enhance registration and session management, policy, and QoS procedures to cope with the characteristics of wireline access networks. The architecture includes trusted network support via a new trusted gateway network function and new or enhanced procedures related to the selection of the gateway, registration, session setup, and policy management. Thus, there is a new support for a wireline access network as well as trusted non-3GPP access network.

Figure 6.9 depicts the architecture for the support of a wireline access network. As can be seen in the figure, the W-AGF serves the AMF via *N1* (access network functionalities) and

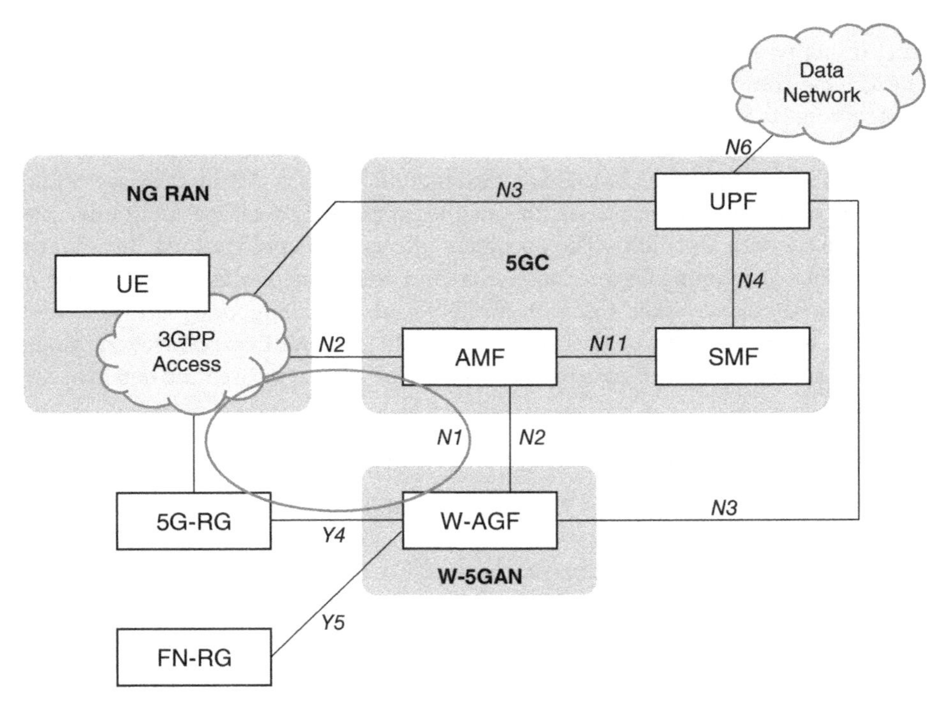

Figure 6.9 Non-roaming architecture of the 5G core network when the Fixed Network Residential Gateway (FN-RG) connects to the Wireline 5G Access Network (W-5GAN) and Next Generation Radio Access Network (NG RAN) of 5G.

N2 (termination) reference points, the UPF (access network) via *N3*, and the 5G Residential Gateway (RG) via *Y4*. In this new architecture, the RG replaces the previously specified UE. The task of the 5G-RG is to support 5G functionalities to connect the new network types to the 5G system. 3GPP TS 23.316 presents the new architecture and the related functions, whereas 3GPP TS 23.501, TS 23.502, and TS 23.503 contain the respective procedures.

The 5G-RG can also connect to the Fixed Wireless Access (FWA) as defined in Release 15. In addition, the 5G-RG is capable of supporting a hybrid mode to connect simultaneously to a 3GPP access and to a wireline access by means of a single access PDU session or Access Traffic Steering, Switch and Splitting (ATSSS) feature. 3GPP specifications TS 23.501, TS 23.502, and TS 23.503 support the ATSSS, whereas 3GPP TS 23.316 supports the interworking with EPC. The W-AGF supports the connectivity of an FN-RG and also the FN-RG that does not have 5G capabilities.

Please note that the Broadband Forum defines the wireline access network, and their respective specifications support the W-AGF functionalities, too.

As a summary, the support of wireline access has the following impacts:

- Global Cable Identifier (GCI) as per the CableLabs definitions to identify the wireline access line that the 5G-RG is using.
- Global Line Identifier (GLI) as per the Broadband Forum definitions that uniquely identifies the line of the 5G-RG within the operator infrastructure.
- Mobility restrictions based on GLI and GCI.
- Modifications to 3GPP TS 23.502 procedures to accommodate the new elements.
- New 5G-RG device that replaces the 5G UE; it supports the needed capabilities of both the 5G system and wireline access as per the Broadband Forum and CableLabs.
- New FN-RG device that acts as UE without the support of 5G capabilities.
- New parameter describing the RG Level Wireline Access Characteristics (RG-LWAC) to cope with QoS control on the wireline access.
- New parameter describing the RG Total Maximum Bit Rate (RG-TMBR) to cope with the QoS model of the wireline networks that uses subscription's maximum aggregate bit rate, including GBR and non-GBR traffic.
- New W-AGF NF introduced for the interworking between the 5GC and legacy wireline access networks.
- SUPI for the FN-RG based on GCI and GLI.
- Support of the Broadband Forum interaction with the Access Configuration System (ACS) related to the provisioning of configuration and remote management of the 5G-RG.
- Support of IPTV multicast over unicast as per 3GPP TS 23.316, Sections 4.9.1 and 7.7.1.

6.3.8.2 Support of Trusted Access Network

In addition to the wireless access scenarios presented in the previous section, support of trusted access network considers the case where the relation between the non-3GPP access network and the 5GC is deeper than in the pure untrusted scenario. Nevertheless, the 3GPP specifications do not define the level of the trust in this respect.

Figure 6.10 depicts the non-roaming architecture for considering the Trusted Non-3GPP Access Network (TNAN) as the access node, i.e., executing the termination of the *N2* and *N3* reference points as well as the relay of the *N1* between the UE.

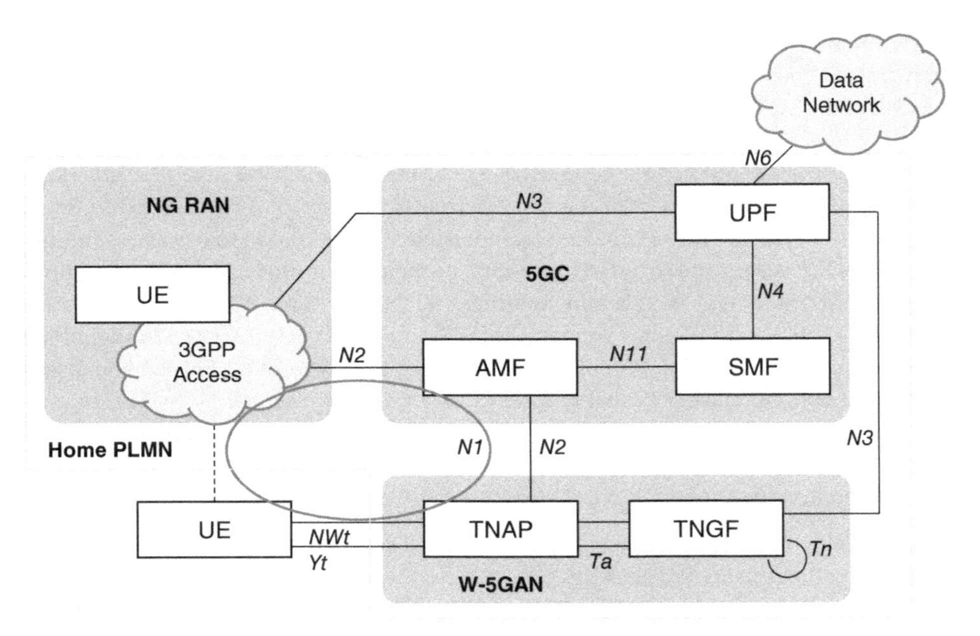

Figure 6.10 Non-roaming architecture for a 5G core network with trusted non-3GPP access.

The TNAN houses the Trusted Network Gateway Function (TNGF) and the Trusted Non-3GPP Access Point (TNAP), which the *Ta* reference point interconnects. The UE connects to the TNG via the *NWt*, which is based on the Internet Key Exchange (IKEv2) in untrusted networks, the difference in this case being the registration procedure. In this model, the Extensible Authentication Protocol 5G (EAP-5G) can be executed between the UE and TNAP on access layers, as well as between the TNAP and TNGF on *Ta*. As for session management and other procedures, the already defined methods of Release 15 for the untrusted non-3GPP access networks are valid in such a way that the TNGF now replaces the N3IWF along with minor procedural modifications.

3GPP Release 16 also specifies a scenario for the trusted non-3GP networks and devices that do not support NAS and that connect to 5G via WLAN. In this case, the Trusted WLAN Interworking Function (TWIF) replaces the TNGF so that the TWIF terminates the *N1* NAS interface, assuming the functions of the UE as per the 5GC.

For more information on this scenario, please refer to 3GPP TS 23.501, TS 23.502, and TS 23.503. The following list of specifications will also help with more information on the generic aspects:

- 3GPP TS 23.203: "Policies and Charging control architecture; Stage 2";
- 3GPP TS 23.316: "Wireless and wireline convergence access support for the 5G System (5GS)";
- 3GPP TS 23.501: "System Architecture for the 5G System; Stage 2";
- 3GPP TS 23.502: "Procedures for the 5G System; Stage 2";
- BBF TR-069: "CPE WAN Management Protocol";

- BBF TR-124 issue 5: "Functional Requirements for Broadband Residential Gateway Devices";
- BBF TR-369: "User Services Platform (USP)";
- BBF WT-456: "AGF Functional Requirements";
- BBF WT-457: "FMIF Functional Requirements";
- BBF WT-470: "5G FMC Architecture";
- CableLabs WR-TR-5WWC-ARCH: "5G Wireless Wireline Converged Core Architecture";
- CableLabs WR-TR-5WWC-ARCH: "5G Wireless Wireline Converged Core Architecture";
- CableLabs WR-TR-5WWC-ARCH: "5G Wireless Wireline Converged Core Architecture."

6.3.8.3 Radio Aspects

As summarized in 3GPP RP-200678 and RP-190999, Release 16 introduces the "NG interface usage for WWC (Wireless Wireline Convergence)." It enhances the NG interface protocols of 3GPP TS 29.413 and 38.413 allowing trusted non-3GPP access and wireline access connectivity with the 5GC as described in Release 16 of TS 23.316, TS 23.501, TS 23.502, and TS 23.503.

This item adds NG protocol support for communications between the trusted non-3GPP access network and the 5GC, as well as between the wireline 5G access network and the 5GC.

6.3.8.4 Charging Aspects

As summarized in 3GPP SP-200525, the respective work item specifies the charging aspect on Wireless and Wireline Convergence for 5G system architecture (5WWC). 5WWC is specified in 3GPP TS 23.501, TS 23.502, TS 23.503, and TS 23.316. Enhancement to the charging aspect for 5WWC is considered part of this series' specifications for this 5WWC.

The following charging scenarios are included in the charging aspect of 5WWC:

- UE connects to 5GC via trusted non-3GPP access;
- 5G-RG connects to 5GC via NR-RAN and W-5GAN;
- FN-RG connects via W-5GAN.

This work item specifies charging requirements, procedures related to charging, charging information, and related triggers for chargeable events for 5WWC scenarios.

The specifications related to 5WWC charging include TS 32.255, TS 32.291, and TS 32.298. The subscriber's identifiers and Permanent Equipment Identifier (PEI) in 5G-RG and FN-RG scenarios specified in TS 23.501 and TS 23.361 are used in charging information. The procedures and related triggers in 5WWC charging scenarios are also specified in the charging aspect for 5WWC. The related changes to OpenAPI are specified in 3GPP TS 32.291.

6.3.8.5 Time-Sensitive Networks

There are many industry areas benefiting from 5G in their daily tasks such as remote operation and control. 5G also tackles some very special aspects such as direct communication between devices, which is beneficial in, say, the mining industry, which needs underground communications. Relying on a local 5G network can reduce cabling costs in industrial sites wanting remote connectivity. 5G can also support autonomously moving units such as factory carts in a logistics center.

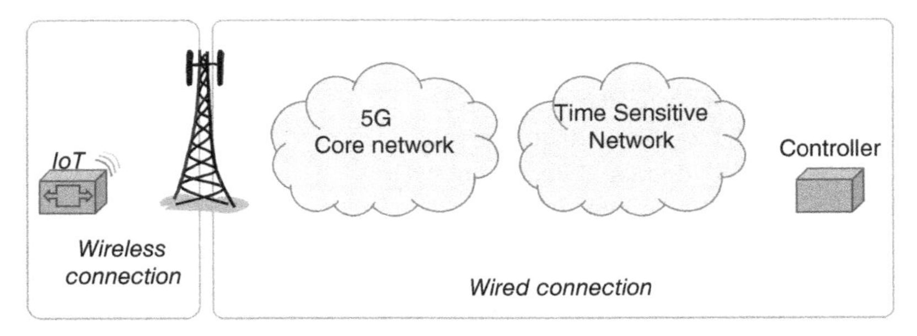

Figure 6.11 The merging of 5G and TSNs can benefit industrial verticals.

There is an industry trend to bring TSN and the wireless world closer together. TSN refers to the evolution of wired Ethernet (based on IEEE 802.3 and IEC/IEEE 60802 standards) making it more reliable and faster while reducing latency. TSN can be applied to industrial automation, vehicle networks, and many other environments requiring enhanced performance.

Thanks to its low latency and ultra-reliable characteristics, 5G is sufficiently capable of being integrated as part of this concept to connect wireless devices such as intelligent industrial sensors, thus reducing the need for cable installations (Figure 6.11). The specifications will further enhance 5G performance along with 3GPP Release 16 [26].

The IEEE TSN specifications thus enable low-latency communication in the factories of the future. Respectively, 5G time-sensitive communication supports deterministic or isochronous communication combined with high reliability and availability and low latency, where end systems and intermediator nodes are synchronized.

The 3GPP supports TSN time synchronization related to the whole end-to-end communications chain, complying with the IEEE 802.1AS Time-Aware System. The 5G point of contact with the TNS system includes a TSN Translator (TT) that supports the IEEE 802.1AS requirements, while 5G components such as the UE, gNB, and UPF, as well as the Network-Side TSN Translator (NW-TT) and Device-Side TSN Translator (DS-TT) components, are synchronized with the 5G GM (the 5G internal system clock providing the reference time; Figure 6.12).

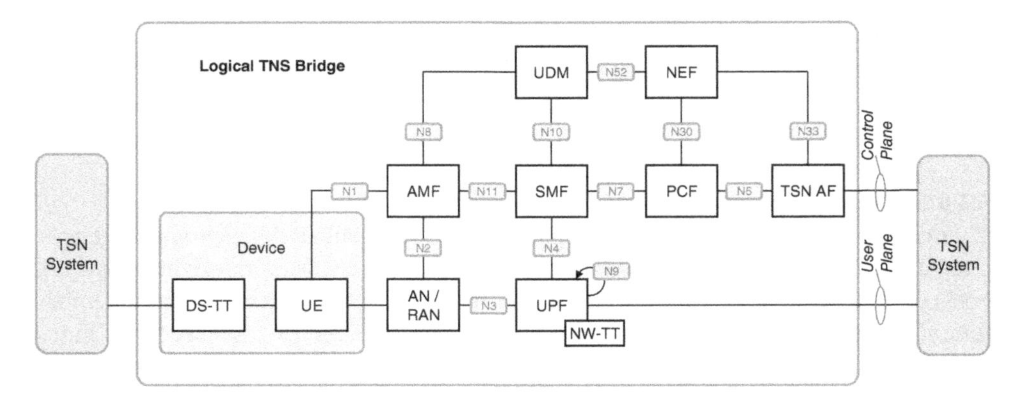

Figure 6.12 The logical TSN bridge as interpreted from 3GPP TS 23.501.

For more information on the TSN, please refer to 3GPP TR 23.725 V16.2.0 and TS 23.501, Release 16.

6.3.9 Location-Based Services

As 23.501, Section 4.4.4, states, the 5G architecture to support the Location Services feature is optional and applicable to both regulatory services and commercial services in Release 16. If the operator decides to support the functionality, the architecture for non-roaming and roaming scenarios is defined in Clause 4.2 of TS 23.273, whereas Clause 4.4 of TS 23.273 defines the respective reference points, and the service-based interfaces to support Location Services are defined in Clause 4.5 of TS 23.273.

6.3.9.1 Positioning Techniques in 5G

The Location-Based Service (LBS) feature is detailed in 3GPP TS 23.502, Chapter 4.13.5, while 3GPP TS 38.305 details the Stage 2 functional specifications of UE positioning in NG-RAN as of Release 15. This specification allows the NG-RAN to utilize one or more positioning methods at a time to locate the UE. Positioning is based on signal measurements and a position estimate. There is also optional functionality to estimate velocity of the UE.

The UE, serving ng-eNB or gNB, performs the measurements, which may include E-UTRAN LTE and other general radio navigation signals, e.g., via the Global Navigation Satellite System (GNSS). Estimation of the position may be performed by the UE or by the Location Management Function (LMF). Positioning may rely on UE-based, UE-assisted, and LMF-based or NG-RAN node-assisted methods.

The 5G system's NG-RAN includes the following standard positioning techniques for estimation of the UE's location:

- Network-assisted GNSS;
- Observed Time Difference of Arrival (OTDOA);
- Enhanced Cell ID (E CID);
- Barometric pressure sensor positioning;
- WLAN positioning;
- Bluetooth positioning;
- Terrestrial Beacon System (TBS);
- Hybrid positioning combining some of the above-mentioned methods; and
- Standalone mode, which refers to an autonomous functionality of the UE without assistance from the network, based on one or more of the above-mentioned methods.

5G supports an integrated LBS, which provides tools for many new use cases [27]. Figure 6.13 depicts the 5G positioning architecture for the NG-RAN and E-UTRAN as described in 3GPP TS 38.305. The main element for the 5G LBS is the LMF. The RAN may include a set of evolved 4G ng-eNBs, 5G gNBs, or both.

TP is a Transmission Point, which refers to a set of geographically co-located transmit antennas for a single cell, part of a single cell, or a single PRS-only (Positioning Reference Signal for radio network) TP. The TPs can be formed by antennas of ng-eNB or gNB, remote radio heads, remote antenna of a base station, or antenna of a PRS-only TP. In practice, each transmission point may correspond to a cell.

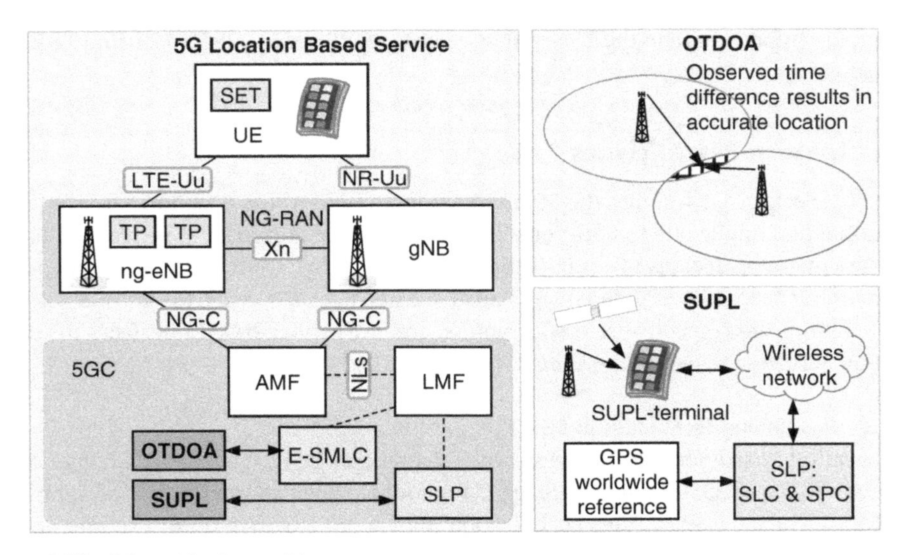

Figure 6.13 5G positioning architecture.

Either the Gateway Mobile Location Center (GMLC) or the target UE itself may trigger a location-based request. The network routes the request to the AMF (detailed in Chapter 5).

Also, the AMF is able to trigger the LBS request for target UE in special situations such as when a user dials an IMS emergency call.

The AMF sends a location service request, accompanied by any optional assistance data, to the LMF, which processes the data. The LMF sends the resolved UE's location back to the AMF.

As depicted in Figure 6.13 and summarized in Table 6.2, the LMF may signal with an Evolved Serving Mobile Location Centre (E-SMLC), SUPL Location Platform (SLP), or with both.

For more information on the LBS, please refer to TS 38.305 (UE positioning) and TS 29.516 (E-SMLC).

Table 6.2 The comparison of E-SMLC and SLP.

E-SMLC	SLP
The E-SMLC provides the LMF with access to OTDOA information. The OTDOA reasons the user's location based on signaling from a set of base stations. The more signals from different base stations the UE receives, the more accurate the positioning.	The SLP is based on SUPL, which provides the user's positioning over the data link. The SLP contains an SUPL Location Center (SLC) and SUPL Positioning Center (SPC), which calculate the UE's position from satellites and their positioning database, Wi-Fi, and/or cellular networks.

6.3.10 Mission Critical Services

As stated in 3GPP 23.501 and TS 22.280, a Mission Critical Service (MCX Service) is a communication service reflecting enabling capabilities provided to end-users from mission critical organizations and mission critical applications for other businesses and organizations (e.g., utilities, railways). In practice, an MCX Service is Mission Critical Push-to-Talk (MCPTT) based on 3GPP TS 23.379, Mission Critical Video (MCVideo) as per 3GPP TS 23.281, or Mission Critical Data (MCData) as defined in 3GPP TS 23.282.

Furthermore, as specified in TS 22.261, "MCX Users require 5GS functionality that allows for real-time, dynamic, secure and limited interaction with the QoS and policy framework for modification of the QoS and policy framework by authorized users."

MCX Services are valid in home networks as well as in roaming scenarios. Also, priority access is valid as per 3GPP TS 22.261.

Some users of MCX Services may be entities related to railway and maritime communications.

6.3.11 Public Warning System

3GPP TS 23.041 defines the Public Warning System (PWS) architecture for 5GS, and is summarized in 3GPP TS 23.501, Chapter 4.4.1, and TS 23.502, Chapter 4.13.4.

6.3.12 Streaming and TV

Streaming is defined in 3GPP TS 26.114. As stated in Ref. [28], the Coverage and Handoff Enhancements for Multimedia (CHEM) feature is designed to help users consume more efficiently the resources for the media. This feature assists networks in delaying handovers and reduces the occurrence of their procedures for the Multimedia Telephony Service for IMS (MTSI). This function is capable of providing additional information to the eNB and gNB indicating the acceptable, increased packet loss values for media configurations.

In practice, there is an extended value set for handover thresholds located at the eNB and gNB that facilitates the delayed staying of the MTSI terminal within a desired sector, cell, access technology, and domain (packet switched or circuit switched) regardless of the increased traffic channel's packet losses. This makes it possible to balance between the experienced QoS and the increased packet losses due to coverage and handovers.

The CHEM feature is based on functionality and optional elements as defined in 3GPP TS 26.114, and related information can also be found in 3GPP TR 26.959 (study on enhanced VoLTE performance) and 3GPP TS 26.114 (multimedia telephony's media handling and interaction).

6.3.13 Cloud-Based Functions and EDGE

The 5G architecture is largely based on data center infrastructure, formed by centralized and regional clouds. The related EDGE clouds are essential for supporting the lowest latency applications.

Edge computing is not only applicable for the latency of the communications link, but also provides the possibility to offload processing capacity from the terminals to the edge hardware. One example benefiting from edge and low-latency concepts and offload processing is VR/AR. To work fluently, they require very low latency for communications and data processing. Typically, such applications rely on powerful and expensive hardware located at the site. As 5G is based on a novelty architecture model of NFV virtualization, and can take advantage of edge computing for offloading data processing and ensuring fast delivery, the same infrastructure can also be applied to external services needing highly demanding performance such as VR/AR applications.

By applying AR/VR data rendering to handle optimal sharing of data processing in the edge cloud instead of the user device, it is possible to reduce the handset processing requirements – which, in turn, can provide the possibility to offer advanced services with lower-cost multifunctional devices. In fact, this opens new cooperative business models for MNOs designing advanced and new added value services, which were not feasible within the older mobile communications infrastructure.

6.3.14 Virtual Reality, Augmented Reality, and Extended Reality

The enhanced performance of 5G will benefit many new applications that were impossible to use with the legacy mobile communication systems. Some of the new applications benefiting from 5G are related to AR and VR. The increasing number of modern services and applications benefits from enhanced network performance of 5G to keep providing users with the most fluent experiences. Only time will show how fast 2G, 3G, and 4G will lose consumers' interest when the more capable and spectrally efficient 5G takes over. One of the indications of the forthcoming popularity of 5G can be seen from the recent GSMA study, which forecasts that 5G is to account for 15% of global industry by 2025.

As an example, the GSMA is investigating and promoting 5G cloud technology for VR/AR applications via the recently established Cloud XR (Extended Reality) Forum [29]. The aim of the forum is to accelerate the delivery and adoption of 5G cloud-based AR and VR technologies.

6.3.15 SON

Another technology familiar from previous systems is the Self-Optimizing Network (SON) concept. It is a constantly evolving enabler, which will also be useful in 5G. One of the related scientific fields is ML, which could assist in the optimal adjustment and fault recovery of 5G networks. The SON is defined by the 3GPP in TS 38.300 and refers to self-configuration and self-optimization. This work item was finalized in June 2018 to be included in 5G as of Release 15. The specification includes definitions for UE support for SON, self-configuration by dynamic configuration of the *NG-C* interface, dynamic configuration of the *Xn* interface, and Automatic Neighbor Cell Relation (ANR) function. The latter is important as it is designed to ease the task of the operator from manually managing Neighbor Cell Relations (NCR). There will thus be definitions for intra-system–intra-NR and intra-system–intra-E-UTRA ANR function, and intra-system–inter-RAT and inter-system ANR function, among various other functions the SON can provide.

3GPP TS 38.300 identifies the following items for the SON:

- Support for mobility load balancing to distribute load evenly among cells and among areas of cells, or to transfer part of the traffic from congested cells or from congested areas of cells, or to offload users from one cell, cell area, carrier, or RAT to achieve network energy saving.
- Support for mobility robustness optimization. For analysis of connection failures, the UE makes the Radio Link Failure (RLF) Report available to the network. The UE stores the latest RLF Report, including both LTE and NR-RLF report until the RLF Report is fetched by the network or for 48 hours after the connection failure is detected.
- Support for Random Access Channel (RACH) optimization. RACH optimization is supported by UE reported information made available at the NG-RAN node and by Physical Random Access Channel (PRACH) parameters exchange between NG-RAN nodes.
- UE History Information from the UE. The source NG-RAN node collects and stores the UE History Information for as long as the UE stays in one of its cells. The UE may report the UE History Information when connecting to a cell of the NG-RAN node.

Please refer to 3GPP TS 38.300, Section 15.5, for more details on the SON in 5G Release 16.

6.3.16 Support for Energy Saving

As stated in 3GPP TS 38.300, the aim of the energy saving function is to reduce operational expenses through energy savings. The function, in a deployment where capacity boosters can be distinguished from cells providing basic coverage, optimizes energy consumption enabling the possibility for an E-UTRA or NR cell providing additional capacity via single or dual connectivity to be switched off when its capacity is no longer needed and to be reactivated on a need basis.

6.3.17 Enablers for Network Automation Architecture for 5G

3GPP TR 23.791 V16.2.0 (2019-06), study of enablers for network automation for 5G, presents use cases related to Release 16 network automation.

6.3.18 5G Voice

5G supports native voice calls via VoNR. While the 5G system comprises 5G gNB components connected to 4G core, i.e., when the NSA architectural model in used, VoLTE can be applied.

For a more detailed description of the 5G voice service, please refer to the respective section in Chapter 5.

6.3.19 Sidelink

3GPP TS 36.300 specifies V2X sidelink communication detailing the respective NG-RAN support via the *PC5* interface as depicted in Figure 6.14.

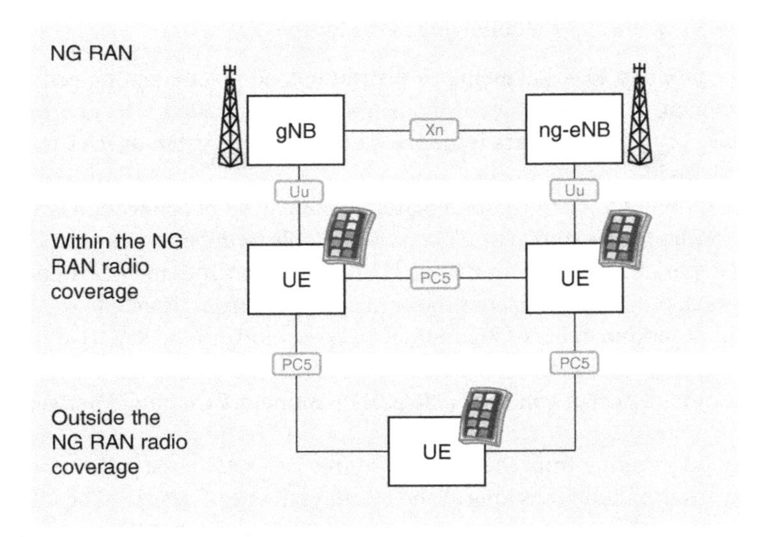

Figure 6.14 NG-RAN architecture supporting the *PC5* interface as interpreted from 3GPP TS 38.300 [30].

Both NR sidelink communication and V2X sidelink communication support V2X services via the *PC5* interface. Please note that NR sidelink communication may also be used to support services other than V2X services such as unicast transmission, groupcast transmission, and broadcast transmission.

6.3.20 Verticals Support

3GPP TS 38.300, Section 16, presents 3GPP-specified service aspects for verticals support. The following summarizes the key statements of this specification:

- **Emergency Services**: Emergency Services facilitate NG-RAN support for Emergency Services directly or via fallback to E-UTRA. System information broadcasting indicates the emergency services support (3GPP TS 38.331).
- **IMS**: The IMS voice service via the NG-RAN includes IMS's ability to support QoS flows with the 5QoS Identifier (5QI) for voice and IMS signaling (3GPP TS 23.501), or through 4G EPC system's fallback, and UE's capability to support IMS voice over PS as defined in 3GPP TS 24.501.
- **Network slicing**: Network slicing scenarios contain (1) NG-RAN for 5G NR that interconnects with 5GC; and (2) 4G E-UTRA connected to 5GC. A network slice consists of RAN and core network segments, and different PDU sessions take care of the traffic for different slices. NSSAI (Network Slice Selection Assistance Information (NSSAI) includes one or a list of Single NSSAI (S-NSSAI) components and the S-NSSAI identifies the network slices as defined in 3GPP TS 23.501.
- **PNI-NPN**: Public Network Integrated Non-Public Network is a network deployed for non-public use. It relies on NFs provided by a PLMN. In PNI-NPN, Closed Access Group

(CAG) identifies a group of subscribers that can access one or more CAG cells. The CAG identifier is part of System Information Block #1 (SIB1) broadcasting.

- **PWS**: PWS includes an NR component that interconnects with 5GC. The NR schedules, pages, informs UE about, and broadcasts warning messages within the defined area. PWS use cases include the Earthquake and Tsunami Warning System (ETWS) and Commercial Mobile Alert System (CMAS). The ETWS use case complies with the regulatory requirements for warning notifications on earthquakes and tsunamis as defined in 3GPP TS 22.168. The ETWS can be of primary notification for short information, and secondary notification for detailed information. The CMAS use case is capable of delivering multiple, concurrent warning notifications as defined in 3GPP TS 22.268.
- **Sidelink**: 5G V2X sidelink communication relies on the NG-RAN architecture and the *PC5* interface as specified in 3GPP TS 36.300. Sidelink transmission and reception can take place when the sidelink-capable UE is within or outside NG-RAN coverage, regardless of the UE Radio Resource Control (RRC) state.
- **SNPN**: Standalone NPN is a network deployed for non-public use. It does not rely on network functions of a PLMN. The PLMN ID and Network Identifier (NID) broadcast on SIB1 present the SNPN. When a UE supporting SNPN operates in the SNPN access mode, it only selects and registers with SNPNs; otherwise it selects PLMNs.
- **TSC**: Support for Time-Sensitive Communications, as defined in TS 23.501 [3], is a communication service for deterministic and/or isochronous communication that has high reliability and availability. I-IoT and Cyber-Physical Control (CPC) applications belong to TSN as described in 3GPP TS 22.104.
- **URLLC**: Vertical support includes URLLC service aspects.

6.3.21 Non-public Networks

A 5G NPN, also known as a private network, is capable of providing 5G network services to isolated organizations. The 5G NPN can reside in the physical premises of the organization, e.g., within a factory or campus area.

The benefits of NPN deployment include personalized QoS, potentially enhanced network performance thanks to the load that is easier to predict, dedicated security, and protection from the rest of the users and infrastructure via physical isolation. The NPN may also ease maintenance and operation.

As stated in Ref. [31], an NPN enables deployment of the 5G system for private use. An NPN may be deployed as an (1) SNPN that an NPN operator manages and that does not rely on NFs provided by a PLMN; or (2) as a PNI-NPN that is a non-public network deployed with the support of a PLMN. Identification of the NPN can take place in the following way:

- **SNPN**: Combination of PLMN ID and NID. An SNPN-enabled UE uses SUPI and credentials for each subscribed SNPN.
- **PNI-NPN**: The network has PLMN ID, while CAG ID identifies the CAG cells.

These two options result in practical scenarios of isolated deployment of SNPN and NPN in conjunction with public networks. As presented by 5G Alliance for Connected Industries and Automation (5GACIA) [32], the latter breaks down into three scenarios especially in the I-IoT environment: (1) shared RAN; (2) shared RAN and control plane; and (3) NPN hosted by the public network:

- **SNPN**: NPN is an isolated and independent network separated from the public network with all NFs residing inside the organization's premises. Communication between the NPN and the public network takes place via a firewall.
- **Shared RAN**: The NPN and the public network share part of the RAN. Communication of the NPN stays within the premises. 3GPP TS 23.251 defines the RAN sharing options.
- **Shared RAN and control plane**: The NPN and the public network share the RAN for the defined premises, while the public network executes the network control tasks and the NPN traffic remains within the premises. Network slicing is one way to deploy this scenario. Network slicing provides a means to set up logically independent networks within a shared physical infrastructure. The network slice identifiers serve to separate the public and private networks. In addition to network slicing, the Access Point Name (APN) solution of the 3GPP works as an alternative. It denotes the target network for the routed data flow by means of differentiation between traffic shares.

For more information on NPN, please refer to the latest versions of 3GPP TS 23.251 (network sharing; architecture and functional description), TS 22.104 (service requirements for cyber-physical control applications in vertical domains, stage 1), and TS 23.501 (system architecture for the 5G system, stage 2).

6.4 Release 17 and Beyond

There are already active discussions on the use cases and verticals that 5G would be capable of handling beyond the capabilities of Release 16. This section presents some of the most concrete ones.

6.4.1 Drones (Unmanned Aerial System)

3GPP TS 22.125 V17.1.0 (2019-12) describes UAS support in the first phase of Release 17. Furthermore, 3GPP TR 23.754 describes a study on supporting UAS connectivity, identification, and tracking as a Release 17 item.

As stated in 3GPP TS 22.125, there is strong interest in using cellular connectivity to support UAS. Some of the benefits of cellular systems include ubiquitous coverage, high reliability and QoS, robust security, and seamless mobility. These very aspects are also important for UAS Command and Control (C2) functions. As part of the development, there are studies on safety and performance so that standard development organizations can define private and civil UAS ecosystems, which could coexist with commercial air traffic, public and private infrastructure, and the general population.

The 3GPP-based ecosystem serves to provide both control and user plane communication services for UAS. Some UAS-related use cases include data services for C2, telematics, UAS-generated data, remote identification, and authorization.

UAS combines UAV, such as drones, a UAV controller, and their respective communications link that uses, e.g., 3GPP cellular systems. The UAV is in practice an aircraft without a human pilot onboard. An operator can control the UAV by relying on the UAV controller, and the UAV may have autonomous flight capabilities.

There are a variety of physical UAV types and capabilities in terms of size and weight and their use, e.g., for recreational purposes or commercial applications. The respective regulation varies depending on the types and areas of use.

UAS requirements also cover C2 as well as the two-way data link between the UAV and the connected network and their servers.

6.4.2 MBMS

3GPP Release 15 did not include the MBMS concept as a native service in the 5G system, nor did Release 16. In a parallel fashion, the MBMS supported by LTE evolves via the joint efforts of broadcasters and the mobile industry to develop further LTE MBMS features to meet the requirements of 3GPP TR 38.913, scenarios, and requirements for next generation access technologies [33].

Prior MBMS development has resulted in Evolved MBMS (eMBMS) and Further Evolved MBMS (FeMBMS) as per LTE Release 14. The 3GPP decided to leave the potential inclusion of the item as a candidate feature of 5G in Release 17.

6.4.3 Machine Learning and Artificial Intelligence

As 5G takes off, there have already been concrete steps taken in the development of AI, which can assist in many areas on top of 5G services. Its evolution along with 5G deployments is moving "traditional" models relying on autonomic network management and cognitive network management. The transition includes ML abilities combined with self-aware, self-configuring, self-optimizing, self-healing, and self-protecting networks.

6.4.4 Use Cases of 6G

The ITU NET-2030 considers the future needs and requirements of 6G, which will be a reality during the 2030–2040 decade.

References

1 3GPP, "3GPP TR 22.891 V14.2.0, Feasibility Study on New Services and Markets Technology Enablers, Stage 1, Release 14," 3GPP, September 2016.
2 CAR 2 CAR Communication Consortium, "Overview of the C2C-CC System," C2C CC, August 2007.
3 5GAA, "V2X," 5GAA. [Online]. Available: http://5gaa.org/5g-technology/c-v2x. [Accessed 5 July 2019].
4 3GPP, "SA1 Completes Its Study into 5G Requirements," 3GPP, 23 June 2016. [Online]. Available: https://www.3gpp.org/news-events/1786-5g_reqs_sa1. [Accessed 27 September 2020].

5 sdxcentral, "The Top 5G Use Cases," sdxcentral, 17 November 2017. [Online]. Available: https://www.sdxcentral.com/5g/definitions/top-5g-use-cases. [Accessed 28 July 2020].
6 GSMA, "Generic Network Slice Template Version 3.0," GSMA, 22 May 2020.
7 5G Americas, "Network Slicing for 5G and Beyond," 5G Americas, 2016.
8 SDX Central, "How 5G SDN Will Bolster Networks," 2019. [Online]. Available: https://www.sdxcentral.com/5g/definitions/5g-sdn.
9 O-RAN Alliance, "O-RAN Use Cases and Deployment Scenarios," O-RAN Alliance, February 2020.
10 GSMA, "VoWiFi and VoLTE Entitlement Configuration, Version 2.0," GSMA, 3 October 2018.
11 GSMA, "Road to 5G: Introduction and Migration," GSMA, April 2018. [Online]. Available: https://www.gsma.com/futurenetworks/wp-content/uploads/2018/04/Road-to-5G-Introduction-and-Migration_FINAL.pdf. [Accessed 5 July 2019].
12 Hoglund, A., Lin, X., Liberg, O. et al. "Overview of 3GPP Release 14 Enhanced NB-IoT," *IEEE Network* 31(6): 16–22, November 2017.
13 GSMA, "Mobile IoT in the 5G Future – NB-IoT and LTE-M in the Context of 5G," GSMA, 14 May 2018. [Online]. Available: https://www.gsma.com/iot/mobile-iot-5g-future. [Accessed 5 July 2019].
14 3GPP, "3GPP TR 38.824: Study on Physical Layer Enhancements for NR Ultra-reliable and Low Latency Case (URLLC), Release 16," 3GPP, 2019.
15 Chee, F.Y., "EU Opens Road to 5G Connected Cars in Boost to BMW, Qualcomm," Reuters, 4 July 2019. [Online]. Available: https://www.reuters.com/article/us-eu-autos-tech/eu-opens-road-to-5g-connected-cars-in-boost-to-bmw-qualcomm-idUSKCN1TZ11F. [Accessed 5 July 2019].
16 Shi, Y., "LTE-V: A Cellular-Assisted V2X Communication Technology," Huawei, Beijing, 2015.
17 Global Market Insights, "Vehicle to Vehicle (V2V) Communication Market Size," Global Market Insights, 2018. [Online]. Available: https://www.gminsights.com/industry-analysis/vehicle-to-vehicle-v2v-communication-market. [Accessed 26 July 2018].
18 5GAA, "V2X," 2020. [Online]. Available: https://5gaa.org/5g-technology/c-v2x.
19 3GPP, "TR 38.913: Study on Scenarios and Requirements for the Next Generation Access Technologies (Release 15)," 3GPP, 2019.
20 KeySight, "Satellites Bolster 5G through (Whitepaper)," KeySight, USA, 22 June 2020.
21 3GPP, "TR 38.811 V15.3.0 (2020-07), Study on NR to Support Non-terrestrial Networks, Release 15," 3GPP, July 2020.
22 3GPP, "TR 22.822 V16.0.0 (2018-06), Study on Using Satellite Access in 5G, Stage 1, Release 16," 3GPP, June 2018.
23 3GPP, "3GPP SA2 Meeting 120 Documents," 3GPP, 21 March 2017. [Online]. Available: https://www.3gpp.org/ftp/tsg_sa/WG2_Arch/TSGS2_120_Busan/Docs. [Accessed 6 August 2020].
24 3GPP, "3GPP TR 21.916, Summary of Rel-16 Work Items (Release 16)," 3GPP, July 2020.
25 3GPP, 12 March 2020. [Online]. Available: https://www.3gpp.org/ftp/tsg_sa/TSG_SA/TSGS_87E_Electronic/Docs. [Accessed 6 August 2020].
26 Ericsson, "5G Meets Time Sensitive Networking," 18 December 2018. [Online]. Available: https://www.ericsson.com/en/blog/2018/12/5G-meets-Time-Sensitive-Networking.

27 Tampere University of Technology, "Positioning and Location-awareness in 5G Networks," Tampere University of Technology, [Online]. Available: http://www.tut.fi/5G/positioning. [Accessed 3 July 2019].

28 3GPP, "TR 21.916: Release 16 Description; Summary of Re-16 Work Items (Release 16)," 3GPP, March 2020.

29 GSMA, "Future Networks," GSMA, 29 November 2018. [Online]. Available: https://www.gsma.com/futurenetworks/digest/gsma-launches-new-forum-to-support-5g-cloud-based-ar-vr. [Accessed 5 July 2019].

30 3GPP, "TS 38.300 V16.2.0 (2020-07), NR and NG-RAN Overall Description, Stage 2, Release 16," 3GPP, July 2020.

31 Chandramouli, D., "5G for Industry 4.0," 3GPP, 13 May 2020. [Online]. Available: https://www.3gpp.org/news-events/2122-tsn_v_lan. [Accessed 22 September 2020].

32 5GACIA, "5G Non-Public Networks for Industrial Scenarios," 5GACIA, July 2019.

33 Nugent, M., "ITU Forum, Towards 5G Enabled Gigabit Society: 5G – Shaping the Future of Broadcasting," 11–12 October 2018. [Online]. Available: https://www.itu.int/en/ITU-D/Regional-Presence/Europe/Documents/Events/2018/5G%20Greece/Session%205%20Nuget%20Shaping%20the%20future%20of%20Broadcasting.pdf. [Accessed 23 September 2020].

7

Security

7.1 Overview

7.1.1 5G Security Architecture

The 5G system of the 3GPP brings important enhancements for security. Figure 7.1 depicts the 5G security architecture of the 3GPP Release 16 network as interpreted from 3GPP Technical Specification (TS) 33.501, Release 16 [1]. This architectural model has not changed since its introduction in Release 15.

The 3GPP 5G system consists of evolved physical radio and core networks and their logical functions. The security architecture can be presented as security domains that consist of Access Stratum (AS), Home Stratum (HS), Serving Stratum (SS), and Transport Stratum (TS). Figure 7.1 outlines the functional elements within these stratums and their security relations. Table 7.1 summarizes the interfaces of Figure 7.1.

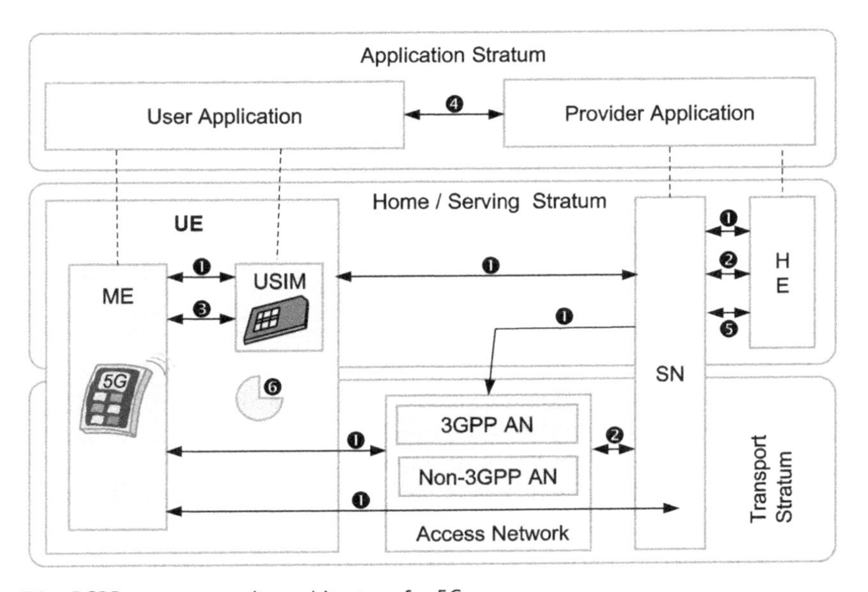

Figure 7.1 3GPP system security architecture for 5G.

5G Second Phase Explained: The 3GPP Release 16 Enhancements, First Edition. Jyrki T.J. Penttinen.

Table 7.1 The interfaces of 3GPP security architecture [1].

Interface	Function
1	**Network access security** features for User Equipment (UE) authenticate and access services securely via the network. These features protect the radio interface, whether it is 3GPP or non-3GPP radio access, and deliver the security context of Serving Network (SN) to Access Network (AN) for access security.
2	**Network domain security** features provide a means for network nodes to exchange user data and signaling in a secure manner.
3	**User domain security** features secure user access to Mobile Equipment (ME).
4	**Application domain security** features provide a means for applications of users and providers to exchange messages securely. 3GPP Release 16 TS 33.501 does not define actual application domain security, though.
5	**Service-Based Architecture (SBA) domain security** features provide secure communication within the serving network domain and other network domains. The features include network element registration, discovery, authorization security, and protection for service-based interfaces.
6	**Visibility and configurability of security** features provide a means to inform users if a security feature is in operation.

7.1.2 Security Functions

3GPP TS 33.401 defines Long Term Evolution (LTE) security, and TS 33.501 defines 5G security. Comparing their definitions, 5G brings completely new SBA domain security [1–3]. Figure 7.2 depicts the 5G network security functions in a non-roaming scenario. The 5G network functions that contribute to security are the following:

AUSF: The *Authentication Server Function* replaces Mobility Management Entity/ Authentication, Authorization and Accounting (MME/AAA) of the 4G system. It terminates requests from the Security Anchor Function (SEAF) and interacts with the Authentication Credential Repository and Processing Function (ARPF). The AUSF and the ARPF could be co-located and form a general Extensible Authentication Protocol (EAP) server for EAP Authentication and Key Agreement (AKA) and EAP-AKA'.

ARPF: The *APF* is co-located with Unified Data Management (UDM). It stores the long-term security credentials such as the user's key *K*. Based on these, it executes cryptographic algorithms and creates authentication vectors.

IPUPS: *Inter-PLMN UP Security* is a new 5G functionality introduced in Release 16. It is located at the perimeter of the Public Land Mobile Network (PLMN) for protecting User Plane (UP) messages. It is a User Plane Function (UPF) functionality that enforces GPRS Tunneling Protocol for the UP (GTP-U) security between UPF elements of the visited and home PLMN via the *N9* interface. Please note that it is possible to activate the IPUPS with other functionalities in a UPF, and it can be activated in a UPF dedicated for IPUPS functionality [4].

SCMF: The *Security Context Management Function* retrieves the key from the SEAF, which is used to derive further keys. The SCMF may be co-located with the SEAF in the same Access and Mobility Management Function (AMF).

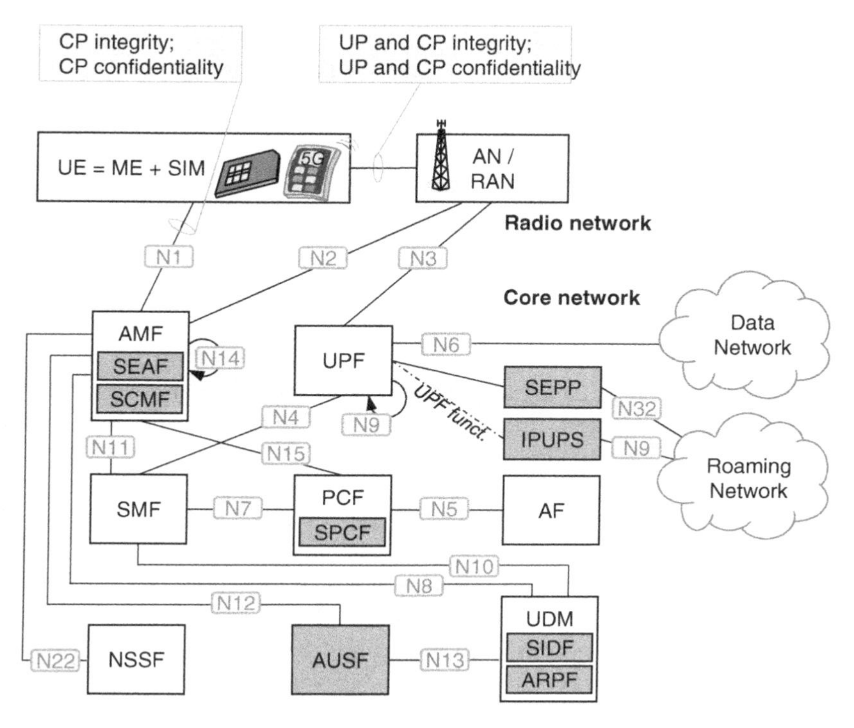

Figure 7.2 5G network functions related to security (highlighted). Release 16 adds the IPUPS component to the 5G security architecture.

SIDF: The *Subscription Identifier De-Concealing Function* is a service offered by the UDM network function of the home network of the subscriber. It de-conceals the Subscription Permanent Identifier (SUPI) from the Subscriber Concealed Identifier (SUCI).

SEAF: The *SEAF* forms, as an outcome of primary authentication, the unified, common anchor key K_{SEAF} for all the access scenarios. K_{SEAF} protects the communications of the UE and the serving network, and resides in the visited network in a roaming scenario. There may be separate K_{SEAF} keys for the same UE connected to 3GPP and non-3GPP (such as Wi-Fi) ANs. SEAF is co-located with the AMF in 3GPP Release 15.

SEPP: The *Security Edge Protection Proxy* is a new functionality compared to previous mobile generations, and Release 15 introduced its basic form. It protects control plane messages at the perimeter of the PLMN, and enforces inter-PLMN security via the *N32* interface [1]. In the contexts of SEPP, the 3GPP also defines a Consumer's SEPP (cSEPP) that is a SEPP residing in the PLMN where the service consumer network function is located, while Producer's SEPP (pSEPP) is the SEPP residing in the PLMN where the service producer network function is located. Furthermore, an *N32-c* connection refers to a Transport Layer Security (TLS)-based connection between a SEPP in one PLMN and a SEPP in another PLMN, whereas an *N32-f* connection is a logical connection that exists between a SEPP in one PLMN and a SEPP in another PLMN for exchange of protected Hypertext Transfer Protocol (HTTP) messages.

The current security specifications state that the SEPP solution must support application layer mechanisms for addition, deletion, and modification of message elements by intermediate nodes with some specific message elements. An example of such a case is Internet Protocol Packet Exchange (IPX) providers modifying messages for routing purposes [1].

Please note that GSMA Fraud and Security work groups currently design further interoperability recommendations of the SEPP to ensure common principles for secure roaming in environments involving Mobile Network Operators (MNOs), IPX, and value added service providers.

SPCF: The *Security Policy Control Function* provides policies related to the security of network functions such as AMF, SMF, and UE. The Application Function (AF) dictates the elements involved for each policy scenario. The SPCF may be co-located with the PCF, or it can be a standalone element. The SPCF contributes to the confidentiality and integrity protection algorithms, key lifetime and length, and selection of the AUSF.

Thanks to mobile edge computing and virtualization, 5G network functions and contents are moving closer to the consumer. The functions and contents are often replicated and exposed in potentially less protected environments. To make contents available to the user with reduced latency, the edges need to cache them via a third party content provider, so there needs to be adequate security measures in place, respectively.

Authentication between network entities ensures that the edge is authorized to receive the network function instance or the content. In addition, the data that is exchanged between those entities need to be protected at rest and in transit. The security mechanisms in the edges will be based on virtualization technologies involving hypervisors and isolation.

7.1.3 Enhanced 5G Security

Release 16 of the 5G system contains the following aspects for security procedures [1]:

- **Primary authentication and key agreement** in 5G establish mutual authentication between the UE and the network. For consequent security procedures between the UE and the serving network, they also provide keying material, such as an anchor key K_{SEAF}. The home network's AUSF provides it to the SEAF of the serving network.
- **Initiation of authentication and selection of authentication method** refers to the ability of the SEAF to perform authentication with the UE. This can take place during any signaling procedure. The 3GPP requires that the registration request of the UE is based on an SUCI or 5G Global Unique Temporary Identifier (5G-GUTI).
- **Authentication procedures**: Intermediate key K_{AUSF} results in anchor key K_{SEAF}. The AUSF can securely store K_{AUSF}. The home operator dictates the utilization of K_{AUSF}.
- **Linking increased home control to subsequent procedures** along with increased home control as a result of 5G authentication and key agreement protocols and their enhanced security compared to Evolved Packet System (EPS) AKA in EPS.

7.2 5G Network Security Procedures

7.2.1 Keys in 5G

The security of 5G has been redefined. In fact, only the initial authentication procedure is still the same with legacy systems, whereas the 5G key derivation procedure is completely new. The 3GPP has defined a 5G key hierarchy as depicted in Figure 7.3. The 5G key lengths are by default 128 bits in the initial phase, while the network interfaces support key lengths of up to 256 bits when needed for future purposes. Table 7.2 summarizes the principle of the 5G keys.

Figure 7.3 also depicts the 5G core elements, which take part in a rather long chain of key derivation.

Figure 7.3 shows the security link between either the Universal Subscriber Identity Module (USIM) or ME and the respective 5G network function. For the security-related network functions presented in Figure 7.3, please refer to Section 7.1.2. The other components present in this figure are the following:

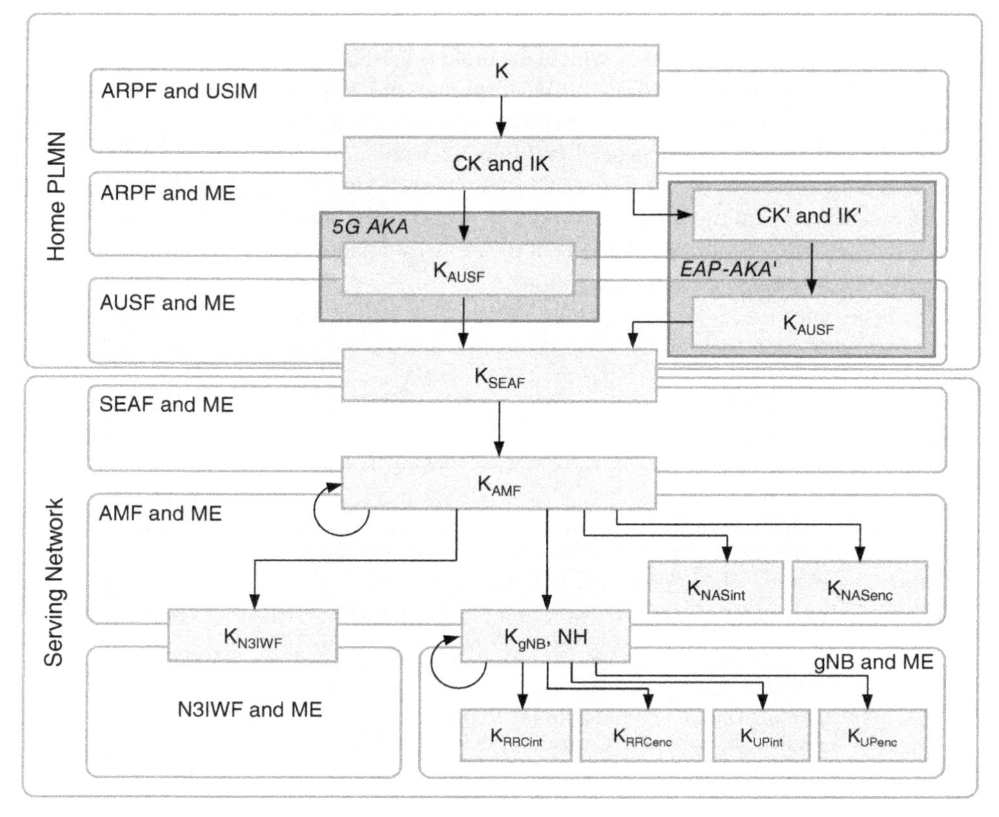

Figure 7.3 The 5G key hierarchy as interpreted from 3GPP TS 33.501 [1]. Release 16 does not change the original definitions of the key hierarchy of Release 15.

The **USIM** is the Subscriber Identity Module (SIM) application residing in the hardware-based Universal Integrated Circuit Card (UICC). The telecom ecosystems oftentimes generalize the UICC and USIM as a "SIM card," as consumers are familiar with that term since the introduction of Global System for Mobile Communications (GSM). The traditional removable card is still valid in 5G, too. Nowadays, the device manufacturer can embed the UICC permanently into the device, soldering it as part of the electrical components of the device. Alternatively, USIM functionality can also reside within a 5G chipset such as baseband module in an integrated form. In any of these cases, the USIM securely stores the subscriber's permanent key K and private key of the Public Key Infrastructure (PKI) key pair, and processes cryptographic tasks together with the ME to protect communications. In practice, the only standardized embedded UICC Form Factor (FF) up to now is the M2M Form Factor (MFF2), while the other embedded and integrated variants are physically still proprietary solutions of different vendors.

ME is the hardware of the 5G device. The ME and UICC (or any of its variants storing the USIM) together form the UE. The hardware of the device can be of many forms such as a consumer's smart device or autonomously functioning Internet of Things (IoT) device, depending on the device manufacturer's design.

N3IWF (Non-3GPP Interworking Function) manages ANs that are not 3GPP-defined 5G RANs (Radio Access Networks). A typical example is Wi-Fi.

gNB (Next Generation NodeB) is the 5G "base station." This receives the ciphering and integrity keys from the key derivation procedures and protects the radio interface.

In addition to the keys summarized in Table 7.2, there are yet more intermediate keys. K_{gNB*} is one of these; the ME and the 5G gNB obtain it via the Key Derivation Function (KDF). Please note that the combination of ME and UICC and its USIM application jointly form UE. The UICC/USIM handles part of the security-related tasks such as storing the K, while the ME takes care of K_{gNB*} derivation.

K'_{AMF} is yet another 5G key. Both the AMF and ME derive it via the KDF when the UE moves from one AMF to another. This case is called inter-AMF mobility scenario.

Table 7.2 The 5G keys.

Key	Description	Domain
CK', IK'	Authentication keys. The 5GC derives them from CK and IK if the Wi-Fi-based access uses EAP-AKA′.	Authentication
K, CK, IK	The long-term, user-specific, unique key K is stored securely in the USIM application of hardware-protected UICC of the user. It is also stored in the ARPF, and does not leave the ARPF or USIM under any circumstances. Instead, both form short-term ciphering key CK and integrity key IK for authentication of the 5G subscriber and as a base for further key derivation.	Authentication
K_{AMF}	Key for the AMF in a serving network. The SEAF and ME derive K_{AMF} from K_{SEAF}. The ME and a source AMF derive K_{AMF} further in the horizontal key derivation procedure.	Key for the AMF in a serving network
K'_{AMF}	Intermediate key. The AMF and ME can derive it upon UE moving between AMFs during inter-AMF mobility using a KDF.	Intermediate key

Table 7.2 (Continued)

Key	Description	Domain
K_{AUSF}	Key for AUSF in a home network. AUSF generates K_{AUSF} from the authentication material (*K*, *CK*, and *IK* for 5G AKA, or from *CK'* and *IK'* in case EAP-AKA′ is used for, e.g., Wi-Fi access). K_{AUSF} derivation thus has two scenarios: (1) if the AUSF receives EAP-AKA′, *CK'*, and *IK'* from the ARPF, the ME and AUSF derive K_{AUSF} from *CK'* and *IK'*; and (2) if the AUSF receives 5G AKA and K_{AUSF} from ARPF, the ME and ARPF derive K_{AUSF} from *CK* and *IK*.	Key for AUSF in home network
K_{gNB}	Key for 5G gNB. The AMF and ME derive it from K_{AMF}. The source gNB and ME obtain it for performing horizontal or vertical key derivation. K_{gNB} is used as K_{eNB} between ME and ng-eNB.	Key for NG-RAN
K_{N3IWF}	Key for non-3GPP interworking functions. The AMF and ME derive it from K_{AMF}. Please note that K_{N3IWF} is not forwarded between N3IWFs. The N3IWF receives the K_{N3IWF} key from the AMF, and uses it for Internet Key Exchange Protocol Version 2 (IKEv2) between the UE and N3IWF in procedures for untrusted non-3GPP access such as Wi-Fi.	Key for non-3GPP access
K_{NASenc}	Encryption key for Non-Access Stratum (NAS) signaling, i.e., on access-agnostic domain. The AMF and ME derive it from K_{AMF}. It applies an encryption algorithm to protect NAS signaling.	Key for NAS signaling
K_{NASint}	Integrity key for NAS signaling. NAS signaling refers to dialogue between the ME and core network nodes, and it is access-agnostic. The AMF and ME derive it from K_{AMF}. It protects NAS signaling by applying an integrity algorithm.	Key for NAS signaling
$K_{NG\text{-}RAN^*}$	Intermediate key. The NG-RAN (either gNB or ng-eNB) and ME derive it when performing a horizontal or vertical key derivation using a Key Derivation Function (KDF). Please note that 5GC performs all key derivations using the KDF.	Intermediate key
K_{RRCenc}	Encryption key for Radio Resource Control (RRC) signaling. The gNB and ME derive it from K_{gNB}. It protects RRC signaling with an encryption algorithm.	Key for RRC signaling
K_{RRCint}	Integrity key for RRC signaling. The gNB and ME derive it from K_{gNB}. It protects RRC signaling with an integrity algorithm.	Key for RRC signaling
K_{SEAF}	Anchor key. The AUSF and ME derive it from K_{AUSF}. It is possible to derive further K_{SEAF} keys for multiple security contexts via a single authentication round, e.g., for the use of both native 5G access as well as untrusted non-3GPP access. AUSF provides K_{SEAF} to the SEAF of the serving network.	Key for AUSF in a home network
K_{UPenc}	Encryption key for UP traffic. The gNB and ME derive it from K_{gNB}. It protects UP traffic between ME and gNB with an encryption algorithm.	Key for UP traffic
K_{UPint}	Integrity key for UP data traffic. The gNB and ME derive it from K_{gNB}. It protects UP traffic between the ME and gNB with an integrity algorithm.	Key for UP traffic
NH	Intermediate key referring to Next Hop. The AMF and ME derive it for providing forward security.	Intermediate key

5G has adapted the widely used PKI scheme for identifier protection based on the X.509 Certification Authority (CA) procedure. The PKI is a set of rules, policies, and procedures to create, manage, distribute, use, store, and revoke digital certificates.

The PKI makes the public key encryption, and protects 5G communications. As stated in Ref. [5], the user's home network provisions the public key. It is securely stored in the user's USIM, whereas the ARPF houses the respective private key.

The home MNO has control over the subscriber's privacy, provisioning, and updates of the public key.

The authentication keys of 5G include *K*, *CK*, and *IK*. If EAP-AKA′ is applied, *CK* and *IK* derive further *CK*′ and *IK*′.

The 5G system's key hierarchy, as depicted in Figure 7.3, involves K_{AMF}, K_{AUSF}, K_{gNB}, K_{N3IWF}, K_{NASenc}, K_{NASint}, K_{RRCenc}, K_{RRCint}, KS_{EAF}, K_{UPenc}, K_{UPint}, and *NH* as summarized in Table 7.2.

3GPP 33.501 describes the key-related requirements of the 5GC and NG-RAN in Section 5.1.3. Please note that Release 16 has brought standalone non-public networks. The key hierarchy for standalone non-public networks when an authentication method other than 5G AKA or EAP-AKA′ is used is given in Annex I.2.3 of 3GPP TS 33.501 [1].

5G also defines dual connectivity as described in Chapter 3, and there is a key derived in that scenario, too. This happens in such a way that when the Master Node (MN) establishes a security context for the first time between secondary node and the device (UE) for an AS security context, the MN forms a secondary node key (K_{SN}) and delivers it to the secondary node over the *Xn-C* interface. The MN forms the K_{SN} by associating a Secondary Cell Group (SCG) counter along with the AS security context. The MN sends the SCG counter value to the UE via RRC signaling the UE needs to generate a new K_{SN}.

7.2.1 5G Identifiers

There are new identifiers in the 3GPP 5G network. These are SUPI and SUCI.

SUPI is a primary identifier in 5G and forms the foundation for all the key derivations together with the subscribers' unique *K* key. The serving network authenticates the SUPI during the authentication and key agreement procedure between UE and network. Furthermore, the serving network authorizes the UE through the subscription profile obtained from the home network based on the authenticated SUPI.

The concealed variant of the SUPI is referred to as the SUCI. As defined in 3GPP TS 33.501, the SUCI is a one-time use subscription identifier, which contains the concealed subscription identifier, which may be, e.g., the Mobile Subscriber Identification Number (MSIN). The SUCI is an optional mechanism managed from the UICC. Its aim is to provide further security by hiding the permanent user identification information. It is a privacy-preserving identifier and it contains the concealed SUPI.

The SUCI is a one-time use subscription identifier, which contains the concealed subscription identifier such as the MSIN portion of the SUPI, and additional non-concealed information needed for home network routing and protection scheme usage. The principle of the SUCI is that the UE generates it using a protection scheme with the raw public key that has already been securely provisioned beforehand in control of the home network. Based on indication of the USIM, dictated by the MNO, calculation of the SUCI can be done either by the USIM or the ME. The UE then builds a scheme input from the part containing subscription identifier of the SUPI and executes the protection scheme. The UE would not conceal the home network identifier, though, such as Mobile Country Code (MCC) or Mobile Network Code (MNC). Please note that there is no requirement for protecting the SUPI in the case of an unauthenticated emergency call.

In the 3GPP-defined 5G system, subscription credentials refer to a set of values stored in the USIM and the ARPF, consisting of at least one or more long-term keys and the subscription identifier SUPI. These credentials are in use to uniquely identify a subscription and to mutually authenticate the UE and 5G core network [1].

5G identifier protection relies on the PKI scheme. As stated in ETSI TS 133.501, V. 15.1.0, Chapter 5.2.5 (Subscriber Privacy), the home network provisions the home network's public key and it is stored in the USIM, while the private key is stored in the ARPF. The home MNO controls the provisioning and updating of the public key. The home network operator also controls subscriber privacy.

The SUPI is not transferred over the radio interface in clear text. The only exception to this rule is the related routing information such as MCC and MNC, which need to be visible.

In case the USIM does not yet contain the provisioned public key, SUPI protection in initial registration is not possible. This scenario uses a null scheme instead, and the ME performs it.

In 5G, the UE consists, as has been in previous generations, the ME and secure element, which may be the traditional UICC or evolved variant of it, including embedded and integrated UICCs. Together, the UICC and the ME form the UE. 3GPP TS 33.501 details the security aspects of the UE.

7.2.2 Network Key Storage and Procedures

In the 5G security architecture, the UE has a corresponding key for every network entity key. Table 7.3 summarizes the procedures related to 5G keys as interpreted from 3GPP TS 33.501 [1].

7.2.3 5G Key Derivation

As stated in 3GPP TS 38.300 and TS 33.501, the following principles apply to 5G New Radio (NR) connected to 5GC security.

The AMF houses the termination points for ciphering and integrity protection of NAS signaling, whereas the gNB terminates ciphering and integrity protection of RRC signaling as well as the UP data.

The 5G security functions provide *integrity protection* as well as *encryption* between user device and multiple network functions. As for the UP, "ciphering" provides user data confidentiality, whereas "integrity protection" provides user data integrity for user data (Data Radio Bearers, DRBs). For RRC signaling (Signaling Radio Bearers, SRBs), "ciphering" provides signaling data confidentiality and "integrity protection" signaling data integrity. Integrity protection is always active for RRC signaling, whereas the configuration of ciphering and integrity protections are optional.

Table 7.3 5G keys and their respective storage and procedures

Element	Keys and processes	Entity
AMF	After successful authentication, key setting takes place as soon as the AMF is aware of the subscriber's identity (5G-GUTI or SUPI). 5G AKA or EAP AKA′ results in a new K_{AMF} that is stored in the UE and the AMF. The AMF receives K_{AMF} from the SEAF or from another AMF. The AMF then generates K_{gNB} to be transferred to the gNB, NH to be transferred to the gNB, and possibly an NH key to be transferred to another AMF. The AMF also generates K_{N3IWF} to be transferred to the N3IWF. The AMF generates keys K_{NASint} and K_{NASenc} dedicated to protecting the NAS layer, and the AMF also generates AN-specific keys from K_{AMF}.	Network
ARPF	The ARPF stores, along with the USIM, the 5G subscription credentials that are the subscription identifier SUPI and long-term keys *K*. They serve as a base for security procedures taken for identifying subscription and for the mutual authentication of the UE and 5G core network. The ARPF processes the *K*, which is either 128 or 256 bits long, and other sensitive data in its secure environment. The ARPF also stores the home network private key that the SIDF uses to de-conceal the SUCI and to reconstruct the SUPI. As stated in 3GPP TS 33.501, during an authentication and key agreement procedure, the ARPF derives *CK′* and *IK′* from *K* in case EAP-AKA′ is used, and it derives K_{AUSF} from *K* in case 5G AKA is used. The ARPF forwards the derived keys to the AUSF.	Network
AUSF	In the AUSF an intermediate key K_{AUSF} is generated during authentication. It can be left at the AUSF as this optimizes the signaling in case the UE registers to different serving 3GPP or non-3GPP networks. This procedure is comparable to fast re-authentication in EAP-AKA′, but the subsequent authentication provides weaker guarantees than an authentication directly involving the ARPF and the USIM. The AUSF generates the anchor key K_{SEAF} from the authentication key material received from the ARPF during an authentication and key agreement procedure. As stated in 3GPP TS 33.501, in case EAP-AKA′ is used for authentication, the AUSF derives a key K_{AUSF} from *CK′* and *IK′* for EAP-AKA′. K_{AUSF} can be stored in the AUSF between two subsequent authentication and key agreement procedures to optimize and speed up the process.	Network
ME	The ME has various tasks related to the keys, and it acts as the counterparty of the respective network functions:The ME generates K_{AUSF} from the *CK* and *IK* it receives from the USIM, and the UE stores K_{AUSF}. In case the USIM supports 5G parameters storage, K_{AUSF} is stored in its non-volatile memory instead.The ME generates K_{SEAF} from K_{AUSF}. If the USIM supports 5G parameters storage, it stores K_{SEAF}. Otherwise, the ME's non-volatile memory stores K_{SEAF}.The ME generates K_{AMF}. If the USIM supports 5G parameters storage, it stores K_{AMF}. Otherwise, the ME's non-volatile memory stores K_{AMF}.The ME also generates the rest of the keys deriving them from K_{AMF}. The ME deletes 5G security context, K_{AUSF} and K_{SEAF} from its memory if the USIM is removed from the ME. This removal can happen when the ME is in power-on state, or if the USIM is missing or swapped upon powering up the ME.	Device
N3IWF	The N3IWF receives K_{N3IWF} from the AMF. The N3IWF uses K_{N3IWF} as the key for IKEv2 between UE and N3IWF for untrusted non-3GPP access procedures.	Network

Table 7.3 (*Continued*)

Element	Keys and processes	Entity
NG-RAN	The NG-RAN refers to the gNB or ng-eNB. The NG-RAN receives K_{gNB} and NH from the AMF. The ng-eNB uses K_{gNB} as K_{eNB}. The NG-RAN generates further AS keys from K_{gNB} and NH.	Network
SEAF	The SEAF can initiate an authentication with the UE during any procedure. This takes place upon signaling establishment with the UE. The UE uses the SUCI or 5G-GUTI in the registration request procedure. When the SEAF so wishes, it can invoke an authentication procedure with the AUSF. Either the SUCI (3GPP TS 33.501) or SUPI (3GPP TS 23.501) is used during authentication. The primary authentication and key agreement procedures bind anchor key K_{SEAF} to the serving network. The SEAF receives K_{SEAF} from the AUSF upon a successful primary authentication procedure in each serving network. The SEAF then generates K_{AMF} and hands it to the AMF. The SEAF never transfers K_{SEAF} outside the SEAF, and the SEAF deletes K_{SEAF} as soon as it has derived K_{AMF}. The SEAF is co-located with the AMF.	Network
SIDF	The SIDF manages the de-concealing of the SUPI based on the SUCI. For this, the SIDF relies on the PKI concept's private key, which is securely stored in the home operator's network. The actual de-concealment takes place at UDM. The SIDF is protected via access rights, so that only the home network's element may present the request to the SIDF.	Network
UDM	UDM receives authentication requests. The SIDF de-conceals the SUCI to obtain the SUPI for UDM to process the authentication request. UDM/ARPF selects the authentication method based on the SUPI and subscription data.	Network
USIM	The USIM stores the long-term key K equally as the ARPF does. The USIM generates key material from K during the authentication and key agreement procedure, and the USIM forwards it to the ME. The USIM also stores the home network public key that conceals the SUPI.	Device

The operator can configure ciphering and integrity protection per DRB. There is a dependency between the DRB and PDU session, though, as all the DRBs that belong to the same PDU session need to be integrity protected in the UP if so indicated in the User Plane Security Enforcement, as stated in 3GPP TS 23.502 [6].

Furthermore, any entity processing clear text related to key management and data handling must stay protected from physical attacks, and they are located in a secure environment. The keys relate to the AS of the gNB, which are isolated cryptographically from the NAS keys of the 5G core network, and the procedures rely on separate AS- and NAS-level Security Mode Command (SMC) procedures.

As a highly relevant rule in 5G security deployment, a sequence number (COUNT) is used as input to ciphering and integrity protection and a given sequence number must only be used once for a given key (with the only exception of identical retransmission) on the same radio bearer in the same direction. If the device manufacturer and operator fail to apply this rule, it may leave the network vulnerable to malicious attacks, as has been proved for the case of LTE voice calls by attacking the connections that reutilize the COUNT (network resetting it back to zeros), which has resulted in a ReVoLTE demo that reuses keystream in Ref. [7]. This specific vulnerability is a result of part of the vendor community's misinterpretation of the 3GPP security specification. To implement it adequately, the stream cipher has to generate a unique keystream for each call to prevent the issue related to keystream reuse.

As 3GPP TS 38.300 states, primary authentication performs mutual authentication between the UE and the network. It also provides the anchor key K_{SEAF}. Posteriorly, K_{AMF} builds upon K_{SEAF} as part of primary authentication. K_{AMF} can also materialize from NAS key rekeying and key refresh events. K_{AMF}, in turn, is the basis for K_{NASint} and K_{NASenc} is a result of the NAS SMC procedure.[1]

As soon as an initial AS security context is ready between the UE and gNB, the AMF and the UE derive K_{gNB} and an NH parameter from K_{AMF}. Each K_{gNB} and NH parameter associates to an NH Chaining Counter (NCC). Every K_{gNB} relates with a derived NCC corresponding to the respective NH value.

At the beginning, K_{gNB} is a result of K_{AMF} derivation. It associates with the NH parameter and NCC value. The UE and the target gNB use K_{gNB*} in the handover procedure, derived from the previous K_{gNB} (horizontal key derivation) or from the NH parameter (vertical key derivation).

Finally, K_{RRCint}, K_{RRCenc}, K_{UPint}, and K_{UPenc} are derived based on K_{gNB} after a new K_{gNB} is derived as depicted in Figure 7.4.

This key derivation procedure provides both backward security (i.e., the gNB is not capable of figuring out previous K_{gNB*} values) and prevents gNB to predict future K_{gNB} values.

In addition to the NAS SMC procedure, the 5G AS SMC procedure handles RRC and UP security algorithms negotiation and RRC security activation so that the AMF provides the UE 5G security capabilities to the gNB, which, in turn, selects an appropriate ciphering algorithm. Furthermore, the gNB selects an appropriate integrity algorithm. The indication of these algorithms is integrity protected. Posteriorly, RRC encryption and decryption starts gradually after respective verification procedures.

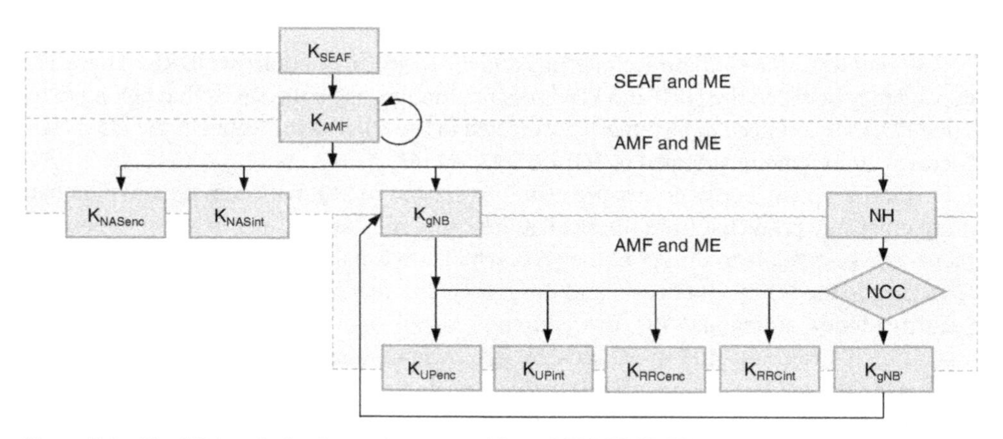

Figure 7.4 The 5G key derivation as interpreted from 3GPP TS 33.501, Release 16.

[1] The NAS SMC procedure consists of a set of messages between the network and the UE. The network sends the NAS SMC to the UE, whereas the UE sends a response with the NAS Security Mode Complete message. The NAS SMC procedure establishes an NAS security context between the UE and the network.

Please refer to 3GPP TS 23.502, TS 38.331, and TS 33.501 for more detailed procedure descriptions and the rest of the functions such as key refresh and key rekeying, state transitions, and mobility scenarios [1, 6, 8].

7.2.4 Security Aspects of Network Slicing

The 3GPP-defined network slice refers to a logical end-to-end network. An operator can create highly dynamic network slices to provide different service types to a set of customers. The network slice can involve the user and control plane of the 5G core network, the RAN, and the interworking functions to other non-3GPP ANs. The AMF manages the network slices, connecting the UE in one or more network slices. The UE is capable of connecting a maximum of eight parallel slices.

Network slicing is an essential function of standalone 5G networks. It provides a means for operators to adjust their service level per vertical in terms of attributes such as data speed, latency, and capacity. Another important aspect of network slicing functionality is its capability to isolate slices to provide adequate security levels between users.

3GPP TS 23.502 defines the network slice-specific authentication and authorization procedure in its Section 4.2.9.1. The 5G system triggers it for Single Network Slice Selection Assistance Information (S-NSSAI) requiring network slice-specific authentication and authorization with an AAA Server (AAA-S). Either a Home PLMN (H-PLMN) operator or third party can host the server; in the latter scenario, the third party needs to have a business relationship with the H-PLMN by relying on an EAP framework [1]. If the AAA-S belongs to a third party, the H-PLMN operator can apply an AAA Proxy (AAA-P).

As stated in 3GPP TS 23.502, the AMF triggers the network slicing procedure during a registration procedure as soon as any network slices require slice-specific authentication and authorization. This triggering can also take place when the AMF determines that network slice-specific authentication and authorization is required for an S-NSSAI in the allowed NSSAI; this scenario happens, e.g., in the event of a subscription change. Lastly, the procedure takes place when the AAA-S that authenticated the network slice triggers a re-authentication.

The AMF works as an EAP authenticator communicating with the AAA-S via the Network Slice-Specific Authentication and Authorization Function (NSSAAF).

7.3 SIM in the 5G Era

7.3.1 Background

The original hardware of the SIM card, referred to as the UICC, will maintain its relevance also in the 5G era as a tamper-resistant, hardware-based secure element to store keys and other confidential information. This is logical, as there is a considerable legacy device base and established ecosystem that serves the ecosystem to manufacture the cards and to manage the logistics and personalization of the SIM profiles per each MNO. The tamper-resistant UICC provides adequate security-level compliance with the international accreditation requirements as dictated by the GSMA SAS and payment institutions.

With rapid developments in the IoT and the need for built-in security for IoT devices, the market will benefit from a deeper integration of USIM functionality into these devices, too. An ultra-small device would not be able to house a traditional UICC FF due to its physical size. Thus, the ecosystem will see an increasing number of small UICC FFs, including a variety of both Embedded UICCs (eUICCs) and Integrated UICCs (iUICCs).

As has been the principle in 3GPP releases, backwards compatibility is essential and will continue to be with the new variants, so the impacts of 5G on the original UICCs are expected to be minimal.

There are 5G-specific files included in the new UICCs as of Release 15 and Release 16. Nevertheless, the 5G file structure remains the same as in previous versions of the USIM. In addition, the pre-existing UICC FFs are still valid, referring to the standardized 2FF, 3FF, 4FF, as well as to the MFF2 used in M2M applications.

7.3.2 UICC Profiles in 5G

3GPP defines the 5G subscription credentials as a set of values in the USIM and the ARPF. 5G subscription credentials consist of a long-term key or set of keys *K*, unique for each user, and the SUPI, which uniquely identifies a subscription. The *K* and SUPI function to mutually authenticate the UE with the 5G core network.

The subscription credentials are processed and stored in the USIM. In addition, the 3GPP has confirmed that the USIM still resides on a UICC in the 5G era.

3GPP TS 31.102 outlines the characteristics of the USIM application [9]. In Release 15, there is one new file that defines the configuration parameter for handover between WLAN and the EPS of 4G. This file is a new service in the Universal SIM Toolkit (UST). There is also a modification to the specification for enhanced coding of Access Technology in EFPLMNwAcT (user controlled mobile network selector with access technology file) to accommodate 5G Systems (5GS).

3GPP TS 31.103 outlines the features of the IP Multimedia Services Identity Module (ISIM) application. In Release 15, there is one new file defined in this specification with a configuration parameter for handover between WLAN and the EPS. There is also an updated file ID for EF-XCAPConfigData (IMS configuration file) in the ISIM, to match the value used in the USIM. Please note that Mission Critical Push-to-Talk (MCPTT) corrections already considered in Release 14 are updated in Release 15. In addition, the Estimated_P-CSCF_Recover_Time timer is removed, which requires an update on data in the EFIMSConfigData file.

3GPP Release 16 brings additional files. Release 16 of 3GPP TS 31.102, Section 4.4.11.1, describes the files that are specific for 5GS. This specification dictates that the "DF5GS" file is present at the ADFUSIM level if any of the following services are available in EFUST (USIM Service Table):

- Service no. 122 5GS Mobility Management Information;
- Service no. 123 5G Security Parameters;
- Service no. 124 Subscription identifier privacy support;
- Service no. 125 SUCI calculation by the USIM;

- Service no. 126 UAC Access Identities support;
- Service no. 127 Control plane-based steering of UE in VPLMN;
- Service no. 128 Call control on PDU session by USIM;
- Service no. 129 5GS Operator PLMN List;
- Service no. 130 Support for SUPI of type NSI or GLI or GCI;
- Service no. 132 Support for URSP by USIM;
- Service no. 133 5G Security Parameters extended;
- Service no. 134 MuD and MiD configuration data;
- Service no. 135 Support for trusted non-3GPP access networks by USIM.

For more information on the USIM and ISIM of Release 16 5G, please refer to the latest versions of 3GPP TS 31.103 and TS 31.103, respectively.

7.3.3 Changes in Authentication

The same authentication principles apply in 5G as have been defined prior to 5G for both the file structure and actual files. The 5G USIM houses the new 5G-specific files and the initial subscription-specific keys, while the ME, i.e., the hardware and software of the device beyond the UICC, takes care of functionality specific to the 5G system. As an example, if there is a network roaming priority list file *OPLMNwACT* present, and it contains 5G network content, which is not yet available in practice, there are no issues as the device passes through the remaining options for prioritizing the available accessed networks.

As has been the case previously, the USIM shall reside on a UICC also in 5G. The UICC may be removable or non-removable. For non-3GPP access networks, USIM applies in the case of a terminal with 3GPP access capabilities. If the terminal supports 3GPP access capabilities, the credentials used with EAP-AKA and 5G AKA for non-3GPP access networks shall reside on the UICC.

3GPP TS 33.501 also specifies the requirements for secure storage and processing of subscription credentials. However, per the interim agreement in 3GPP Technical Report (TR) 33.899, the two accepted solutions are UICC and Smart Secure Platform (SSP). Furthermore, 3GPP TR 31.890, V. 15.1.0, Section 8.3, states that it is expected that the UE might be able to access the 5G system with the existing UICCs that are found already in the commercial market. It may thus happen that the UE will try to access the 5G system even if the respective UICC is not capable of recognizing the 5G network. As the 3GPP specifications aim to provide a good level of backwards compatibility in general, it is important to ensure the functionality of this scenario by updating the specifications where appropriate and feasible.

For further information on the subject, the reference document 3GPP TS 33.501 also presents examples of additional authentication methods with the EAP framework. Advanced subscription management valid for 5G can be found in the GSMA white paper titled "eSIM: The What and How of Remote SIM Provisioning" (March 2018), and the latest remote SIM provisioning documents of the GSMA can be found at their web page, including the architecture and technical specifications [10].

7.3.4 SIM Evolution

3GPP TS 33.501 defines the 5G security architecture and procedures, and includes statements related to the 5G SIM, too. 5G will still rely on "traditional" SIM cards. Nevertheless, there will also be embedded and integrated variants in the consumer and IoT markets of 5G. The removable UICC, commonly known as SIM card, is based on the ISO/IEC 7816 smart card standard. It was adopted by the GSM and other 3GPP systems, and is used nowadays in some non-3GPP systems, too. It will maintain relevance as a tamper-resistant, hardware-based secure element to store keys and other confidential information also in 5G. This is logical as there is an important base of legacy devices and established UICC ecosystems for the manufacturing, logistics, and personalization of the operators' UICC profiles. The UICC provides adequate security-level compliance with the strict international requirements of the GSMA SAS and payment institutions.

All the currently utilized ETSI-specified FFs as depicted in Figure 7.5 (size in mm) are physically compatible with 5G, although some of the previous UICC hardware variants cannot necessarily handle the increased processing of 5G. 5G merely needs some changes to the UICC's contents (file structure) as there are new files defined for 5G functionalities in the standalone mode.

With fast development of IoT and the need for respective built-in security, the market will benefit from a deeper integration of USIM (which forms a file structure within the UICC)

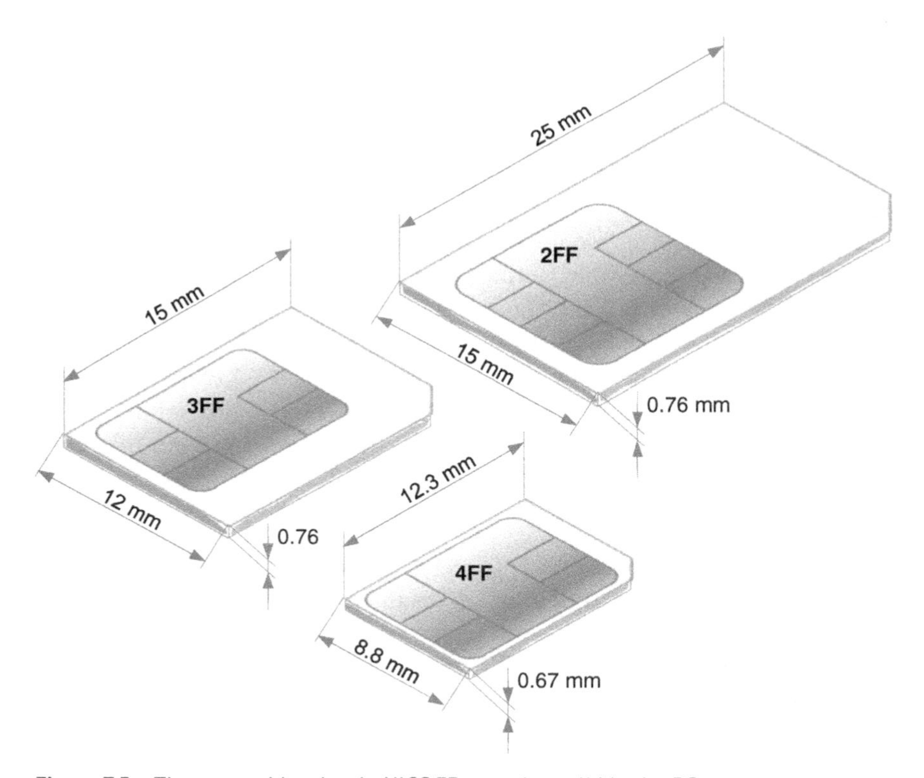

Figure 7.5 The removable, plug-in UICC FFs are also valid in the 5G era.

functionality into these devices. An ultra-small device would not be able to house traditional UICC FFs. Thus, 5G markets are expected to use smaller UICC FFs of embedded and integrated UICCs. New over-the-air provisioning technologies are standardized, too, to support the embedded and integrated variants as they are soldered permanently into the device.

The European Telecommunications Standardisation Institute (ETSI) facilitates the development of new interoperable variants of the UICC. ETSI refers the evolved UICC to an SSP. Apart from this new model, the old UICC continues to exist, while the SSP is expected to be gradually more popular. Meanwhile, the MFF2 of ETSI will also still be valid in 5G together with any proprietary variants of eUICC, that is, permanently soldered UICCs in the device. Figure 7.6 summarizes some examples of these options.

For both eUICC and iUICC, referred to as eSSP and iSSP, respectively, according to the new ETSI terminology, the GSMA has developed Embedded SIM (eSIM) specifications for a new way of managing the MNO profiles. Commonly known as Remote SIM Provisioning (RSP), the new eSIM ecosystem is being developed jointly with GlobalPlatform and SIMalliance.

The 3GPP defines the new 5G subscription credentials as a set of values in the USIM and the ARPF. These credentials refer to a long-term key or set of keys *K*, unique for each user, and the SUPI, which uniquely identifies a subscription. The *K* and SUPI are designed to mutually authenticate the UE with the 5G core network.

The subscription credentials are processed and stored in the 3GPP-defined USIM. In addition, the 3GPP has confirmed that the USIM still resides on the UICC in the 5G era (Figure 7.7). In general, the same principles apply as previously defined for both the file structure and actual files throughout prior 3GPP generations.

The eSIM principles designed for UICC profile management apply to 5G, too, and the eSIM can serve as a security anchor to other stakeholders such as device manufacturers

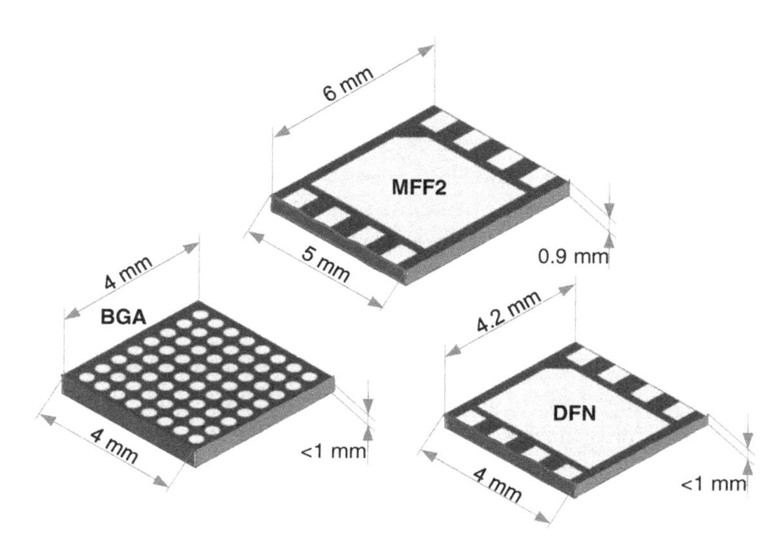

Figure 7.6 Some examples of eUICC elements.

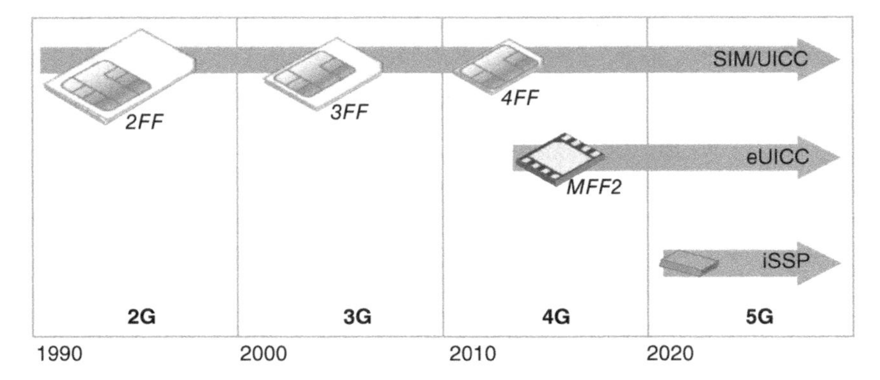

Figure 7.7 UICC variants and respective commercial timelines.

and application providers requiring security in the endpoints. Moreover, the evolution of 5G subscription management is planned to have a convergence path to support both consumer and M2M devices by a common platform.

7.3.5 eSIM

Additional UICC variants will complement the "traditional" SIM FFs in the 5G era. For the eUICC and iUICC variants, there is a need to manage the elements over the air. ETSI facilitates the new interoperable variants of the UICC.

ETSI refers to the evolution of the UICC by the new term SSP. The SSP is not equivalent to the UICC, though, and the UICC continues to exist in a parallel fashion with SSP variants such as eSSP and iSSP.

For all the variants, the remote management of the MNO profiles and subscription data can be based on the GSMA eSIM definitions, which are supported by the ecosystem that also includes GlobalPlatform and SIMalliance. This approach minimizes fragmentation in the offering of remote management solutions of different vendors [11].

eSIM refers to the extension of the "traditional" SIM that was introduced in the GSM. The original SIM has been evolving ever since to cope with the new requirements of 3G, 4G, and 5G in the form of a UICC and USIM (operating system and files) that the hardware houses. Physically, the eSIM can be embedded into the device, e.g., in the form of ETSI-specified MFF2, or it can be integrated even deeper into the processor. Nevertheless, eSIM functionality can also be housed in traditional, removable FFs of the SIM (FF2, FF3, FF4). In all these cases, the same physical SIM can be utilized for managing the subscription, including the change of MNO.

As stated by the GSMA, eSIM is a global specification by the GSMA, which enables remote SIM provisioning of any mobile device. eSIM now allows consumers to store multiple operator profiles on a device simultaneously, and switch between them remotely, though only one can be used at a time. The specification now extends to a wider range of devices beyond the single companion device made possible with the first release. Manufacturers and operators can now enable consumers to select the operator of their choice and then securely download that operator's SIM application to any device [12, 13].

Remote provisioning means much smaller devices can be supported. The first products have already come to market, and we can expect to see many further launches in the future: it is now easier to extend mobile connectivity to devices such as tablets, smart watches, fitness bands, portable health systems, and various other devices.

eSIM is the only globally-backed remote SIM specification for consumer devices. This universal approach will grow the IoT by allowing manufacturers to build a new range of products for global deployment based on this common eSIM architecture.

Consumer benefits include:

- Simpler device setup without the need to insert or replace a SIM card;
- Devices that can operate independently of a tethered smartphone, with their own subscriptions;
- A range of new, enhanced mobile-connected devices.

The GSMA has developed the architectural and technical specifications needed to manage the eSIM. These definitions form the core of the service, while the underlying connectivity of the related wireless and cellular networks is based on respective standards-setting organizations such as the 3GPP. In addition, there are related test specifications, compliance specifications, and security-related materials for the eUICC Security Assurance Scheme, eUICC for consumer device protection profile, and eUICC PKI certificate policy. ETSI recognizes the eUICCs and iUICCs in their SSP standards, and the 3GPP will refer to them accordingly to include remote SIM provisioning aspects. As an example, as defined by ETSI, an iSSP is an integrated SSP confined in a dedicated subsystem within a System on Chip (SoC). The SoC is usually soldered in the terminal and so the SSP is an integral part of the terminal.

The latest versions of the GSMA eSIM architectural and technical specifications are the following:

- eSIM Architecture Specification SGP.21 V2.2, published 1 September 2017 [12].
- eSIM Technical Specification SGP.22 V2.2.2, published 5 June 2020 [13].

7.3.6 eSIM Architecture

One of the key elements of the eSIM architecture is the Local Profile Assistant (LPA). The LPA can be stored either in the device (LPAd) or in the eUICC (LPAe). Figures 7.8 and 7.9 summarize the components, roles, and interfaces associated with remote SIM provisioning and management of the eUICC for consumer devices depending on the LPA's storage:

- SM-DP+ is Subscription Manager Data Preparation. It is used by the operator to order profiles for specific eUICCs as well as other administrative functions.
- LPA is a functional element in the device or in the eUICC that provides the Local Profile Download (LPD), Local Discovery Services (LDS), and Local User Interface (LUI) features. When LPAs are located in the device, they are called LPAd, LPDd, LUId, and LDSd, respectively. When LPAs are located in the eUICC, they are called LPAe, LPDe, LUIe, and LDSe, respectively. Where LPA, LPD, LDS, or LUI are used, they apply to the element independent of its location in the device or in the eUICC.
- eUICC is embedded Universal Integrated Circuit Card.

- SM-DS is a Subscription Manager Discovery Server. It is responsible for providing addresses of one or more SM-DP+(s) to an LDS.
- CI is Certificate Issuer. It is an entity that is authorized to issue digital certificates.
- EUM is an eUICC Manufacturer.
- DLOA is Digital Letter Of Approval.
- DLOA Registrar is a role that stores DLOAs and provides an interface to enable the management system to retrieve them.
- Management System is any authorized system, e.g., an MNO backend system or an SM-DP+, interested in verifying the level of certification, evaluation, approval, qualification, or validation of a component (e.g., eUICC platform).

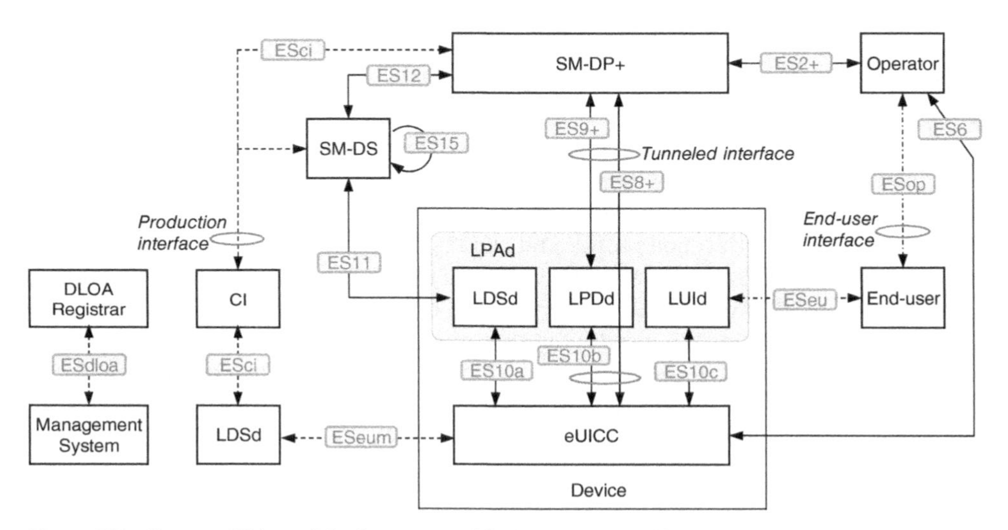

Figure 7.8 Remote SIM provisioning system, LPA in the device [12].

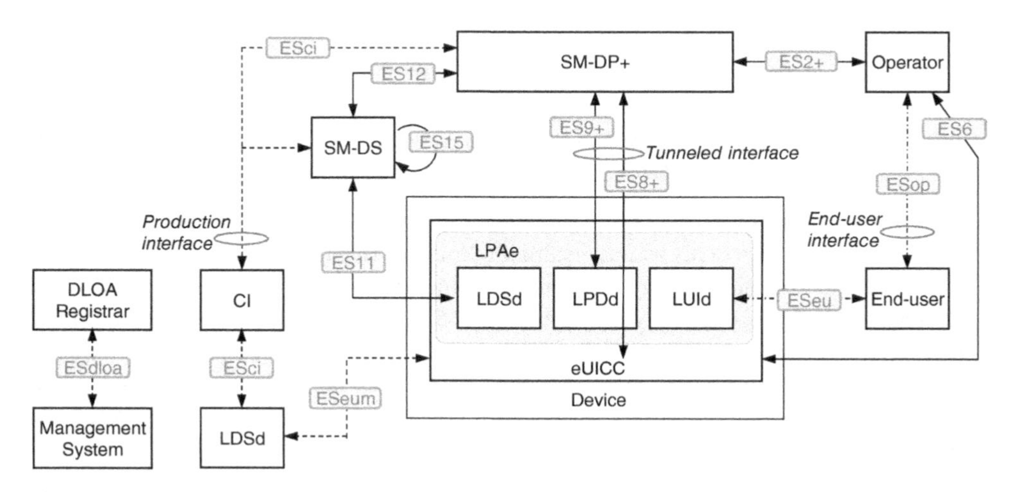

Figure 7.9 Remote SIM provisioning system, LPA in the eUICC [12].

In addition, a Root Subscription Manager Discovery Service (SM-DS) is a globally identified central access point for finding events from one or more SM-DP+ components.

7.3.7 Technical Solution

TS SGP.22 details the procedures for the eSIM. The key procedures are remote provisioning, local profile and eUICC management, device and eUICC initialization, notifications, and SM-DS-related actions. SGP.22 also details the data elements, functions, and interfaces of remote SIM provisioning [13].

The following sections summarize the key provisioning and management procedures.

7.3.7.1 Profile Download

The download initiation procedure consists of the following subprocesses: contract subscription process, download preparation process, contract finalization process, and optional subscription activation process. Profile download can happen in three ways:

- Activation code. In this option, the end-user enters an activation code to the LPAd manually or by scanning a QR code upon the device capabilities.
- SM-DS. In this option, the LPAd retrieves an SM-DP+ address and event ID from the SM-DS.
- Default SM-DP+. In this option, the LPAd retrieves the default SM-DP+ address from the eUICC.

SGP.22 details the steps of different cases, including errors and profile lifecycle, in its Section 3.1 [13].

7.3.7.2 Local Profile Management

Section 3.2 of GSM SGP.22 details the procedures when an end-user initiates the local profile management procedure using the LUI. As specified in GSMA SGP.21, user intent (which can be either simple or authenticated confirmation) is required for procedures directed to operational profiles. These procedures are the following:

- Enable profile. This procedure is used to enable a profile already upon download and installing it on an eUICC. The profile can be operational, test, or provisioning profile. For the operational profile procedure, the end-user can see a list of installed profiles within the eUICC to enable and disable them.
- Disable profile. This procedure disables an already downloaded, installed, and enabled profile on the eUICC.
- Delete profile. This procedure deletes a profile on the eUICC.
- List profiles. This procedure provides the LPAd with a means to list profiles and their states in a human readable format.
- Add profile. This procedure provides the end-user with a means to add a profile, which can then be enabled, disabled, and deleted via the respective procedures. Network connectivity is needed to add the profile. The download can be triggered via an activation code, or by retrieval of a pending profile download event from the SM-DS or the default SM-DP+.
- Set/edit nickname. This procedure adds and modifies a downloaded profile nickname associated to a profile.

Chapter 3.3 of GSMA SGP.22 details the local eUICC management procedures such as eUICC Identifier (EID) retrieval, eUICC memory reset, and SM-DP+ address setting [13].

7.3.7.3 Discovery Service

As stated in Ref. [12], Section 4.12, the role of the SM-DS is to provide mechanisms that allow an SM-DP+ to inform the LDS within any device that an SM-DP+ wishes to communicate with it about a pending event.

In a simple deployment, only the root SM-DS is configured on the eUICC, with a unique root SM-DS address. The LDS in the target device polls the root SMDS using the same logical location. When the root SM-DS has an event-ID for the target device, it responds with the SM-DP+ address.

In a deployment involving cascaded SM-DS components, the SM-DP+ sends an event registration to an alternative SM-DS, which may not be configured as the root SM-DS on the eUICC. This SM-DS cascades the event registration to the root SM-DS. The LDS of the target device polls the root SM-DS until it receives the address of the alternative SM-DS to request the event from there. The alternative SM-DS then returns the SM-DP+ address.

7.3.8 Security Certification of 5G SIM and Subscription Management

The SAS of the GSMA assesses the security of the UICC and eUICC suppliers and eSIM subscription management service providers [14]. The GSMA SAS has been divided into two schemes:

- SAS-UP refers to UICC Production. This scheme has been deployed since 2000. It means that the production sites and processes of UICC manufacturers undergo a thorough security audit. The compliant sites are granted a security accreditation for a 1-year period, which is extended to two additional years upon each successful renewal. SAS for UICC Production is described in GSMA documents FS.04 SAS-UP Standard V8 and FS.05 SAS-UP Methodology V7.
- SAS-SM refers to Subscription Management. This scheme has been adapted to create confidence of the remote provisioning for eSIM products, extending the SAS-UP model. SAS for SM is described in the GSMA documents on the SAS for Subscription Management, FS.08 (SAS-SM Standard) and FS.09 (SAS-SM Methodology).

In addition to the above-mentioned documentation, the GSMA has also produced a common document applicable to both schemes under the title FS.17 SAS Consolidated Security Requirements.

7.4 Other Security Aspects

7.4.1 Security Certification of Data Centers

The current telecom and overall data service business trend relies increasingly on virtualized data center networking and software-defined networks. In the professional environment, certificates are applied for both the personnel operating the data centers as well as the premises; the latter is becoming increasingly critical, as the data will also include more sensitive information. 5G architecture includes critical elements such as data repository and authentication that are housed by default within the MNO cloud infrastructure; nevertheless, there are no technical restrictions to rely on functional elements managed by third service providers for such tasks.

The personnel of data centers need to master management of cloud data considering security aspects. The certification of data center personnel thus typically requires knowledge of networking technologies, including virtualization and cloud technologies ensuring adequate safety aspects. There are many certificates designed for experts in this domain.

There exists a variety of certificates for the data centers, too. For the security of data centers, some of the commonly recognized entities managing the certificates include HIPAA, PCI, Uptime Institute, and Colocation America.

The certified data center brings additional confidence for parties relying on the center. For critical functions, including those that 5G requires, there must be a guarantee for security. It is not solely a matter of business partners trusting each other but the statutory certifications are required by law. An example of this is the HIPAA and requires auditing prior to operations. Standard certifications, on the other hand, are a set of requirements of certain authority entities dictating criteria for performance operations. Unlike stationary certification, these tier standards are not required by law; nevertheless, they do have an important role in data center business. The standard certified refers to other types of principles as the certification is based on data center owner compliance plans considering expenses and specific needs of potential customers. This type of certification process is more a matter of reputation, which may be challenging to prove if the data center is about to start operations.

Certified Public Accountants (CPA) has set up a Statement on Standards for Attestation Engagements 16 (SSAE 16) certificate. It is in practice a set of guidelines for the level of controls at a service organization. The guidelines are applied to safe storage of data in the servers, and for the secure transfer of the data within and between data centers and external entities.

Another example of security certificates is the Service Organization Control (SOC) reporting framework. This contains three reporting standards, which are SOC 1, SOC 2, and SOC 3. SOC refers to a reporting standard for a business's financial reports.

7.4.2 GSMA Security Controls

GSMA document FS.31 presents baseline security controls [15]. The latest version, V2.0 published in February 2020, presents a set of practical aspects of business and technological controls, including controls for UE and ME, UICC and eUICC management, IoT, RAN, roaming and interconnect, core network management and network operations, and security operations. The document serves as a guideline for a specific set of security controls that the mobile telecommunications industry should consider deploying. It is important to note that the presented security controls do not override local regulations or legislation in any territory, but merely supplement security levels within the mobile telecommunications industry.

References

1 3GPP, "TS 33.501 V16.3.0, Security Architecture and Procedures for 5G System, Release 16," 3GPP, July 2020.
2 3GPP, "TS 33.401 V16.3.0, 3GPP System Architecture Evolution (SAE), Security Architecture," 3GPP, July 2020.
3 SIMalliance, Technical Whitepaper, 5G Security – Making the Right Choice to Match your Needs, October 2016.
4 3GPP, "TS 23.5-01 V16.5.1, System Architecture for the 5G System, Stage 2, Release 16," 3GPP, August 2020.
5 ETSI, "TS 133.501, V. 15.1.0, Chapter 5.2.5: Subscriber Privacy," ETSI, 2018.
6 3GPP, "TS 23.502 V16.5.1, Procedures for the 5G System, Stage 2, Release 16," 3GPP, August 2020.
7 Rupprecht, D., Kohls, K., and Holz, T., "Call Me Maybe: Eavesdropping Encrypted LTE Calls with ReVoLTE," Ruhr University Bochum & New York University Abu Dhabi. [Online]. Available: https://revolte-attack.net. [Accessed 7 September 2020].
8 3GPP, "TS 38.331 V16.1.0 (2020-07), NR Radio Resource Control Protocol Specification, Release 16," 3GPP, July 2020.
9 3GPP, "TS 31.102 V16.4.0, Characteristics of the Universal Subscriber Identity Module (USIM) Application, Release 16," 3GPP, June 2020.
10 GSMA, "eSIM Specification," GSMA, 2020. [Online]. Available: https://www.gsma.com/esim/esim-specification. [Accessed 30 September 2020].
11 GSMA, "eSIM Whitepaper: The What and How of Remote SIM Provisioning," GSMA, March 2018. [Online]. Available: https://www.gsma.com/esim/wp-content/uploads/2018/12/esim-whitepaper.pdf. [Accessed 30 September 2020].
12 GSMA, "RSP Architecture, Version 2.2," GSMA, September 2017.
13 GSMA, "SGP.22 RSP Technical Specification, Version 2.2.2," GSMA, June 2020.
14 GSMA, "Security Accreditation Scheme (SAS)," GSMA. [Online]. Available: https://www.gsma.com/security/security-accreditation-scheme. [Accessed 23 September 2020].
15 GSMA, "Baseline Security Controls Version 2.0," GSMA, 5 February 2020. [Online]. Available: https://www.gsma.com/security/wp-content/uploads/2020/02/FS.31-v2.0.pdf. [Accessed 25 September 2020].

8

5G Network Planning and Optimization

8.1 Network Design Principles

8.1.1 Introduction

One of the key tasks of the Mobile Network Operator (MNO) is to balance deployment costs and service level. If the operator compromises coverage, capacity, or quality, then capital and operating expenditure are lower. On the other hand, the churn increases accordingly due to unhappy customers.

On the other hand, an overdimensioned network results in an excess of capital expenditure without adequate return on investments (RoI). In an optimally dimensioned network, there is a sufficient level of capacity and radio cell coverage to comply with the quality expectations of the subscribers.

5G brings a variety of new aspects that also influence network planning. Some of the key points for 5G network design compared to the previous generations include the following:

- Artificial intelligence for more efficient network functionality and performance;
- Device-to-Device (D2D) communications to optimize latency and increase reliability of connectivity in, e.g., the Vehicle-to-Everything (V2X) environment;
- Enhanced energy utilization to provide benefits for, e.g., remotely located Internet of Things (IoT) devices;
- Adequate evolved layer 1 and 2 techniques for high frequencies, low latency, and massive IoT (mIoT) capacity;
- Evolved location-based services;
- Evolved radio interface multiple access scheme;
- Evolved self-optimizing network functions;
- Evolved transmission via integrated backhauling and increased use of optical switching;
- Exposure of network functions and resources to third parties to create more open ecosystems;
- Higher radio frequencies and wider bandwidths via new carrier aggregation options to provide more capacity and data speeds, combined with small cell environment;
- Interworking that also includes non-3GPP access networks such as Wi-Fi, including security management;

5G Second Phase Explained: The 3GPP Release 16 Enhancements, First Edition. Jyrki T.J. Penttinen.

- Network slicing to provide verticals with optimized and personalized performance;
- Service-Based Architecture (SBA), in a form of Network Functions Virtualization (NFV) and Software Defined Networking (SDN), to minimize network layers and optimize network resource utilization;
- Control and User Plane Separation (CUPS).

5G is a result of new and developed solutions that enhance radio, transmission, and core network performance, and provide flexible deployment options. Radio Access Network (RAN) development also brings cloud-based network models and a variety of advanced solutions such as direct communication between terminals and more variable heterogeneous network deployment models. The possibility of deploying 5G in a highly flexible and modular manner is a result of virtualization. One of the most important aspects of the 5G system is network slicing, which provides multiple performance figures for different user types by relying on a single logical architecture while the network resources are orchestrated on demand per slice.

As stated in Ref. [1], 5G specifications introduce low, mid-, and high bands, and support both Long Term Evolution (LTE) and 5G New Radio (NR) with their respective dual connectivity for simultaneous connectivity via LTE and NR, which results in a gradual increase in performance.

The 5G system is based on Orthogonal Frequency Division Multiple Access (OFDMA) in both uplink (UL) and downlink (DL), and includes optional Single Carrier Frequency Division Multiple Access (SC-FDMA) for UL. Release 15 includes an initial base for Ultra-Reliable Low Latency Communications (URLLC) by physical layer frame structure and support of its numerology, whereas Release 16 has enhanced URLLC further.

Other enhancements or additions of Release 15 include support for massive Multiple In, Multiple Out (MIMO) and beamforming (5G beamforming is applicable for data, control, as well as broadcast channels), support for Frequency Division Duplex (FDD) and Time Division Duplex (TDD), and scalable Orthogonal Frequency Division Multiplexing (OFDM) subcarrier spacing to adapt to many radio channel types up to 400 MHz per component carrier. Release 15 also supports Carrier Aggregation (CA) modes for up to 16 NR carriers, and aggregation bandwidths of up to 1 GHz.

Furthermore, 5G supports Low-Density Parity Code (LDPC)-based error correction, enhancing LTE's turbo codes at high data rates, whereas the 5G control channels use polar codes. 5G specifications include support for Cloud RAN (C-RAN) and its split model for the Radio Link Control (RLC) and packet data convergence protocol layers.

Release 16 adds further definitions to URLLC, and brings unlicensed spectrum operation below 7 GHz, which can be based on Licensed Assisted Access (LAA). Release 16 also introduces for the first time integrated access and backhaul. Furthermore, Release 16 defines native NR-based Cellular V2X (C-V2X) and radio access for non-terrestrial networks such as satellites. Release 16 also supports radio bands above 52.6 GHz, and has dual-carrier, CA, and mobility enhancements, as well as User Equipment (UE) power consumption enhancements, and brings non-orthogonal multiple access.

All of the above-mentioned aspects also have an impact on network planning and optimization. The following sections present some of the key aspects that need to be taken into account in the design.

8.1.2 Base Architectural Models

The initial phase of the 5G network typically relies on the already deployed 4G infrastructure, if the operator has it. In this scenario, Figure 8.1 outlines the hybrid 4G and 5G architecture.

Based on the technical means to interconnect 4G and 5G, 3GPP Release 15 specifications provide MNOs with a variety of options to deploy 5G in a gradual manner. Following the standard terminology, a full and solely 5G-enabled network refers to Option 2 as depicted in Figure 8.2.

The 5G-specific base station is called gNB. It communicates with the 5G UE, which is the user's mobile phone or other 5G device such as a sensor.

Prior to Option 2 deployment, operators may construct 5G capacity and coverage by relying on the previous 4G infrastructure as described in the NSA connectivity scenarios. The 3GPP has designed many possibilities from which Options 3, 4, and 7 can be assumed to be the most popular ones.

Option 1: This option represents the SA 4G network, comprising solely LTE eNB and Evolved Packet Core (EPC) components.

Option 2: This option represents the SA 5G network, which is the ultimate goal of 5G operators. It consists solely of high-performing 5G radio and core systems.

Option 3: 5G UE relies on both 4G and 5G radio networks, and solely on 4G core. 4G core is, in fact, not even aware of the new 5G radio signaling interface in this scenario, and the old LTE base stations (evolved NodeB, eNB) serve as anchors to interconnect the 5G gNB elements and 4G core. The user data can be delivered between the core and only LTE

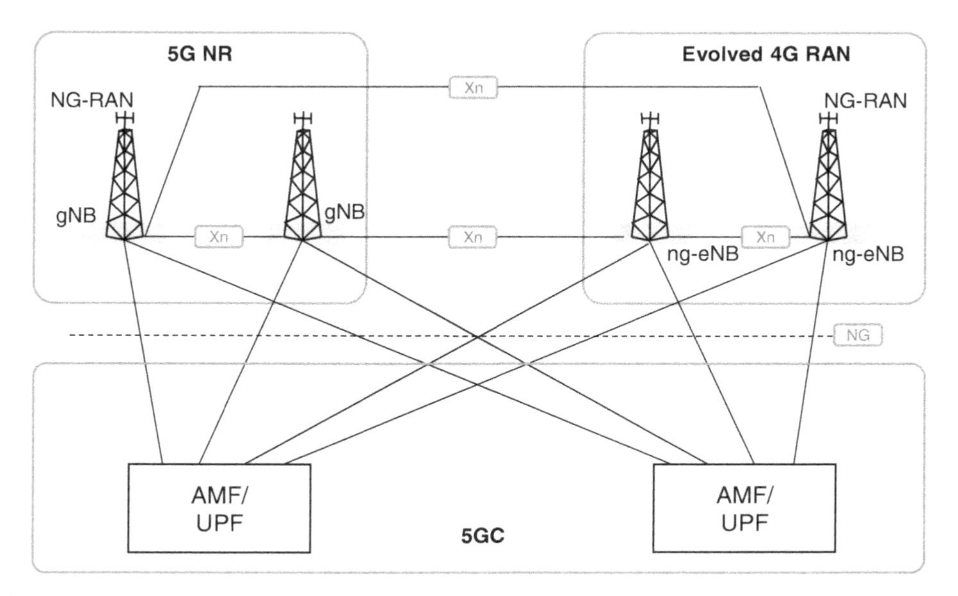

Figure 8.1 Release 15 5G architecture. Radio access can take place via next generation NodeB (gNB), which is the native 5G NodeB (base station), and ng-eNB, which is the evolved 4G NodeB (E-UTRAN base station).

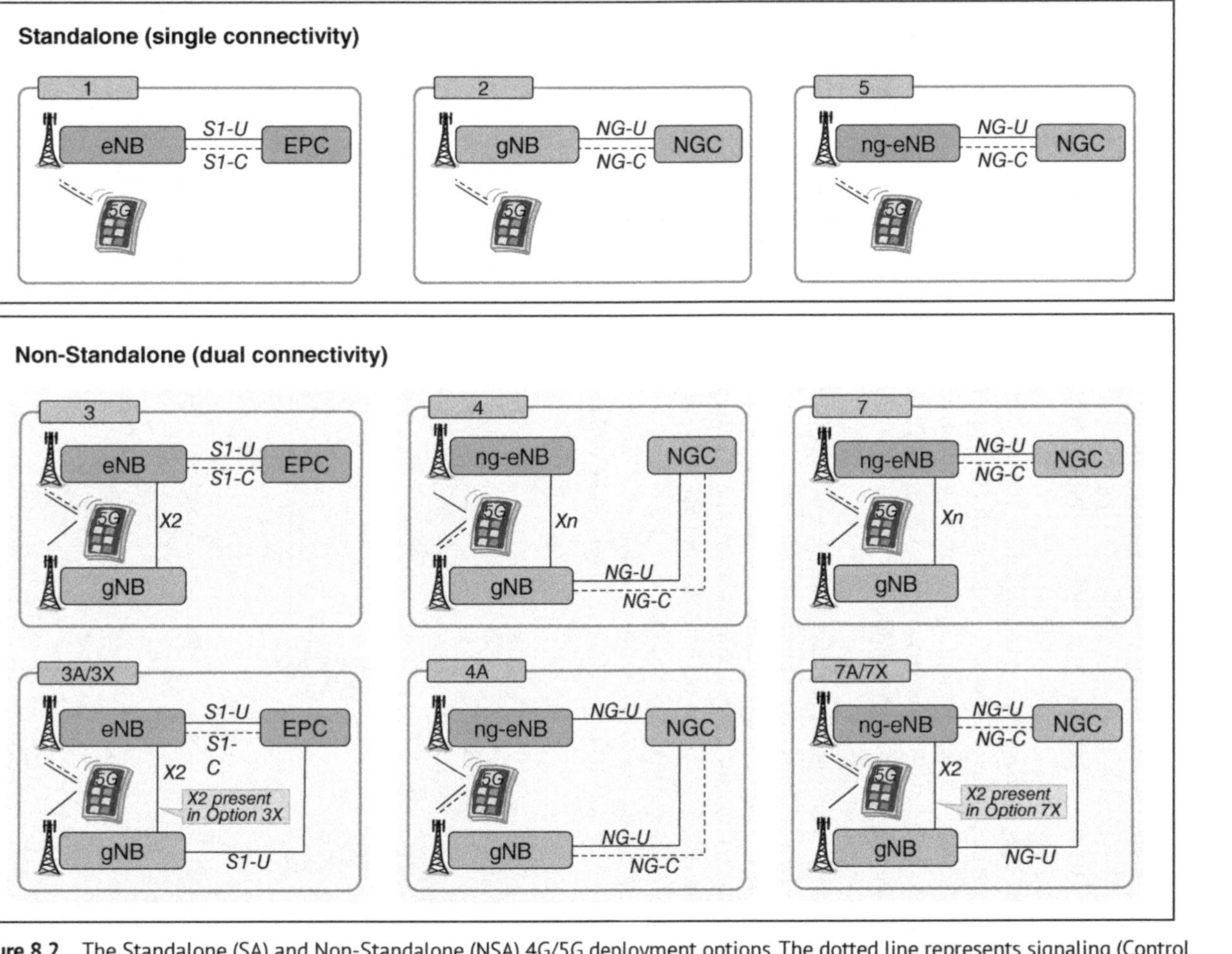

Figure 8.2 The Standalone (SA) and Non-Standalone (NSA) 4G/5G deployment options. The dotted line represents signaling (Control Plane (CP)) and the solid line means data connectivity (User Plane (UP)). Please note the notation: eLTE eNB refers to the ng-eNB.

eNB, or via both eNB and 5G gNB. In Option 3A, there is no *X2* interface. This option also has a variant called 3X, which is marked in Figure 8.2 with the *X2* interface. In Option 3X, traffic flow is converged at the 5G gNB and divided from there to the 4G eNB, while the 5G NR takes care of the majority of the traffic. Whenever 5G NR has lower coverage performance, the traffic split mechanism can offload more traffic to the eNB. This facilitates the optimized bandwidth in the *X2* interface. Option 3X is thus a highly practical candidate for initial 5G deployments.

Option 4: The evolution of 5G can lead into an adjusted, intermediate step of the NSA, NR-assisted architecture that relies on the 5G Next Generation Core (NGC). In this option, the 5G gNB acts now as the anchor for connecting the data directly and from the further evolved 4G eNB elements (eLTE eNB) between the 5G Core (5GC).

Option 5: This option uses 5GC and LTE ng-eNB access, with the enhanced interface between them to support 5G functions such as network slicing and network virtualization still relying on previous 4G radio access and respective performance.

Option 6: This option does not exist in the current specifications.

Option 7: Complexity-wise, this scenario is between Options 3 and 7. It is similar to Option 3, but the 4G eNB elements are upgraded to better cope with the delivery of 5G performance; hence, the updated term of the eLTE eNB.

In Release 15, as interpreted from, e.g., 3GPP TS 23.501 and TS 38.501, the 3GPP has thus defined various architectural options that allow the UE to connect to the network relying on 4G LTE, evolved 4G eLTE, and 5G NR access that are interconnected to 4G EPC or 5GC networks [2, 3].

As described in Chapter 5, LTE/eLTE and NR can also use Dual Connectivity (DC) by deploying different combinations of the master and secondary nodes.

Table 8.1 summarizes the key aspects of the respective deployment options [4].

These examples show that there are many ways that MNOs may deploy their 5G networks in the initial and more mature phases. In the practical environment, MNOs may also consider the following deployment scenarios as interpreted from the *5G Deployment Considerations* document of Ericsson [5].

Co-located, same frequencies: NSA 5G network, 4G and 5G radio using low-band and/or mid-band set in sub-6 GHz. The coverage areas of 4G and 5G are comparable ensuring seamless user experience whenever the terminal performs handover between 4G and

Table 8.1 Summary of 4G and 5G deployment options as defined in 3GPP TS 23.501 and TS 38.401 [2, 3].

Option	Core	Master radio	Secondary radio
1: SA LTE	4G EPC	4G LTE	N/A
2: SA NR	5GC	5G NR	N/A
3: EN-DC	4G EPC	4G LTE	5G NR
4: NE-DC	5GC	5G NR	4G ng-eLTE
5: eLTE	5GC	4G ng-eLTE	N/A
7: NGEN-DC	5GC	4G ng-eLTE	5G NR

5G. In this scenario, 5G provides high capacity in dense city centers and urban areas, especially for the Evolved Mobile Broadband (eMBB) and fixed wireless access users, whereas 4G can be used as a "gap filler" for the rest of the areas to serve the less demanding applications and IoT devices. In this scenario, the radio equipment of 4G and 5G may be co-located into the same sites, making it straightforward to reuse antenna towers, transmission, power supplies, and site shelters.

Partially co-located, different frequencies: In this scenario, 4G may provide a basic coverage layer on the sub-6GHz bands, while NSA 5G serves customers on the mmWave bands. Customers of the latter case benefit from the highest capacity and data rates while enjoying the lowest possible latency. In this scenario, the physical 4G sites can be reused as such for 5G, but the denser 5G gNB network also requires additional sites. This scenario is also feasible for enhancing the already existing 4G network by co-locating 4G radio equipment to the new 5G sites.

5G focused: In this scenario, SA 5G, as defined in Option 2, operates on all the possible bands of the operator, including low, mid-, and high bands. The low bands provide large coverage for the basic 5G services, while the mid-band is useful especially in urban areas, balancing radio coverage and data speeds. The mmWave provides users with the highest data speeds for the most demanding data services, especially in the dense urban environment.

The deployment of 5G will be gradual, and as network coverage enhances over the years, there will be an increasing number of diverse devices capable of taking full advantage of the NR of 5G. The year 2019 was the time for NSA deployments, until Release 16 and gradual SA networks started facilitating the full potential of 5G.

The high-level 5G scenarios can thus be categorized into NSA and SA deployments, which are described in more detailed in Annex J of 3GPP Technical Report (TR) 23.799. Furthermore, 3GPP TR 38.801 presents practical deployment scenarios summarized in the following.

Non-centralized deployment: This scenario refers to a set of 5G gNB elements equipped with a full protocol stack. The scenario is suitable especially in macro cell and indoor hotspot deployments. The gNB elements can be connected with other gNB elements or LTE eNB and eLTE eNB elements.

Co-sited with E-UTRA: In this scenario, 5G NR functionality is co-sited with 4G E-UTRA. This deployment scenario is suitable for many cell types, including the urban macro, which provides the largest coverage. Load balancing can be used in this scenario to optimize the 4G and 5G radio access spectrum resources.

Centralized deployment: 5G NR supports centralization of the upper layers of the NR radio protocols, and various gNB elements can be attached to the centralized unit via a transport network. The non-co-located and co-located deployment scenarios with 4G E-UTRA are applicable in this scenario.

Shared RAN deployment: This scenario refers to an environment where various hosted core operators are present.

Heterogeneous deployment: This scenario refers to the heterogeneous RAN service areas. It can be an optimal solution for indoor deployments to ensure a fluent interworking between shared RAN and non-shared RAN.

As stated in Ref. [6], Release 15 was introduced to accelerate initial availability of specifications for the first deployments. One reason for this "early drop" was the intention to realign the ecosystem as there were some proprietary 5G networks already deployed prior to the publishing of the technical specification set of the 3GPP.

The philosophy of 5G is remarkably different from previous generations due to virtualization of the network functions. This results in new, more optimized ways of utilizing the resources in a dynamic manner when needed via network slices. Network slice refers to a logical network that provides specific network capabilities and network characteristics. The MNO can have a number of these network slices that can be configured, each being optimized to certain use cases and environments such as broadband communications, critical communications, and mIoT communications. Equally, the essential parameters of each slice can be adjusted separately, including the level of security.

As soon as the first 5G services are deployed, there will only be a small number of 5G-capable terminals, and the NR coverage will be limited. For this phase, the solution is to deploy 5G radio coverage gradually in such a way that the NR elements are connected directly to the LTE radio elements as depicted in Figure 8.3. Another option is to provide both interfaces NR-LTE and NR-EPC for relying on the legacy core network of 4G either via LTE radio or by connecting 5G NR directly to 4G EPC.

More information on this topic can be found at 3GPP TS 38.300 (5G new radio), and the GSMA white paper "Road to 5G: Introduction and Migration" [7].

8.1.3 3GPP Split Options

As an additional solution in the interim phase of 4G EPC support for initial 5G deployments, the 4G core network's Serving Gateway (S-GW) and Packet Data Network Gateway (P-GW) functions have been broken down into two parts: UP and CP.

The first set of technical specifications of 5G includes the concept for NSA architecture, which will be the option for DC. The initial DC was included already in LTE 3GPP Release 12. Where CA makes it possible to serve different carriers by the same eNB, DC

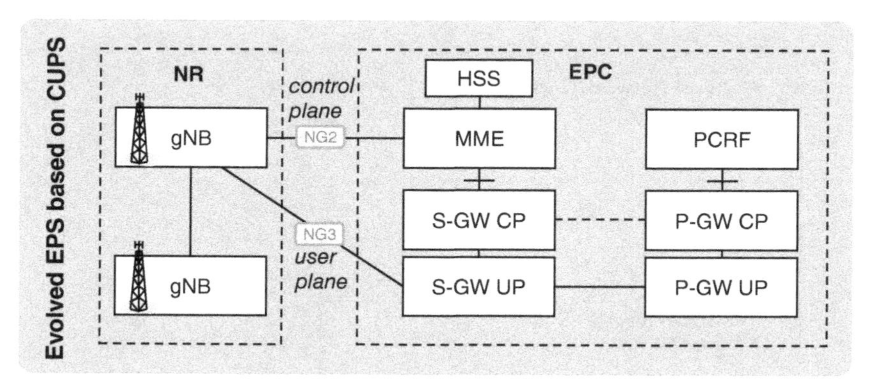

Figure 8.3 4G EPC supports 5G connectivity by dividing the S-GW and P-GW functions into UP and CP.

serves the carriers by different backhauls, which refers to separate eNB elements, or combined eNB and gNB elements.

Details of the selected options are defined in 3GPP TS 38.300 (Stage 2 radio architecture). ng-eNB here refers to a node providing E-UTRA UP and CP protocol terminations towards the UE and connected via the NG interface to the 5GC. gNB, in turn, is a node providing NR UP and CP protocol terminations towards the UE and connected via the NG interface to the 5GC.

Release 15 presents the initial split Option 2 for the 5G system, while Release 16 brought the extended set of Options 1–8. In practice, as Release 16 facilitates the deployment of the rest of the network services and functions, this phase is a logical moment for the transition from intermediate options to consideration of full-scale 5G options.

3GPP TS 38.401 V16.2.0, Annex A, includes informative deployment scenarios of gNB and en-gNB as a basis for centralized and distributed node deployment considerations. Figure 8.4 depicts logical nodes Central Unit (CU), which can be further divided into CP (CU-C), UP (CU-U), and Distributed Unit (DU). These units are internal to a logical gNB and en-gNB. Figure 8.4 also shows the *NG* and *Xn* interface protocol terminations.

The split options of the 5G radio network are thus more diverse compared to 4G as detailed later in this chapter.

8.1.4 Deployment Scenarios of ETSI

ETSI TR 138 913, V14.2.0 (2017–05) presents a study of scenarios and requirements for 5G [8]. These deployment scenarios represent eMBB, Massive Machine Type Communications (mMTC), and URLLC use cases, as well as deployment scenarios related to Enhanced Vehicle-to-Everything (eV2X) services. The TR proposes a high-level description of a variety of deployment scenarios, including carrier frequency, aggregated system bandwidth,

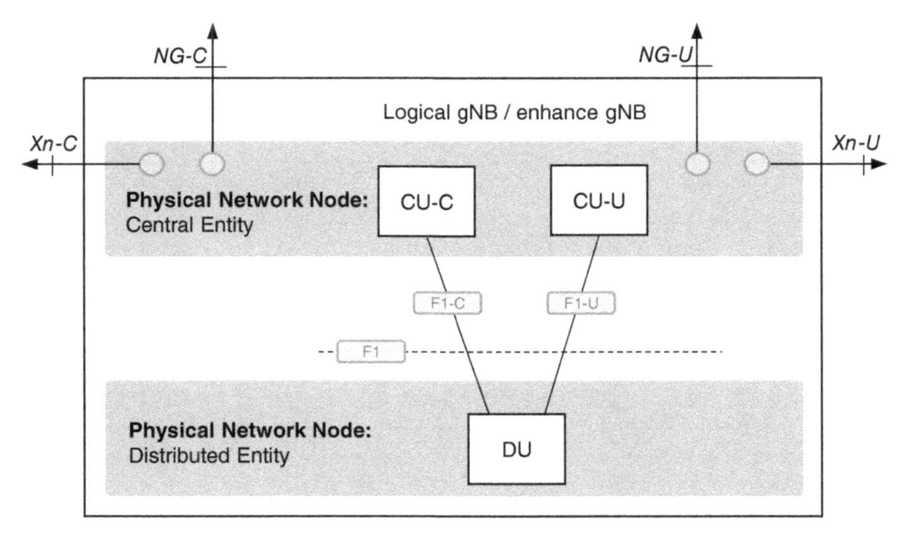

Figure 8.4 Example of deployment of logical gNB/en-gNB as interpreted from Annex A of TS 38.401 [3]. The central and distributed entities are physical network nodes.

network layout, Inter-Site Distance (ISD), base station and UE antenna elements, UE distribution and speed, and service profile.

The outcome of the TR is a list of requirements that were considered in the 5G specification. These considerations are valid for both Release 15 and Release 16 network deployments.

8.2 5G Radio Network Planning

8.2.1 Overview

This section looks at Phase 2 radio network planning aspects, including the radio link budget (RLB) of all known 5G Phase 1 and Phase 2 frequency bands.

Radio network planning balances capacity, coverage, and quality. The core and transport networks, in turn, need to support the generated traffic from the radio network to avoid bottlenecks, while the commercial aim is to optimize the costs.

Radio network dimensioning is based on the RLB and radio wave propagation prediction models. They help operators to estimate the maximum feasible distance from the base station up to where the UE can still be served in different topological environments.

Radio waves attenuate more in dense urban areas due to obstacles such as high buildings. Coverage is largest in open areas as attenuation is lowest for the Line-Of-Sight (LOS) scenarios. Also, higher frequencies attenuate more than the lower ones. The lower bands are thus used typically in large rural and suburban areas; on the other hand, vegetation and trees may decrease their radio propagation range. Radio network planning is thus a complex task that requires theoretical and practical knowledge about radio waves, field measurements, tools, and simulators.

5G radio wave propagation prediction models are based on already existing and new methods. For frequencies above 6 GHz, which is a new area compared to previous generations, the impact of obstructing materials is rather significant. The 60 GHz band is a special one as the oxygen molecules cause an additional 20 dB attenuation peak to radio wave propagation, which further reduces coverage.

The low band refers to sub-1 GHz, mid-band to 1–6 GHz, and high band to frequencies above 6 GHz. In the initial phase of 5G, the high bands on mmWave areas of 24 , 28, and 38 GHz are assumed to be popular in the United States, whereas Europe and China use 26 GHz. There are also many frequency range candidates on lower bands such as 600–900 MHz, 1.5, 2.1, 2.3, and 2.6 GHz to be evaluated for 5G deployments. These bands are especially useful in applications requiring less capacity such as IoT communications as they provide a rather large coverage.

Thanks to the radio wave propagation characteristics, the mid-bands of 3.30–4.20 and 4.40–4.99 GHz are especially suitable for the initial phase of 5G. They provide a feasible compromise for capacity and coverage, and facilitate the expedited 5G deployment schedules.

The same old principles of Radio Frequency (RF) propagation apply to 5G; the higher the frequency is, the smaller the coverage area. The basic rule suggests that when a frequency doubles, the received power level lowers by 3 dB – which equals a 50% reduction of the original power. Thus, thanks to their wider bandwidth, frequencies above 6 GHz are mostly adequate for delivering high radio capacity in small cell environments. These cells are oftentimes limited to very short distances outdoors, or within single floors indoors, which makes them especially suitable for a dense city environment.

Along with the increased complexity of 5G and high dependency on traffic types, geographical topology, and other factors, its RLB is typically based on radio network planning simulators and digital cluster maps. Nevertheless, by applying adequate propagation models, it is possible to make a rough estimate of the expected cell sizes with only basic tools. This type of exercise may be feasible in the nominal planning phase to estimate the rough number of base stations in the planned area and the respective high-level investment.

The RLB takes into account the key aspects of the communication link, such as transmitter power, antenna cable loss, and antenna gain for each beam (which can be highly dynamic in 5G). In the receiving side, sensitivity and noise figure are some of the parameters. It also takes into account fading variations and indoor attenuation in radio reception.

The ultimate goal of the RLB is to dimension the expected cell size (radius) so that the planned services, such as voice calls and data, comply with the designed quality criteria. This happens by balancing both UL (transmission from user device to base station) and DL (the opposite direction) so that the desired services, such as two-way voice call, can be used adequately.

As depicted in Figure 8.5, the closer the user is located to the base station, the higher the 5G data speed. The higher received power helps to lower the data error rate. The final bit rate per device depends on the modulation scheme and the number of the devices communicating simultaneously in the cell area. To prioritize and balance between different services and users, 5G has an advanced Quality of Service (QoS) classification.

Correct dimensioning of the RLB is of utmost importance for MNOs as it indicates how dense the network will be – which, in turn, dictates the initial Capital Expenditure (CAPEX) and longer-term Operating Expenditure (OPEX) of the network. A well-optimized radio network can ensure adequate RoI and sustainable business, while poorly planned radio networks may waste energy, money, and capacity.

As an example, it is important to balance the power levels of the useful and interfering transmissions instead of increasing the power levels in every site. A skillful radio engineer who knows by heart the theories and practices of radio propagation, network planning, optimization, and measurement techniques is one of the most invaluable assets of an MNO or cooperating party planning 5G networks.

Optimized radio network may save lots of money by reducing unnecessary transmitter power as it decreases interference. Figure 8.5 presents the principle of the interference sources in 5G. Interference may be caused by nearby UE to the base stations, or the nearby base station can generate interference to the UE. 5G OFDM is an adaptive multi-subcarrier technique, which tolerates well such interferences, adjusting accordingly the data throughput.

The radio communication links are highly dynamic. The radio network must be optimized constantly based on subscription forecasts, field measurements, and automatized network analytics. As 5G may be based on a significant number of small cells operating in mmWave bands, especially in dense city areas, one option is to deploy radio elements on light poles that are connected to the core network via fiber optics or radio links.

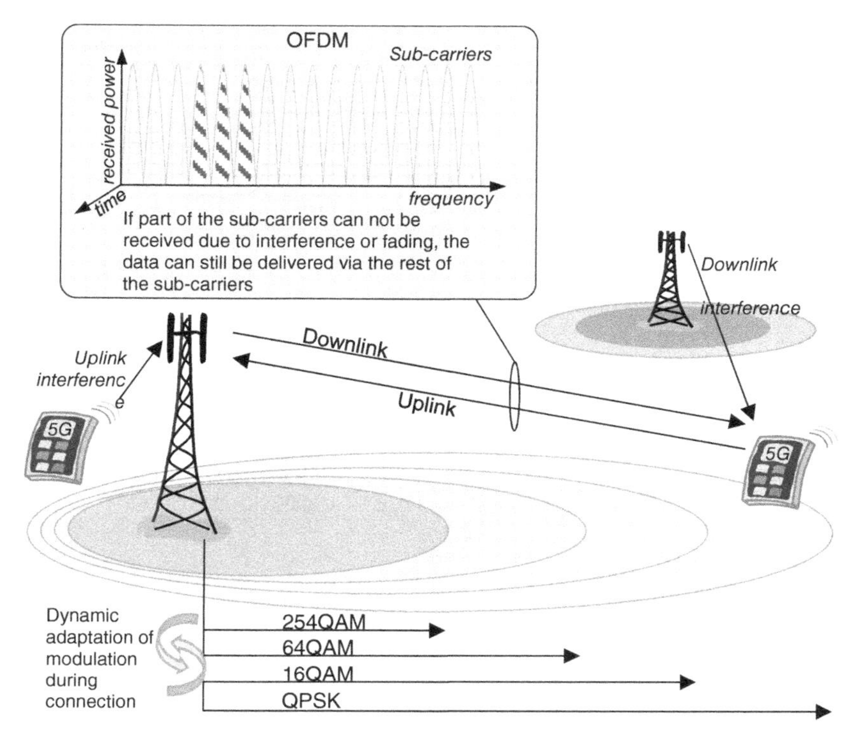

Figure 8.5 The principle of the 5G RLB. The maximum coverage area depends on the modulation scheme, among other factors. 254QAM provides 5G users with the lowest coverage but the highest data speeds.

Evolution of traditional site deployment has lowered the importance of the original, coaxial antenna cable-based installations of the base station site equipment and the respective antennas. The respective cable losses can be largely reduced by deploying a Remote Radio Unit (RRU) installed physically at the base station site, or further away from it by relying on remote cloud as depicted in Figure 8.6. The RRU contains the RF functions, Analogue-to-Digital (A/D) and Digital-to-Analogue (D/A) conversion, up/down converters, operation and management processing capabilities, and a standardized optical interface that connects the RRU and the rest of the base station.

8.2.2 Radio Channel Modeling

For an MNO to plan radio coverage and respective capacity for the 5G network, radio channel modeling continues to be one of the most essential tasks. There are a variety of models developed and under research, and oftentimes there may be set of different models instead of only a uniform one due to the differences in geographical topologies at the global level, and even within a single country. As one example of this variety of strategies, the model might be based on highly accurate three-dimensional map data of the environment with ray tracing principles in the densest city areas, whereas accurate coverage area prediction

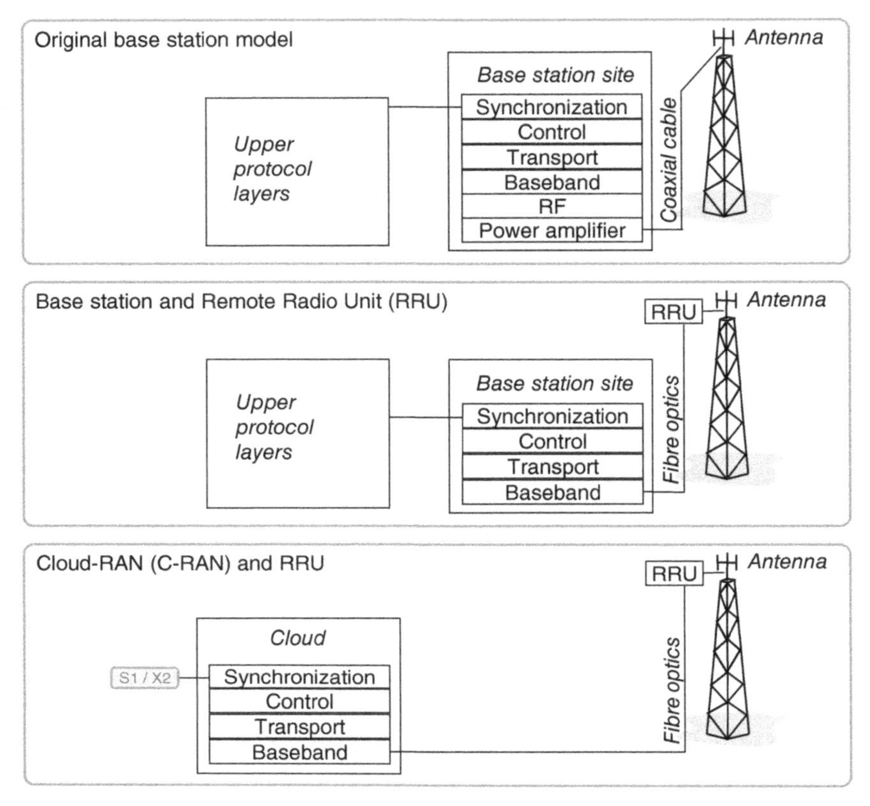

Figure 8.6 The evolution of traditional site deployment relying on a coaxial antenna cable towards the remote radio head in local and remote locations.

is not so essential in the most rural and remote areas. It is a matter of balancing the RoI as the expense of the most accurate map data is superior compared to the basic raster data, which is sufficient for remote locations.

It can be generalized that radio propagation models have been rather established for the practical needs of earlier mobile generations up to 4G. As 5G includes considerably higher frequency bands and may utilize much wider bandwidths and novelty modulation schemes, there have been several research projects investigating suitable models for the new era. Some examples of the research include studies carried out by the 3GPP and ITU-R, as representatives of standardization and regulation. There are also models provided by METIS, mmMagic, MiWEBA, COSTIC1004, IRACON, and the 5G mmWave Channel Model Alliance. One of the recent references investigating these models is found in Ref. [9].

The number of 5G RF bands is expected to be much higher than in previous mobile generations, including multiple mmWave bands, so previous models need to be revised. Massive MIMO antenna technology and hybrid beamforming are important methods in the 5G era to achieve the highest data speeds aimed at a very large number of users within dense areas. The respective array's gain challenges the radio transmission loss of the higher-frequency bands [10].

For commercial products, not all the antenna elements may be supported by separate RF chains. In practice, the arrays may be connected to a Baseband Unit (BBU) by only a small number of RF chains while antenna elements may be divided into subarrays [11]. Thus, each subarray may be composed of several predefined beam shapes. It is expected that the main directions of the beams will cover an angular sector of interest.

3GPP TR 25.996, V14.0.0, contains relevant information as a basis for MIMO radio dimensioning. It is a useful resource for investigating more thoroughly the modeling and impacts of key parameters such as radio path arrival angles to the received power level for suburban macro cell, urban macro cell, and urban micro cell topologies.

8.2.3 5G Radio Link Budget Considerations

The base of the RLB, or Power Link Budget (PLB), is the same as in previous systems; path loss is estimated based on the key parameters, considering the set of gains and losses.

The basis for the coverage area estimate is thus the offered capacity in terms of bit rate, and the respective achievable coverage area. The higher the data rate, based on the adaptive code rate of the OFDM principle, the smaller the coverage area, which means that the highest data rates are achieved close to the base station, while the lowest bit rates are available in the edge area of the cell coverage.

Additional aspects for 5G are a result of novelty techniques such as higher-order MIMO antennas and respective gain per single path. This is a highly dynamic environment as the number of simultaneous users, including a massive set of IoT devices, makes the available capacity fluctuate accordingly.

The goal of the initial phase of 5G radio network planning is to obtain a rough estimate of coverage and capacity within the planning area. This can be done by using RLB. This serves as a simple tool for estimating the achievable path loss values per planned environment between the transmitter and receiver in UL and DL. One of the benefits of this nominal planning phase is to have a high-level estimate of the number of base stations needed to serve the planned area. The more detailed estimate is obtained posteriorly by utilizing a radio network planning tool and applying more accurate models and detailed digital maps. Both nominal and detailed radio network planning are essential phases as the radio network may have a major impact on the expenses of the deployment, and optimization of the base station locations, utilized power levels, antenna heights, and other practical aspects means that the RoI can vary considerably.

In practical network planning, the predesigned base station locations are selected as candidates within certain preferred search areas. This plan may not always be realistic as "site hunting" may be sometimes highly challenging due to restrictions. As an example, the construction of towers or even installation of pure antenna elements on walls may be forbidden in certain environments such as historical city centers. Thus, nominal and detailed plans before the actual deployment may change.

The RLB thus gives a rough estimate of the expected usable cell radius, which is practical especially for the nominal radio network coverage planning phase. For detailed planning, radio network planning tools are typically utilized in such a way that the propagation models and other assumptions are adjusted per cluster type based on the practical field tests.

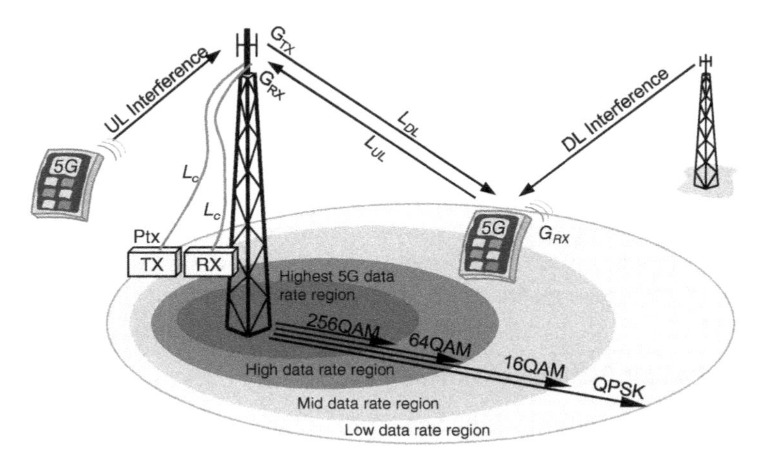

Figure 8.7 Principle of the 5G RLB parameters.

Figure 8.7 shows the principle of the 5G RLB for the estimation of maximum usable path loss between the gNB and 5G UE, which indicates the cell radius in both UL and DL. By applying the path loss value to an adequate radio propagation model, the cell radius can be estimated for different geographical types such as dense urban, urban, suburban, rural, and open area.

Balancing of the UL and DL is considered important for two-way use cases such as Internet Protocol (IP) voice call, which requires similar performance in both directions when aiming to achieve a fluent user experience in the cell edge region (i.e., in all circumstances, both A and B subscribers can hear each other). For the balancing of less critical cases, the data rates in UL and DL depend on the application. Thus, when downloading data such as web pages, the data speed in UL is not the limiting factor, therefore even considerably lower data rates in UL can still provide adequate interpretation of fluent performance.

The DL and UL of 5G use different modulation schemes with each providing different levels of performance, which, in turn, converts into variable cell radius values. In addition to the received useful carrier signal, final radio link performance depends on the level of interfering signals.

The key parameters in the DL direction are the output power level of the gNB transmitter (P_{TX}) and the transmitting antenna gain (G_{TX}).

In the early stage of mobile communications, the cable and connector losses (L_c) were a rather significant issue. Even if it is still a valid deployment scenario to include 5G base station installations along with the development of the Remote Radio Unit (RRU) deployments either at the base station or cloud, cable loss can be assumed to be compensated by the antenna system front-end. This is because the gNB's transmitter is typically located in the antenna module, and the respective transmission from the gNB up to the antenna system is delivered via practically lossless fiber optics.

The further split model of the gNB into RRU, DU, and CU as detailed in Chapter 4 can help to achieve a feasible balance of baseband processing rather far away from the accrual physical site. This is because the respective data can be transported via low-loss fiber optics between the base station and a cloud that processes the radio functions in a centralized way.

5G will also be increasingly based on multi-array, adaptive antennas embedding advanced beamforming technologies. Their increased performance may be taken into account in the 5G RLB, too, as an additional gain that can be estimated by functional modeling and related capacity and coverage simulations.

On the other hand, in an effort to balance the split model's radio signal processing unit's locations, it is important to take into account the considerable amount of data the potentially deployed multi-array-based adaptive antennas produce and therefore would need to be transported outside the physical site location for post-processing.

The radiating power (P_{EIR}, Effective Isotropic Radiating Power) refers to the radiating power level for the omni-radiating antenna. In practice, the mobile communication system typically relies on directional antennas, or in the case of 5G, on further optimized, beamforming antennas with directional individual links so the antenna gain G_{TX} is taken into account accordingly to represent the realistic power level.

At the receiving end of the DL, the key parameters of the 5G UE are sensitivity (S), receiver antenna gain (G_{RX}), and noise figure (NF). In typical, generalized cases, the 5G terminal's antenna gain can be assumed to be either 0 or negative depending on the level of integration of the components and device's antenna size.

In the UL, the key parameters are the transmitting power of the 5G UE, the antenna gain of the terminal and gNB (which are likely the same as in DL if the same antenna types are utilized), the cable and connector losses, and the sensitivity of the gNB receiver.

The maximum allowed path loss values L_{DL} and L_{UL} are calculated for different modulation schemes by assuming the minimum functional received power level per coding scheme, and by subtracting the emitted and received powers. The estimate can be done per geographical area type of interest outdoors, and indoor coverage can be estimated via typical penetration loss values for buildings.

As a rule of thumb, the most robust modulation schemes, Binary Phase Shift Keying (BPSK) and Quadrature Phase Shift Keying (QPSK), provide the largest coverage areas, but at the same time they result in the lowest capacity. The higher-grade constellation Quadrature Amplitude Modulation (QAM) schemes, up to QAM256 of 3GPP Release 16, provide the highest data rates, but at the cost of reduced coverage areas. Also, the least protected radio transmission (highest coding rates) provides the highest data rates but within smaller coverage areas than do the heavier protected codes.

The 5G system has dynamic modulation and coding (Modulation and Coding Scheme, MCS), which provides an optimal combination of modulation and code rate at any given time. Despite the more advanced variant of the 5G radio link, both LTE and 5G NR are based on the OFDM, so the same principles can be used in the simulations to map the minimum required received power level for each modulation scheme and the mapping of the respective data speeds that can be achieved with those combinations.

The well-known radio path loss prediction models can be utilized as a basis for 5G coverage estimations, adjusting them for the support of extended frequency bands and bandwidths of the 5G system.

The link budget calculation is typically based on a minimum throughput requirement at the cell edge. This approach provides the cell range calculation in a straightforward way.

By analyzing the order of defining the throughput requirement prior to the link budget calculation, it is possible to estimate the bandwidth and power allocation values for a single user, which mimics the behavior of realistic scheduling sufficiently well for link budget calculation purposes.

The 5G link budget can be estimated by analyzing the effect of the key parameters as summarized in Tables 8.2 and 8.3. In these examples, a relatively narrow band of 360 kHz is assumed for UL transmission, and an equally non-aggressive 10 MHz band for DL transmission.

The link budget can be planned in the following way, e.g., in the DL direction with the 10 MHz assumption as shown in the example above. The radiating isotropic power

Table 8.2 The principle of the DL RLB.

DL			
Transmitter, eNB	**Parameter**	**Unit**	**Value**
Transmitter power	P_{tx}	W	40.0
Transmitter power (a)	P_{tx}	dBm	46.0
Cable and connector loss (b)	L_c	dB	2.0
Antenna gain (c)	G_{tx}	dBi	11.0
Radiating power (EIRP) (d = a − b + c)	P_{EIRP}	dBm	55.0
Receiver, terminal			
Temperature (e)	T	K	290.0
Bandwidth (f)	f_{BW}	MHz	10.0
Thermal noise	N	dBW	-134.0
Thermal noise ($g = k_b \cdot T \cdot f_{BW}{}^1$)	N_T	dBm	-104.0
Noise figure (h)	N_f	dB	7.0
Receiver noise floor (i)	N_{rx}	dBm	-97.0
Signal-to-Interference-Noise-Ratio (SINR) (j)	$S/(I+N)$	dB	−10.0
Receiver sensitivity (k)	S_{rx}	dBm	−107.0
Interference margin (l)	I	dB	3.0
Control channel share (m)	L_{cc}	dB	1.0
Antenna gain (n)	G_{rx}	dBi	0.0
Body loss (o)	L_b	dB	0.0
Minimum received power (p)	P_{rx}	dBm	−103.0
Maximum allowed path loss, DL			**158.0**
Indoor loss			**15.0**
Maximum path loss for indoors, DL			**143.0**

[1] k_b = Boltzmann constant (1.38×10^{-23} J/K, or 2.08×10^{10} Hz/K), T = temperature in kelvin, and f_{BW} = bandwidth.

Table 8.3 The principle of the UL RLB.

UL			
Transmitter, terminal	**Parameter**	**Unit**	**Value**
Transmitter power	P_{tx}	W	0.3
Transmitter power (a)	L_c	dBm	24.0
Cable and connector loss (b)	L_c	dB	0.0
Antenna gain (c)	G	dBi	0.0
Radiating power (EIRP) (d)	P_{EIRP}	dBm	24.0
Receiver, eNB			
Temperature (e)	T	K	290.0
Bandwidth (f)	f_{BW}	MHz	0.36
Thermal noise	N	dBW	−148.4
Thermal noise (g)	N_T	dBm	−118.4
Noise figure (h)	N_f	dB	2.0
Receiver noise floor (i)	N_{rx}	dBm	−116.4
SINR (j)	$S/(I+N)$	dB	−7.0
Receiver sensitivity (k)	S_{rx}	dBm	−123.4
Interference margin (l)	I	dB	2.0
Antenna gain (m)	G	dBi	11.0
Mast head amplifier (n)	G_{mha}	dB	2.0
Cable loss (o)	L_c	dB	3.0
Minimum received power (p)	P_{rx}	dBm	−131.4
Maximum allowed path loss, UL			**155.4**
Smaller of the path losses:			**155.4**
Indoor loss			**15.0**
	Maximum path loss indoors, UL		**140.4**
	Smaller of the path losses indoors:		**140.4**

(d) can be calculated by taking into account the transmitter's output power (a), the antenna feeder and connector loss (b), and transmitter antenna gain (c), so the formula is: d = a − b + c.

The minimum received power level (p) of the UE can be calculated as p = k + l + m − n + o, utilizing the terminology of the link budgets shown above. The noise figure of the UE depends on the quality of the model's hardware components. The minimum Signal-to-Noise Ratio (SNR) or more practically, SINR value j is a result of the simulations. The sensitivity k of the receiver depends on the thermal noise, the noise figure of the terminal, and SINR in such a way that k = g + h + j.

The interference marginal l of the link budget represents the average estimate of the non-coherent interference originating from the neighboring eNB elements. The control channel proportion m degrades slightly the link budget. In the link budget calculations, the effect of the antenna of the terminal can be estimated at 0 dB if no body loss is present near the terminal. In the case of the external antenna, the antenna gain increases respectively, but the logical estimate of the average terminal type is a low-gain in-built antenna.

The effect of data speed can be estimated at a rough level by applying a rule of thumb that assumes a path loss of 160 dB in UL using a 64 kb/s data rate. Whenever the bit rate grows, the maximum allowed path loss drops respectively. A simple and practical assumption is that the doubling of the data rate increases the path loss by 3 dB. This can be assumed to work sufficiently well for a selected channel coding and modulation scheme.

As stated in 3GPP TS 38.300, informative Annex B, to improve UL coverage for high-frequency scenarios, the operator can configure an additional Supplementary Uplink (SUL) to enhance the UL and DL balance at the cell edge. Details of SUL can be found in 3GPP TS 38.101, but as a generic principle, with SUL, the UE is configured with two ULs for one DL of the same cell as depicted in Figure 8.8.

The manually calculated RLB is a rather simplified way to understand roughly the impact of the high-level radio parameter values on the achievable coverage. It also works as a first approach to estimate the feasibility of scenarios and the total number of sites within a given geographical area. For precise 5G radio network planning, computerized simulations are essential, relying on digital three-dimensional maps and adjusted parameter values based on a variety of field test results.

For the manual, first stage exercise, the link budget can be carried out by estimating the total gains and losses in the chain between transmitter and receiver, both in UL and DL. The result can be, e.g., the achievable received power level as a function of the distance between the transmitting and receiving antennas of the base station and device. One way, as presented in Ref. [12], is to estimate the received signal level of the UE and compare it with the sensitivity figure of the receiver to understand if it still suffices for adequate reception of the data, being at least equal to or better than the planned received power level value obtained from the RLB.

For area coverage considerations, there are various propagation models developed throughout the existence of the mobile communication generations. To select an adequate model, it is necessary to consider the model's suitability for the frequency range, base station height, and area cluster support (e.g., rural, suburban, urban, dense urban).

For the most accurate radio network planning, simulation tools and digital maps (e.g., three-dimensional vector models) are needed.

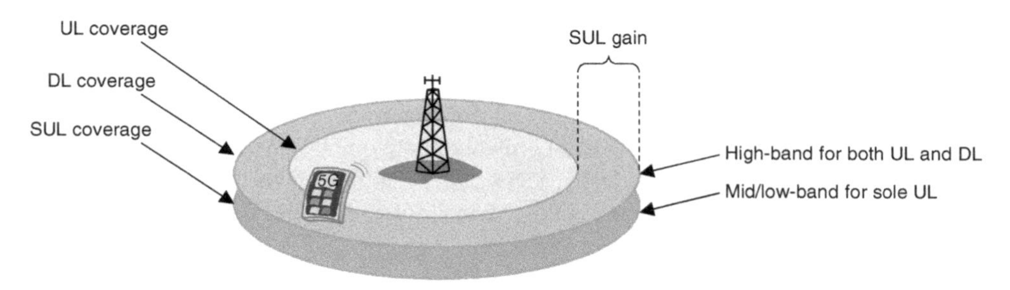

Figure 8.8 The principle of SUL.

8.2.4 5G Radio Link Budget in Bands Above 6 GHz

This section presents a summary of the ITU feasibility study of IMT in bands over 6 GHz [13]. The explored propagation characteristics of these bands provide invaluable information about the expected performance, especially as the highest frequency bands that will be deployed as per the WRC-19 decisions will be a reality.

The first edition of *5G Explained* presented radio link propagation formulas for selected mmWave frequencies based on ITU-R Report M.2376–0 (07/2015) [13]. The frequencies presented are 28, 39, 60, and 72 GHz. Table 8.4 summarizes the observations of these bands. The statements are valid for both Release 15 and 16 5G radio networks.

For more information on radio network planning aspects, please refer to 3GPP TS 43.030 (radio network planning aspects), TR 38.901 (channel modeling study), and ETSI TR 138 913 (5G radio link budget and scenarios). ITU presents various propagation models that apply to the 5G bands, e.g., in Refs. [14] and[16]. ITU also presents fundamental propagation prediction models and principles for low bands, e.g., in Ref. [17].

Table 8.4 Summary of the selected mmWave propagation studies [13].

Frequency band	Summary of observations
28 GHz	In this presented example, the low-power gNB can provide indoor coverage of some tens of meters in dense urban areas (such as New York city center) up to some hundreds of meters in more open Non-Line-Of-Sight (NLOS) environments such as semi-open campus locations. The LOS scenario offers 1 Gb/s up to almost 1 km, and 100 Mb/s up to about 3.5 km. It should be noted that these values can vary greatly depending on the more specific parameters, geographical topology, and antenna heights, among other variables, but the example of Ref. [14] gives a rough understanding on the achievable 5G radio coverage on 28 GHz. As a comparison, Ref. [15] presents a set of study results on 28 GHz, and depending on the parameters (antenna height/type, area type, and LOS/NLOS scenarios), typical cell ranges vary between 14 and 270 m on this band.
39 GHz	In this scenario, the UE's transmit power is 10 dBm, antenna gain 15 dBi, bandwidth 500 MHz, input noise power –80.9 dBm, radio receiver's noise figure 10 dB, and implementation loss 10 dB. The transmitter output power is 19 dBm, with transmit antenna gain +24 dBi provided by beam steering. It should be noted that the model presented in Ref. [14] is simplified, and it does not consider advanced baseband techniques like coding gain effects on the channel model. Nevertheless, this information is valuable to understand the expected data rate as a function of cell range in the 39 GHz band.
60 GHz	In this scenario, the UE's transmit power is 10 dBm and antenna gain is 15 dBi. Unlike in the previous scenario for 39 GHz, a bandwidth of 2 GHz is now applied. The input noise power remains at –80.9 dBm, as well as the radio receiver's noise figure 10 dB and implementation loss 10 dB. The transmitter output power is 19 dBm. Unlike in the previous case, a higher-gain transmit antenna of +29 dBi is applied as there is such commercial potential for highly directive point-to-point installations. Even the frequency range as such causes more attenuated path loss at 60 GHz compared to 39 GHz, applying the above-mentioned parameters, and the coverage area can be greatly enhanced.
72 GHz	In this scenario, the transmitter power is assumed to be +26 dBm, transmitter EIRP +60 dBm, and receiver antenna gain +21 dBm. The resulting coverage at 1 Gb/s is indicated to be around 120 m when NLOS is applied. For LOS conditions, the coverage at 1 Gb/s may be in range of 470 m.

8.2.5 Sidelink Deployment Scenarios

The sidelink provides direct Device-to-Device (D2D) communications under 5G architecture. 3GPP TS 36.300 presents the feasible sidelink scenarios for two user devices (UE) within or outside the radio cell coverage area as summarized in Figure 8.9. Both devices can act as transmitter and receiver in each case. Furthermore, one or more devices can receive the transmission of the UE.

8.3 RAN Deployment

C-RAN provides a common platform for a software-based network solution. According to Ref. [1], C-RAN can be a cost-efficient solution that provides real-time cloud for centralized processing, high bandwidth for optical transport, and distributed, configurable, and

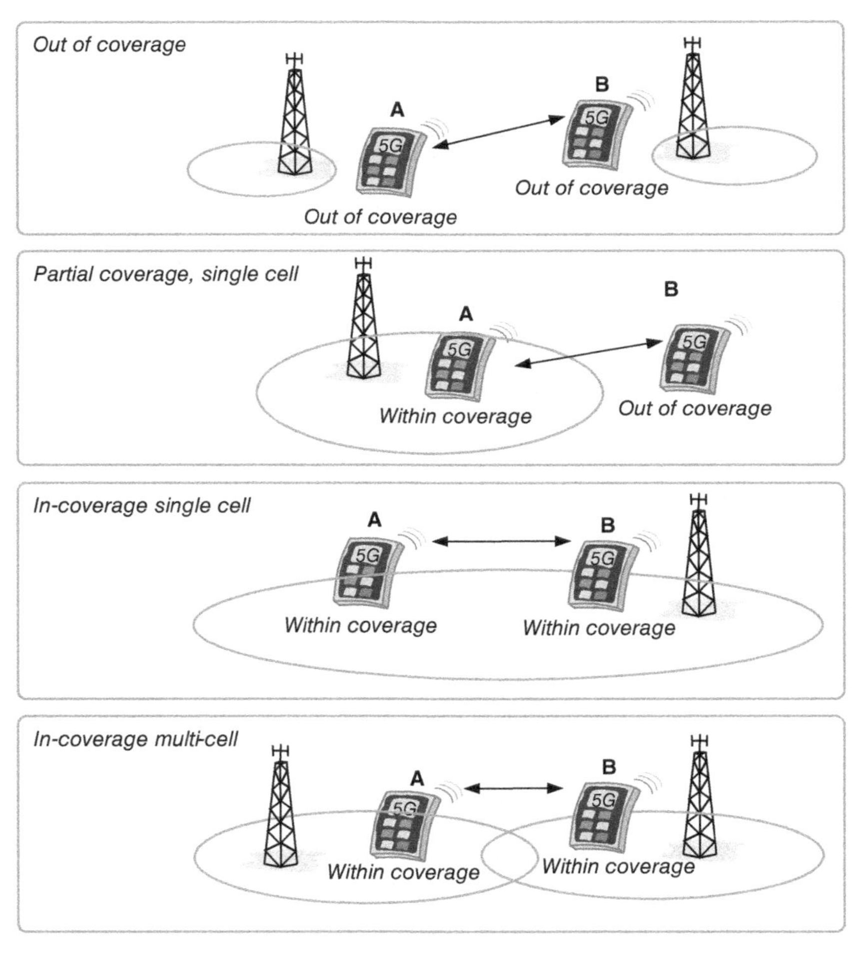

Figure 8.9 The sideline communication scenarios as interpreted from TS 36.300 [18].

wideband RRU. The virtualization of the network provides, among other benefits, a virtual base station pool housing physical and Medium Access Control (MAC) protocol layers, which optimize the resource utilization. Furthermore, the C-RAN facilitates OPEX and CAPEX savings.

The 3GPP as well as other entities such as the O-RAN Alliance present division options that optimize the deployment of the RAN.

8.3.1 O-RAN Deployment Scenarios

The O-RAN Alliance's white paper "O-RAN Use Cases and Deployment Scenarios," Chapter "O-RAN Cloud Native Deployment" [19], presents O-RAN cloud deployment scenarios, which Figure 8.10 summarizes. In this figure, the VNFs are located at the top, and each scenario illustrates how the VNFs are implemented in a proprietary network element, or on the O-RAN-compliant O-Cloud based on COTS servers and decoupled software stack and hardware.

Furthermore, on the notation of Figure 8.10, "O-Cloud" refers to an O-RAN cloud platform that supports the RAN functions via hardware accelerator add-ons per each RAN function, while the software stack is separated from the hardware. The "Proprietary" elements are not O-RAN-defined components that use open interfaces.

In practice, the O-RAN Alliance indicates that in efforts for cloudification, Scenario B has been the initial focus. In it, the proprietary element located at the physical site provides the O-RAN Radio Unit (O-RU) function, while the O-RAN Central Unit (O-CU) and O-RAN Distributed Unit (O-DU) are on Edge O-Cloud, and the Near-RT RIC is on a different, regional O-Cloud.

8.3.2 3GPP Functional Split Options of 5G

As can be seen in Figure 8.11, 5G gNB supports more internal interfaces and split scenarios of core, radio, and transport networks [1].

The CU supports the higher 5G protocol layers, including Service Data Adaptation Protocol (SDAP), Packet Data Convergence Protocol (PDCP), and Radio Resource Control (RRC), while the DU supports the lower protocol layers at RLC, MAC, and physical layer. In case the CU connects with the 4G core, the SDAP is not present, though, as only 5GC supports it.

Each CU can support one or various DUs. In practice, the number of cells formed by DUs per single CU depends on the network vendor's product strategies, and can be in the range of several dozens. As each DU can support one or various radio cells, the total number the CU (and thus gNB) serves can be hundreds.

The 3GPP has standardized the open *F1* interface between the CU and DU in a vendor-agnostic way, so according to the specifications, the DUs and CUs can be deployed from a mix of different vendors.

The possible practical variants of the DU-RU interfaces can be based on the Common Public Radio Interface (CPRI), Enhanced CPRI (eCPRI), and Next Generation Fronthaul Interface (NGFI) as depicted in Figure 8.11.

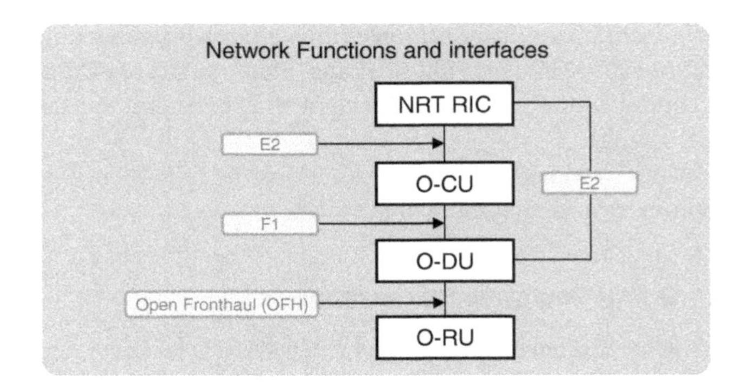

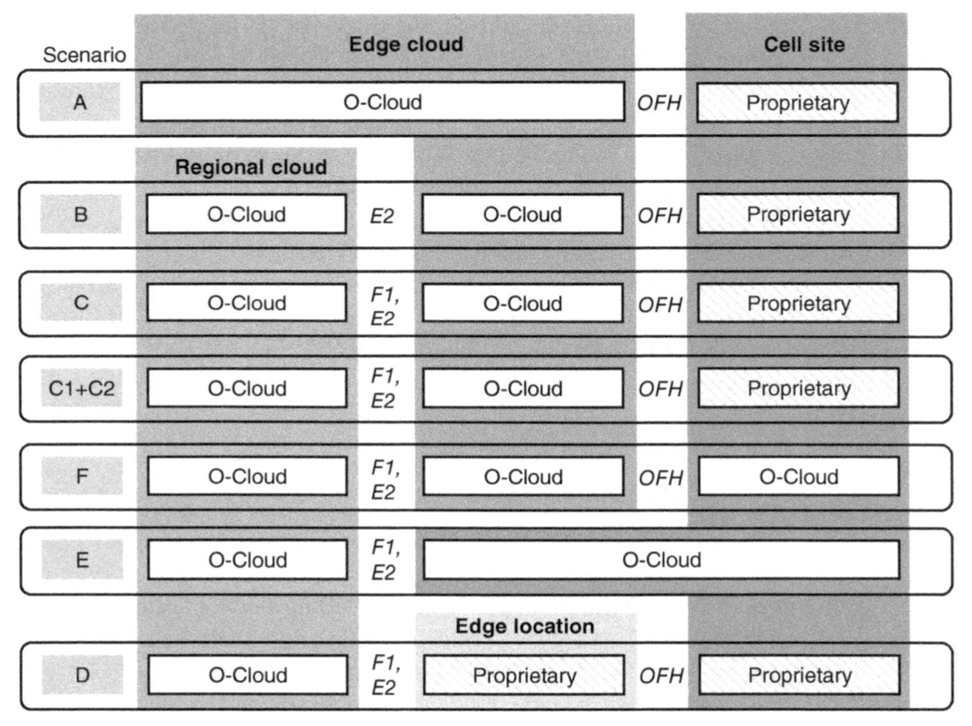

Figure 8.10 Cloud deployment scenarios as interpreted from the guideline of the O-RAN Alliance [19].

The CPRI refers to the industry joint activity for specifying the internal interface of radio base stations between the Radio Equipment Control (REC) and the Radio Equipment (RE) modules [20]. The CPRI line bit rate is flexible, and can be selected from 10 options varying from 614.4 Mb/s (Option 1) up to 24 330.24 Mb/s (Option 10) [21].

The interface between the 5G RU and DU can be the original version of the CPRI, or its evolved variant eCPRI. Also, a new NGFI interface is under consideration [22]. According to Ref. [20], CPRI cooperation defines the latest version 2.0 of the eCPRI specification that enhances support for 5G fronthaul via functionality to support CPRI 7.0 over the Ethernet, which consequently provides a means for CPRI and eCPRI interworking [23]. In relation

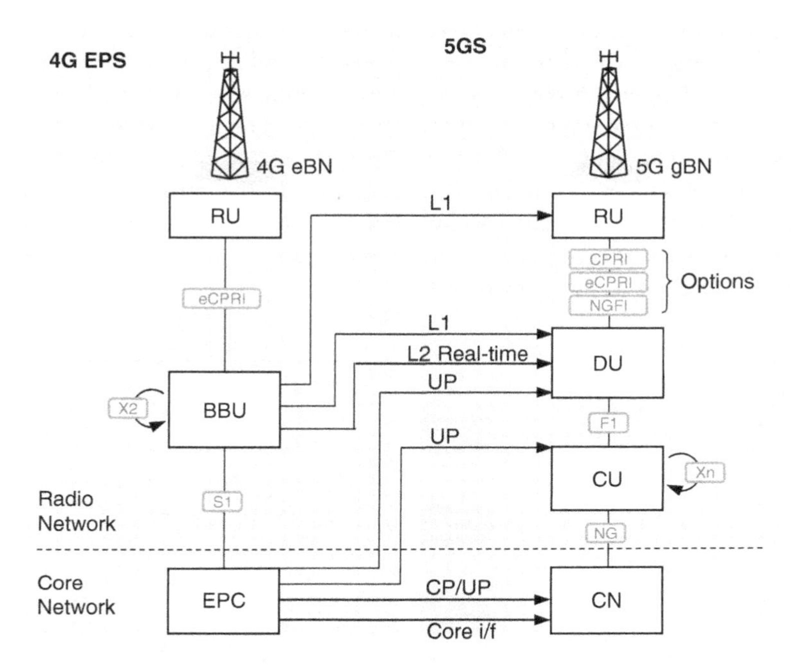

Figure 8.11 5G brings more diverse split options for the network components.

to 4G eNB and 5G NB, the original CPRI concept divides it into two nodes, which are REC and RE. This principle is also valid for the eCPRI, although with renamed modules of eREC and eRE, respectively. One of the differences the eCPRI brings is its more flexible functional split across the two nodes.

As an alternative to the above-mentioned interfaces, the NGFI is an interface between a baseband pool and a set of Remote Radio Head (RRH) components. It is designed to cope with the demanding 5G infrastructure requirements. In the generic C-RAN architecture, the BBU processes the contents and sends them to the RRU in I/Q samples via the CPRI or eCPRI fronthaul interface. According to Ref. [22], the NGFI redefines baseband processing split between BBU and RRU, and the positioning of eNB stack components between BBU and RRU. At the same time, the terminology of the NGFI refers to the BBU as the Radio Cloud Center (RCC), whereas the RRU is referred to as the Radio Remote System (RRS). The NGFI architecture outlines a point-to-multipoint architecture for the RCC-RRU that adds yet another component to the respective architecture called Radio Aggregation Unit (RAU) that communicates with the RCC and performs transport for a variety of RRU components. The IEEE has produced the NGFI P1914-series that standardizes the transport fronthaul interface for future cellular networks [24]. The key NGFI standards of the IEEE are 1914.1-2019 (Packet-based Fronthaul Transport Networks), 1914.3-2018 (Radio over Ethernet Encapsulations and Mappings), P1914.3a (Amendment on Encapsulation Enhancements and Elaborations).

The 5G NR gNB radio functions can thus be modularized via CU, DU, and RU. The division between the physical entity taking care of the DU and RU can be deployed based on the options that the 3GPP and O-RAN Alliance have designed. A base station (group of radio cells) can house a CU that can serve various DUs, while each DU can serve various RUs. In this model, the RU is physically at the base station site, while the CU and DU can reside at the physical base station site or elsewhere, e.g., in regional or central Edge Cloud.

The functional radio split has already been defined for 4G between the BBU and RU, whereas 5G provides further granularity for the division by the introduction of the RU, DU, and CU split. Figure 8.12 depicts the 5G functional split options, and Table 8.5 summarizes their respective main aspects [18, 25, 26]. Table 8.6 summarizes further the key tasks of each protocol layer presented in Figure 8.12.

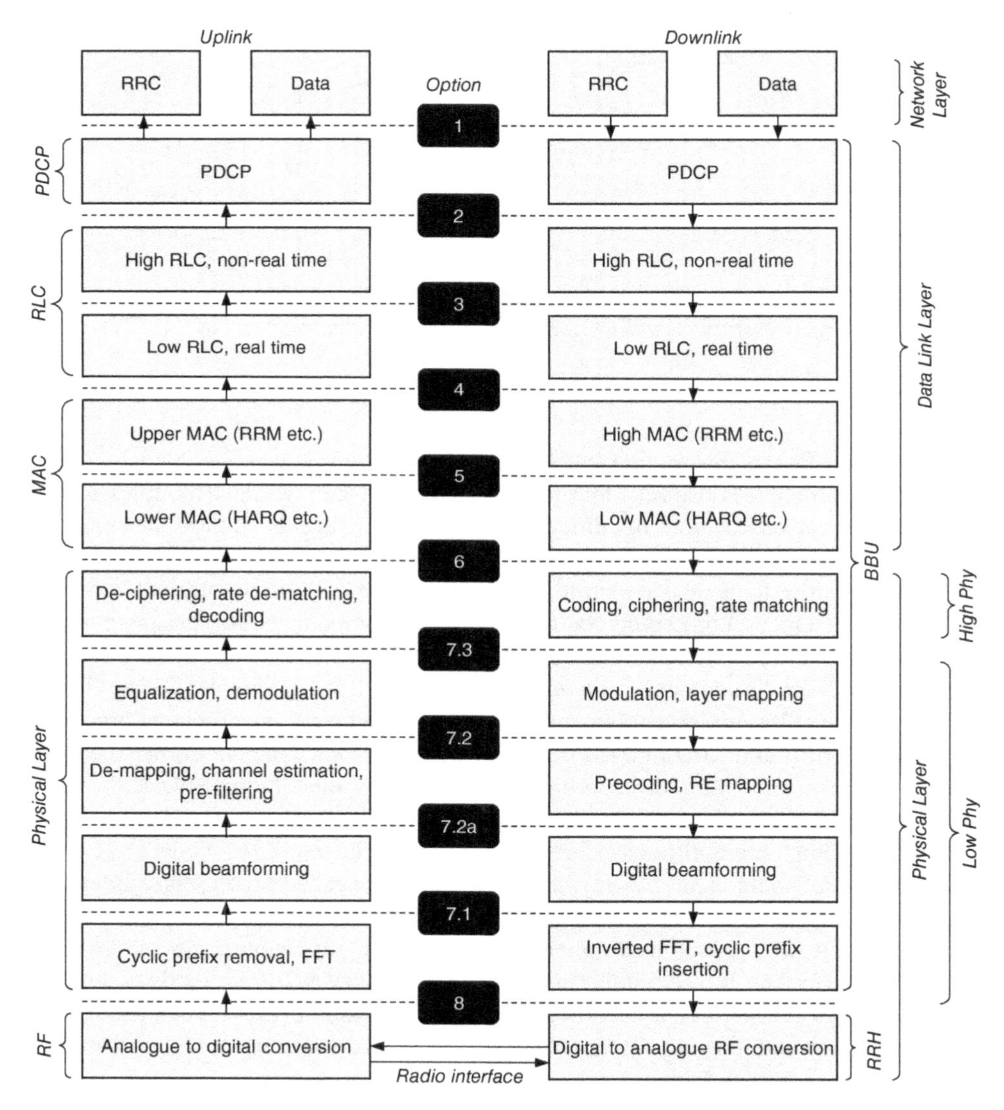

Figure 8.12 The split options of 5G as per the definitions of the 3GPP. Please note that the other entities such as O-RAN present different variants of the division.

Table 8.5 Summary of the 5G split options as interpreted from 3GPP TR 38.801 [26].

Option	Description
1	DC 1A-like split of LTE Release 12. The RRC is located in the CU, whereas the PDCP, RLC, MAC, physical layer, and RF are in the DU.
2	LTE Release 12 DC Option 3C-like split. The CU houses the RRC and PDCP, while the DU houses the RLC, MAC, physical layer, and RF. Release 15 refers to this interface as *F1*.
3	Intra-RLC split. The DU houses the low RLC, i.e., the partial function of RLC, MAC, physical layer, and RF. The CU houses the PDCP and high RLC. In this option, the CU has most of the RLC functions, while the RU takes care of the ARQ-related and real-time functionalities such as aggregation.
4	RLC-MAC split. The DU houses the MAC, physical layer, and RF, while the CU takes care of the PDCP and RLC.
5	Intra-MAC split. The DU houses the RF, physical layer, and part of the MAC layer such as HARQ. The CU takes care of the upper layer tasks such as the scheduling of, e.g., Inter-Cell Interference Coordination (ICIC) and Coordinated Multi-Point (CoMP).
6	MAC-Phy split. The DU houses the physical layer and RF, whereas the CU takes care of the upper layers.
7.1	Option 7 refers to the intra-Phy split. In this option, the DU houses the RF and part of the physical layer, whereas the CU takes care of the upper layers. Option 7.1 refers to the scenario where the DU performs Inverted Fast Fourier Transform (IFFT), Fast Fourier Transform (FFT), and cyclic prefix insertion and removal, whereas the CU takes care of the physical layer. As an example, this interface exchanges I/Q samples of the reserved subcarriers on the frequency domain (time domain samples are not exchanged in this option unlike in Option 8).
7.2	The DU manages precoding and digital beamforming or parts of it. This interface is between functions performing modulation/demodulation and precoding/channel estimation.
7.2a	The DU manages the precoding and digital beamforming or parts of it. This interface is between functions performing precoding/channel estimation and digital beamforming.
7.3	This interface is between the functions managing coding/decoding and modulation/demodulation.
8	Phy-RF split. The DU houses the RF functionality, while the CU manages the upper layer. Option 8 refers to the legacy C-RAN scenario.

As stated in 3GPP TR 38.801, the benefits of the flexible functional split with the division of the NR functions between CU and DU include the production of scalable and cost-effective solutions, as well as efficient load management and real-time performance optimization. It also enables NFV and SDN. As a consequence, the configurable functional split enables adaptation to a variety of use cases.

Table 8.6 The main tasks of the protocol layers of the 5G RAN.

Layer	Description
RRC	Radio Resource Control. Takes care of Radio Management (RM) and Connection Management (CM). It has many tasks depending on the state of the RM and CM (RM Deregistered/Registered, and CM Idle/Connected). The key RRC tasks include network selection, system information broadcast, cell reselection (handover), and paging.
PDCP	Packet Data Convergence Protocol. PDCP is located in the Radio Protocol Stack on top of the RLC layer. Key PDPC tasks include transfer of UP and CP data, header compression, ciphering, and integrity protection. PDCP is specified in 3GPP TS 25.323 (3G), TS 36.323 (LTE), and TS 38.323 (5G). Header compression can be based on IP header compression as per IETF RFC 2507 or robust header compression as per RFC 3095.
RLC	Radio Link Control. The 5G NR RLC communicates with the PDCP and underlying MAC. Transfer of upper layer Packet Data Units (PDUs) in Acknowledged, Unacknowledged, and Transparent Modes. The PDU can consist of control or user data, and has error correction by retransmission of lost RLC PDUs. Segmentation and reassembly of RLC Service Data Units (SDUs), and resegmentation of RLC data PDUs when a complete RLC PDU cannot be transmitted. Reordering of RLC data PDUs.
MAC	Medium Access Control. The 5G NR MAC communicates with the RLC and underlying physical layers. Mapping between logical channels and transport channels. Multiplexing/demultiplexing of MAC SDUs of logical channels into/from Transport Blocks (TBs) delivered to/from the physical layer on transport channels. Scheduling information reporting. Error correction through HARQ. Priority handling, transport format selection, and padding. More information on MAC can be found in 3GPP TS 36.321.
Phy at 6–7.3	Physical layer. The 5G NR physical layer handles a variety of radio interface-related tasks under MAC. The tasks between Options 6 and 7.3 include coding, ciphering, and rate matching
Phy at 7. 3–7.2	Phy, physical layer. The tasks between Options 7.3 and 7.2 include modulation and layer mapping. More information on this layer can be found in 3GPP TS 38.211 (NR physical channels and modulation).
Phy at 7.2–7.2a	Phy, physical layer. The tasks between Options 7.2 and 7.2a include precoding and RE mapping. More information on this layer can be found in 3GPP TS 38.212 (NR multiplexing and channel coding).
Phy at 7.2a–7.1	Phy, physical layer. The tasks between Options 7.2a and 7.1 include digital beamforming. More information on this layer can be found in 3GPP TS 38.215 (NR physical layer measurements).
Phy at 7.1–8	Phy, physical layer. The tasks between Options 7.1 and 8 include FFT, IFFT, and cyclic prefix. More information on this layer can be found in 3GPP TS 38.213 (NR, physical layer procedures for control) and 3GPP TS 38.214 (NR physical layer procedures for data).
RF	Radio Frequency (physical layer). The tasks under Option 8 include D/A and A/D conversion, transmission, and reception of the modulated signals over the air.

As the specifications allow such a large set of options, selection of the most adequate one is an important deployment optimization task for the operator, and it depends on, e.g., the planned services the operator aims to offer to end-users. The selected options have an impact on, e.g., service QoS in terms of latency and data throughput, intended support of communications density and load within a planned area, and performance with the connected transport networks.

The O-RAN provides operators with open interfaces within the RAN and its RU as well as BBU's DU/CU functions, in a finer granularity than has been the case in previous generations. The Network Management System (NMS)/orchestrator takes care of the controlling of the functions in this open environment. Please note that the O-RAN concept is not limited to 5G even though it has been the "driving force," but the same principle can be applied technically to all the currently deployed older generations, too. Some of the benefits of the O-RAN concept are the possibility for bringing new businesses by NR manufacturers that can be assumed to facilitate more cost-efficient deployment options to operators.

As the CU can be placed via the *F1* interface either within the equipment housing the DU or further away to a regional cloud data center, this model opens new business opportunities to data center operators. The operator can also select the split between the RU and DU, from Small Cell Forum's Option 6 to the O-RAN Alliance-defined Option 7.2. The low-level physical interconnection of the fronthaul for the DU and RU can be done via, e.g., CPRI or eCPRI, of which the industry seems to favor the latter thanks to its more up-to-date capabilities to handle low-latency values.

Please refer to Chapter 4 for more details on the split options of 5G.

8.4 5G Core Network Planning

8.4.1 Overall Considerations

For core network planning, the key task is the correct dimensioning of the capacity and reliability of the infrastructure, taking into account the future outlook for the needed traffic.

In any network planning, the core network must ensure adequate capacity and performance for customers. Quality can be measured by the service uptime, which typically requires redundancy to minimize any single point of failures. In the strictest environments, geographical "active-active" redundancy may be used for the critical network elements. This is rather expensive as the elements execute the same tasks in a parallel fashion; if one fails, the redundant element takes over the tasks in real time.

Also, the timing of the investment is important to optimize the costs of materials, storage, and deployment efforts. A good near-term and longer-term forecast for network utilization is thus of utmost importance.

5G core network functionality is predominantly virtualized, and it is based on SBA. Data centers will be an important part of the 5G core network. Their deployment plan considers interconnectivity with the 5G infrastructure.

Data center operations can be outsourced, too, which will bring new considerations for MNOs, such as requirements for data center certification models, and a Service Level Assurance (SLA) for the expected quality using appropriate redundancy and recovery classes.

As has been the case in all previous mobile networks, the essential dimensioning tasks remain the same for 5G, too. The main elements in optimal network planning are related to capacity, coverage, and QoS. The balancing of these parameters depends on the wanted costs; in the optimal network, the RoI is a balance: not too expensive as it wastes capacity, and not too low cost as customers' mix increases along with bad quality.

Along with the expected huge number of IoT devices that will surge within forthcoming years communicating simultaneously in always-on mode, 5G networks will need to support considerably increased traffic capacity. Not only will capacity demand increase but also traffic patterns will diversify as there will be eMBB, critical communications, and mIoT traffic mixed within the same network.

Nevertheless, the 5G system has been designed in such a way that the network slices take care of each traffic type individually, and the hardware resources are only utilized as needed for the network functions. Furthermore, the network slices per use case are designed so that they only include the needed network functions, which further optimizes capacity utilization.

5G networks are remarkably different from previous generations due to the virtualization and SBA model.

In 5G core network dimensioning, some of the practical questions are:

- How can the strategy for the slices be planned, i.e., what network functions are needed in each use case?
- How can enough hardware processing power be reserved?
- How could transmission capacity be dimensioned for peak and average traffic?
- How is it possible to prepare for unexpected traffic peaks?

5G networks are designed for a completely new era with considerably increased data rates, number of simultaneously connected IoT devices, low latency, and a variety of use cases that are possible to optimize via the network slicing concept, differentiating characteristics of the connection [27].

5G challenges for the near future need evolved means for network scalability and flexibility. The key technologies for this are NFV and SDN.

The planning of 5G thus requires renewed dimensioning models for cost-optimized deployments, handling a variety of different traffic demand scenarios. Not only is radio interface dimensioning of utmost importance, but the optimal placement of the core network elements is essential for providing adequate performance for critical communications, with the lowest possible latency values. Thus, the location of physical edge computing elements, i.e., data centers, needs to be considered.

8.4.2 Virtualization

NFV offers the needed flexibility as it removes hardware dependency yet enables a means for fast deployment and service updates. NFV can thus contribute positively to the cost optimization of network deployments and service offering.

SDN, in turn, decouples data and CPs of network functions. It provides an open Application Programming Interface (API), which is in practice based on OpenFlow protocol, for decoupled planes.

Modeling and optimization for NFV and SDN are related to the physical location of SDN controllers and switches, and VNF resource allocation and placement.

As described in Ref. [28], considering the resource allocation and placement of VNFs, a mathematical model can be formed to find an optimal placement for virtual core gateways handling sporadic traffic increase, e.g., when a large crowd event takes place. It also is possible to adapt machine learning techniques to find an optimal placement for VNFs as a function of data center resources. Some examples of such research can be found in Refs. [29–31].

In the initial phase of 5G network deployment, the most typical scenario is to rely on the already existing legacy core network of 4G. In the 4G EPC, the UP and CP functions are based on dedicated hardware and respective software for each function. Examples of these elements are the P-GW, S-GW, Home Subscription Server (HSS), and Mobility Management Entity (MME). In the 5G core network, instead, functions that handle solely the CP such as the MME may be deployed based on VNFs. This refers to the concept that allows the use of a common hardware, and its resources are shared dynamically for all the VNFs on a cloud infrastructure. For functions taking care of both UP and CP, such as S-GW and P-GW, the respective traffic management can rely on SDN-based or NFV-based solutions.

The SDN-based architecture makes the CP of mobile core functions run as VNFs, whereas the gateway functions of S-GW and P-GW are decoupled into SDN controllers and SDN switches. The SDN controllers are deployed at the data center. They configure the SDN switches handling the UP traffic, and they take care of the EPC CP signaling procedures.

In the NFV-based architecture, the CP core network and gateway functions, which are S-GW and P-GW, are executed as VNFs at data centers' common hardware. In this scenario, CP and UP processing of the gateways runs on common cloud servers. This means that the previously utilized hardware of the core network can be replaced by adapting transport switches forwarding control and UP traffic within RAN, data centers, and external networks.

Ref. [28] states that the cost of network load may be improved significantly by distributing the data center infrastructure because as there are more available data centers, more VNFs may also be deployed complying with the requirements for latency to decrease additional SDN control traffic and respective cost of the network load. It should be noted, though, that there is an optimal amount for data centers, and values exceeding four do not improve significantly the load cost optimization.

8.4.3 MEC

The ETSI White Paper No. 24/2018 (MEC Deployments in 4G and Evolution Towards 5G) presents practical aspects on the deployment of Mobile-Edge Computing (MEC) in the 5G system architecture [32]. Along with the renewed SBA of the 5G system (3GPP TS 23.501), it also facilitates flexible deployment of the CP and data plane. One of the benefits of such an approach is the possibility to also natively support edge computing. Thus, the operator can integrate MEC architecture in a straightforward manner into the 5G infrastructure.

Based on Ref. [32], Figure 8.13 depicts an example of MEC mapping to the 5G system architecture. In this scenario, the data plane of the MEC host is mapped to the 5G User Plane Function (UPF) component.

The MEC platform can thus take care of the 5G traffic routing and steering function in the UPF, while the Policy Control Function (PCF) and Session Management Function (SMF), as well as the Application Function (AF) via PCF, serve routing-related influencing. This scenario is similar to 4G traffic influencing within the EPC MEC deployment scenarios.

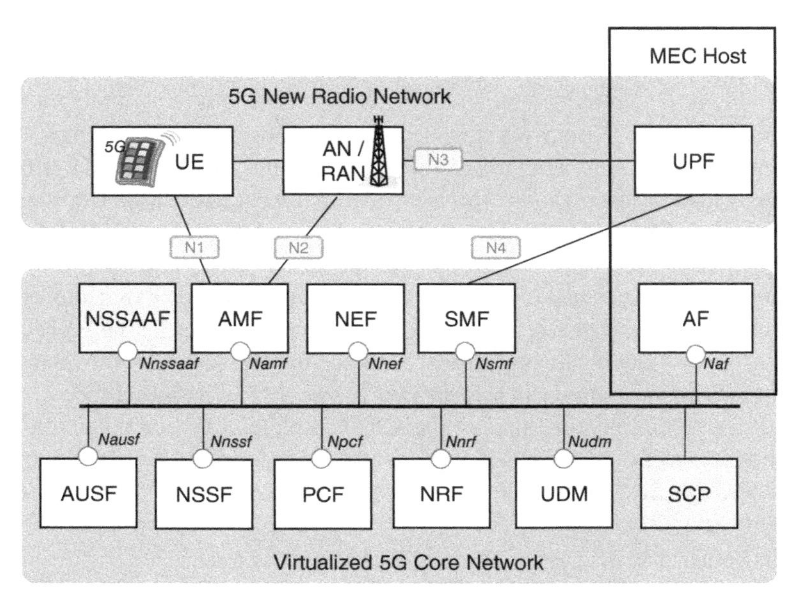

Figure 8.13 MEC mapping with 5G architecture as interpreted from Ref. [32].

The specifications allow operators to decide the practical position of MEC at the edge site. For more details on the model, the ETSI white paper presents some expected migration examples to host MEC, paving the way for 5G deployments while complying with the practical requirements for reusing the edge computing resources, interaction with the 5G CP, and integration with the 5G network [32].

8.4.4 Transport Network Considerations

The transport network connects the radio and core networks. Dimensioning of the transport network becomes increasingly important because the 5G radio interface will be capable of generating considerably more traffic than was the case in previous generations.

The transmission network can be based on fixed and wireless solutions, the fiber optics and radio links being typical technologies. For the radio links, adequate frequency planning maximizes the capacity and minimizes the interference. The transport network may rely on increased intelligence, flexibility, and automation to better cope with the high dynamics of 5G.

For fiber optics deployment, planning must be done well before the expected traffic reaches a critical limit. Overall 5G planning is largely related to the new cloud environment, and edge computing plays a key role in that ecosystem as described in Chapter 5.

8.4.5 Deployment Options of ITU

The ITU-T publication GSTR-TN5G presents the aspects of transport network support of IMT-2020/5G [33]. It also includes RAN deployment scenarios.

The 5G transport network may contain fronthaul, midhaul, and backhaul networks. The ITU-T has identified respectively four RAN deployment scenarios as summarized below:

1. Independent RRU, CU, and DU locations: this scenario includes fronthaul, midhaul, and backhaul networks. The RRU and DU are located 0–20 km apart from each other, while the distance between the DU and CU can be tens of kilometers.
2. Co-located CU and DU: the CU and DU are located at the same site and there is no midhaul.
3. RRU and DU integration: the RRU and DU are close by, e.g., in the shared building hundreds of meters apart. Fiber optics connect RRU and DU eliminating the need for transport equipment and reducing costs. The midhaul and backhaul networks are present in this scenario.
4. RRU, DU, and CU integration: this scenario uses only backhaul, and is feasible for small cell and hotspot deployment.

The O-RAN Alliance builds upon the foundations laid by the 3GPP. As presented in Ref. [34], the 3GPP's 5G C-RAN architecture splits the BBU functionality into two functional units: DU and CU. The DU is responsible for real-time L1 and L2 scheduling functions, whereas the CU handles the non-real-time, higher L2 and L3 protocol layers.

The 5G C-RAN model provides the possibility to host the DU physical layer and software layer in an Edge Cloud data center or central office, whereas the CU physical layer and software can be co-located with the DU or hosted in a regional cloud data center. The Open RAN concept also provides the possibility to use COTS servers for DU and CU software execution that are able to link the RU of other vendors.

Table 8.7 summarizes the roles of the RU, DU, and CU.

The gNB houses the CU. The gNB also has a DU that connects to the CU via the *Fs-C* interface for CP functions and via *Fs-U* for UP functions. A CU can have one or more DUs. In the latter case, the CU controls the DUs over a midhaul interface.

The 3GPP split architecture provides the 5G network operator with an advanced means to use different distribution scenarios for the protocol stacks between CU and DU or multiple DUs, helping with the detailed level of RAN sharing.

The benefit of centralized baseband deployment is the possibility to apply more efficient load balancing between different RUs. In practice, due to some intense processing at the physical site, such as multi-array antenna beamforming, it might be most feasible for the operator to co-locate the DU with the RUs at the base station.

Table 8.7 The roles of the RU, DU, and CU.

Unit	Description
RU	Connects to the front-end and manages the lower part of the physical layer. An example of the functions is beamforming.
DU	Situated close to the RU, and takes care of the higher physical layer functions as well as the RLC and MAC layers. It houses a subset of the eNB or gNB functions according to the selected functional split option. The CU controls the DU.
CU	Executes the RRC and PDCP layer functions.

Baseband processing can be housed at the edge. The edge is especially feasible in the 5G era as it can provide low-latency communication links to carry out local breakout, but still ensure fluent mobility management. The edge can thus further optimize resource utilization of the 5G radio network.

The MNO may consider applying the separation of DU and RU to optimize both the CAPEX and OPEX of the RUs and the overall capacity of the network (a resource pool of RUs managed by the DU handles local traffic peaks more efficiently than a single RU is able to do).

8.4.6 Dimensioning of the Core and Transport

5G network performance can benefit from *programmable transport* as the transport resources can be allocated dynamically when needed and current capacity consumption can be managed. This concept can be applied in locally variable environments for, e.g., managing energy consumption dynamically. Some use cases are network sharing for offering flexible service provisioning, and optimization of the utilized resources via dynamic resource allocation.

Another optimization method is transport-aware radio. This refers to enhanced functionality in the *X2* interface, i.e., between base station elements. Traditionally, the handover could be performed fluently via the *X2* interface, but if the respective transport network is experiencing congestion, there will be service degradation. In fact, handover in such a situation would make the situation worse. Transport-aware radio functionality may thus conclude that the overall performance is better without *X2* signaling while the congestion takes effect, and the network can be advised accordingly.

The third optimization method is dynamic load balancing. This method is beneficial, e.g., in non-predictable weather conditions, which may affect the radio link negatively, lowering the planned data rate due to the attenuation of rain. The offered load can be observed via service orchestration, which analyzes the ongoing level, and the potential impacts on the SLA. As exceeding the agreed SLA levels typically leads to additional expense via non-compliance fees, service orchestration can have embedded intelligence to evaluate the cost impact. In the negative case, service orchestration can analyze the alternative paths and their respective cost impacts to decide dynamically the most cost-efficient option, and to execute an order to change the data flow path accordingly.

8.5 Network Slice Planning

8.5.1 Overview

Network slicing is one of the key features of fully functional and optimized 5G. It provides a means to design and deploy customized communication systems and to integrate services from different verticals. It can be expected that there will be a massive number of simultaneously communicating devices that require sporadic connectivity and small-scale data transfer. The respective required QoS may vary considerably depending on the use case.

mIoT services would thus benefit from a highly dynamic, end-to-end slice representing radio access and core network with enhanced CP and UP. Other use cases may require a higher grade of security, whereas others may want the highest grade of reliability, i.e., lowest possible latency.

The requirements for slices serving different use cases is thus one of the essential dimensioning tasks of the MNO to plan services reflecting practical requirements of each use case. Some of the optimization tasks may be related to reduced CP signaling while maximizing UP resource utilization [35].

8.5.2 Network Slice Ecosystem Roles

The network slicing function provides extended roles for existing and totally new stakeholders. 3GPP TS 28.530 defines the roles within the network slicing ecosystem as summarized in Table 8.8.

Please note that one or more organizations can assume the roles presented in Table 8.8 simultaneously, and furthermore, a single organization can also have multiple roles at the same time. An example of such a multi-role is the CSP that can also be an NOP.

Also, in the case of Network Slice as a Service (NSaaS), the CSP role can be refined into an NSaaS Provider (NSaaSP, or Network Slice Provider, NSP) role, whereas the CSC can be referred to as an NSaaS Customer (NSaaSC) or Network Slice Customer (NSC). The role chain continues in such a way that the NSC can also serve customers in a CSP role, whereas the network slice tenant can assume the role of an NSC.

Table 8.8 The roles in the network slicing ecosystem.

Role	Description
CSC	Communication Service Customer uses communication services.
CSP	Communication Service Provider designs, builds, operates, and provides communication services.
NOP	Network Operator designs, builds, and operates networks to provide network services.
NEP	Network Equipment Provider supplies network equipment to networks. Also a VNF supplier is an NEP, providing virtualized "equipment."
VISP	Virtualization Infrastructure Service Provider designs, builds, operates, and provides virtualized infrastructure services. They offer these services to network operators, and can do so directly to other customers such as communication service providers.
DCSP	Data Centre Service Provider designs, builds, and operates data centers to provide respective services.
NFVI Supplier	NFVI Supplier supplies network function virtualization infrastructure to its customers.
Hardware Supplier	Hardware Supplier supplies network slicing-related hardware.

8.5.3 Network Slice Planning Principles

Network slicing builds dedicated logical networks on a shared infrastructure. A network slice is a set of logical network functions that fulfills the service requirements for communication related to respective use cases.

As a concrete benefit, 5G network slicing provides a tool for the operator to deploy a variety of networks minimizing the OPEX. It also speeds up the time to market, as the setup of the respective virtualized networks is much faster than in the traditional network deployment scenarios. Network slicing provides the slice tenants with a rich set of management capabilities.

The key benefit of network slicing is the use of a shared physical infrastructure for various parallel networks that all house different features and their performance levels. These differentiating features can include the varying QoS and security levels of the data plane as well as customized properties of the CP. Furthermore, network slicing allows dynamic locating of functions, e.g., in the edge and based on demand or traffic patterns, depending on the regional needs for the features and the respective capacity. Automated operations manage network slicing to make this dynamic behavior happen in practice.

Network slicing also provides new business models so that the slice owner can control the proper network slice independently from the infrastructure owner. In addition to the original network operator, a third party can also provide and manage the slicing services.

There are three layers in the network slicing concept:

- Service Instance Layer represents the services and their instances that either the network operator or third party provides;
- Network Slice Instance Layer provides the characteristics of the network a Service Instance requires. The Network Slice Instance (NSI) may have subnetwork instances that the NSI may share with other NSIs;
- Resource Layer relates to the actual physical resource, including storage, processing, and transport of data, and to the Logical Resource, which refers to the physical resource partitioning and physical resources grouping related to a network function.

8.5.4 Network Slice Templates of the GSMA

The GSMA has worked with the mobile communications industry to assist network slice providers to map the use cases of network slice customers into generic attributes. The GSMA has thus created guidelines in the form of NG.116, the Generic Network Slice Template (GST). This is a set of attributes that characterizes a type of network slice and respective service. GST is not related to a specific network deployment. Based on the GST, the Network Slice Type (NEST) represents the actual assigned values of the network slice to express a given set of requirements to support a network slice customer use case (Figure 8.14).

In practice, a network slice customer, such as automotive or other vertical, provides the requirements related to the use case it would like to fulfill to the Network Slice Provider (NSP). The NSP then maps the requirements into the attributes of the GST with appropriate values generating an NEST [36].

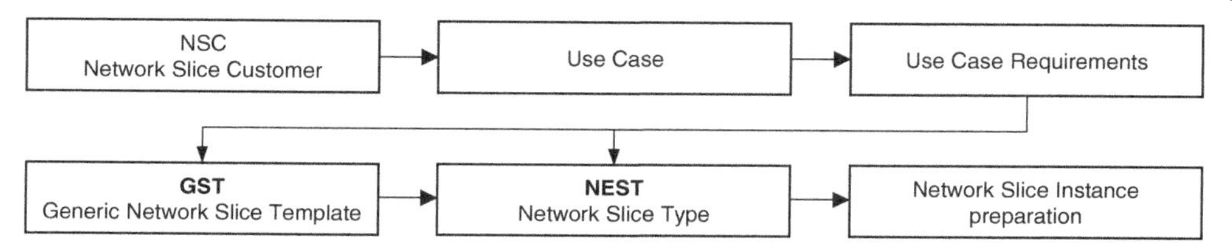

Figure 8.14 The principle of GST and NEST [38].

In the GSMA NG.116 context, which is aligned with the 3GPP terminology, the following applies.

Business customer refers to a tenant of the network slice. These customers represent, e.g., vertical industries. For instance, a business customer could be an enterprise or specialized industry customer (often referred to as "verticals") [37].

Network slice is a logical network that provides specific network capabilities and network characteristics to serve a defined business purpose of a customer. Network slicing allows multiple virtual networks to be created on top of a common shared physical infrastructure. A network slice consists of different subnets, e.g., RAN subnet, core network subnet, and transport network subnet [37].

NSP is typically a telecommunication service provider, which is the owner or tenant of the network infrastructures from which network slices are created. The NSP takes the responsibilities of managing and orchestrating corresponding resources that the network slicing consists of [37].

GST is a set of attributes that can characterize a type of network slice/service. GST is generic and is not tied to any specific network deployment [38].

NEST is a GST filled with values. The attributes and their values are assigned to fulfill a given set of requirements derived from a network slice customer use case [38].

GSMA PRD NG.116 contains a set of predefined attributes and their definitions, such as availability, UL and DL throughput per network slice, area of service, delay tolerance, energy efficiency, NB-IoT support, positioning support, and V2X communications mode.

NG.116 presents examples of the NEST of which Table 8.9 presents a minimum set of attributes needed for the NEST for URLLC Slice/Service Type (SST).

8.5.5 Network Slice as a Service

3GPP TS 28.530 presents the NSaaS model. The CSP can offer such a service to its CSC. This model allows multiple scenarios, such as:

- The CSC can use the network slice as the end-user.
- Th CSC can manage the network slice via the management interface of the CSP.
- The CSC can assume a role of CSP and offer their own services on the network slice the CSP has given. Thus, a network slice customer can also be an NOP, building their own network, which includes the network slice that a CSP gives.
- The ones offering the network slicing in different roles can limit the level of the network slice management capabilities they expose to their customer.

Table 8.9 Example of an NEST for URRLC as per GSMA PRD NG.116 V3.0 [37].

Attribute	Value
Availability	99.999%
DL throughput per UE: maximum DL throughput per UE	100 000 kb/s
UL throughput per UE: maximum UL throughput per UE	100 000 kb/s
Slice quality of service parameters	82[2]
Supported device velocity[3]	2 km/h
UE density	1000 per km^2

The CSP that provides NSaaS commits to a set of capabilities to meet the service-level requirements. These properties may include, e.g., radio access technology, bandwidth, end-to-end latency, reliability, guaranteed and non-guaranteed QoS, security level, etc.

8.5.6 Network Slice Management

3GPP TS 28.530 V16.2.0 (2020-07) presents the management and orchestration of the network slices, as per Release 16 [39]. The tasks include the preparation, commissioning, operation, and decommissioning of the network slice.

The stages of Figure 8.15 are the following:

- Preparation includes network slice design and capacity planning, on-boarding and evaluation of the network functions, and preparation of the network.
- Commissioning takes care of NSI provisioning, which creates the NSI for allocation and configuration of the planned resources.
- Operation includes the activation, supervision, performance reporting, resource capacity planning, modification, and deactivation of an NSI.

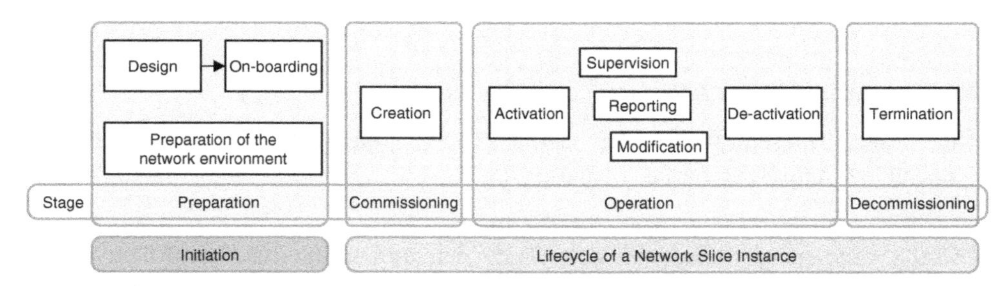

Figure 8.15 The stages of the life cycle for network slicing management.

[2] Refers to the 5QI value 82, indicating the packet error rate of 10^{-5} and packet delay budget of 10 ms.
[3] Maximum speed supported by the network slice at which a defined QoS and seamless transfer between Transmission Reception Points (TRxPs) can be achieved.

- NSI modification may include capacity or topology changes, and, e.g., a report or new network slicing requirement can trigger it.
- Deactivation sets the NSI inactive and terminates the respective communication services.
- Decommissioning refers to NSI decommissioning to terminate the NSI.

8.6 EMF Considerations

8.6.1 Safety Regulation

There have been debates about the potential impacts of wireless technologies on human health since the deployment of the first generation mobile communication systems. Observing the considerable growth of mobile communication systems since the beginning of the wireless era back in the 1980s, we are not able to see any correlation with increasing malignant tumor cases on the population regardless of some sources suggesting the risk.

Related to Electromagnetic Frequencies (EMFs), cellular systems are based on non-ionizing electromagnetic radiation. The proportion of the RF spectrum feasible for wireless communication systems spans from a few kilohertz up to a few hundred GHz, while the spectrum beyond that houses visible light as well as ultraviolet, X, gamma, and cosmic rays as depicted in Figure 8.16. The international and national frequency authorities plan the frequency allocation rules so that all wireless systems can work on their dedicated slots in harmony. The ITU dictates the global principles, whereas the national frequency regulators make the country-specific plans. As an example, in the United States, the Federal Communication Commission (FCC) decides which commercial systems can use certain bands of the RF spectrum.

In addition to the regulation of the principles for system allocations, the frequency authorities also ensure that the transmitted power level and thus the RF radiation of the telecommunications equipment – such as smartphones and base stations – do not exceed the safety values. These safety values are set well below the commonly agreed limits for the exposure of RF radiation that would start causing effects on human beings.

8.6.2 Scientific Understanding

So far, the consensus of the scientific community is that the only proven effect of non-ionizing radiation on humans is the increased temperature of the biological tissue, such as skin, if the power level is high enough. To minimize this thermal effect, the limits for the radiating power of any radio transmitter are set with a generous margin to avoid the occurrence of any thermal effects.

Equally, for authorized telecom maintenance personnel, there are strict rules within the immediate proximity of transmitter antennas, e.g., on a tower; the minimum allowed distance depends on the radiating RF power level, and if there is a need to go closer, the systems within that range need to be switched off.

In high-power sites such as macro cell towers, there may a variety of antennas of different systems, including cellular base station antennas with a typical effectively radiating

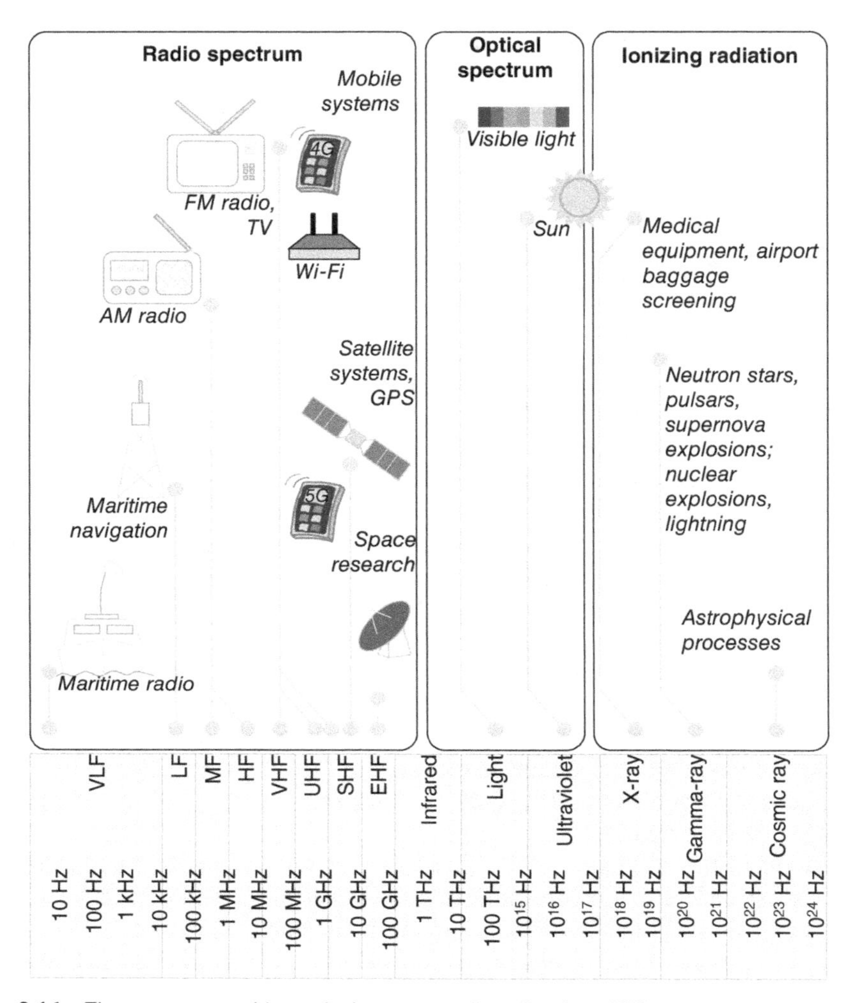

Figure 8.16 The spectrum and its typical sources and applications [40].

power, in front of the antenna of a few tens up to a few hundreds of watts. Depending on the national and international regulations, the highest power systems, such as television broadcast antenna, may result in up to, e.g., 50 kW of radiating power within the front lobe of the antenna.

Regardless of the decades of studies and broadly agreed understanding that RFs within the agreed safety limits are not harmful to humans, there are still questions about the potential health issues of radio communications. This is understandable as we are not capable of observing radio waves or their thermal effects with our limited human senses. To understand the relevancy of the respective studies, we need to understand any limitations and the impact of generalized assumptions, including the inaccuracies and errors due to the statistically limited number of subjects or their differences such as rats and mice. As an example, the US Food and Drug Administration (FDA) nominated RF radiation for study by the National Toxicology Program (NTP) concerning the potential health effects of long-term cellphone use [41]. The activity studied rats and mice that were exposed to full-body RF radiation at frequencies that are used in 2G and 3G systems. In this specific example, the highest levels of radiation exceeded four times the maximum regulatory limit when scaled to humans. The results indicate that there was "clear evidence of tumors in the hearts of male rats," and "some evidence on tumors in the brains and in the adrenal glands of male rats," while it was unclear whether there was correlation between cancer and RF exposure for female rats, and male and female mice. As a comparison, observing the levels that were comparable to the maximum allowed value for humans, the study did not find any health effects. The conclusion of this study states that NTP scientists are not sure why male rats appear to be at greater risk for developing tumors compared to female rats. In addition, the NTP found longer lifespans among exposed male rats and this may be explained by an observed decrease in chronic kidney problems that are often the cause of death in older rats. Continuing this specific example, the source also clarifies that the findings in animals cannot be directly applied to humans because the exposure and duration were greater than what people may receive from cellphones, and because the rats and mice received RF radiation across their whole bodies instead of the localized exposures humans may receive. Neither did the studyfind correlation with the heath issues caused by the RF radiation on humans. This study claims that it has been able to question the assumption that RF radiation is of no concern as long as the energy level is low and does not significantly heat the tissues.

With such unclear and sometimes even contradictory results, it is no wonder that the topic still generates confusion.

Mobile communication uses non-ionizing radiation, which does not have enough energy to impact atoms or molecules. Strong RF radiation originating from high-power radar, e.g., can cause thermal effects on living tissue, though, and this is why regulators have put such demanding limits on wireless equipment.

The mobile industry follows established and strict safety limits set by regulators for RF radiation. As an example, although our home Wi-Fi router can work on the same band as a microwave oven, it would not cause thermal effects as its transmitting power is inferior. Mobile communication specifications limit the maximum radiating power level of devices, including base stations and cellular phones.

8.6.3 Safety Distance

As a result, all mobile device manufacturers need to comply with the maximum radiating power and Specific Absorption Rate (SAR) limits. SAR refers to the amount of energy absorbed by a mass of a biological tissue. The maximum allowed SAR value falls typically in the range of 1.6–2 W/kg. As an example, the FCC has set the limit to 1.6 W/kg in the United States. Equally, based on the globally agreed principles of safety limits, the regulators have applied strict limits for base station radiation.

The principles of physical RF radiation and propagation remain the same as for preceding systems, although 5G networks may also use higher-frequency bands such as mmWaves. As depicted in Figure 8.17, the power level of the transmitter, such as the one used at the 5G base station, decreases quickly as a function of distance. Doubling the distance of a

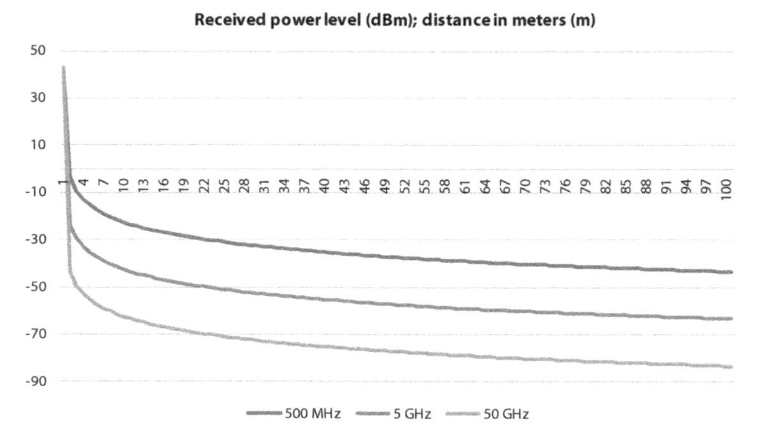

Figure 8.17 The received power from the base station decreases quickly as a function of distance. On the other hand, mobile device radiation is set to a very low level to comply with the SAR limits. This example presents the received power (in dBm) for a 20 W (43 dB) base station in a freely propagation LOS environment. In practice, there are obstacles such as buildings and vegetation that attenuate the signals further.

receiver from a transmitter means that the EMF strength at the respective new location decreases exponentially, as can be noted in the respective RF propagation equation of $P_r = P_t + G_t + G_r + 20 \log ((\lambda/(4\pi d))$ where P_r and P_t are the received and transmitted powers in dBm, G_t and G_r are transmitting and receiving antenna gains in dBi, λ is the wavelength in m, and $\pi = 3.14$.

This example demonstrates the fast attenuation of low, mid-, and high bands in ideal conditions when there are no obstacles within the propagation route, presenting a transmitted power of 20 W (corresponding to 43 dB) on 500 MHz, 5GHz, and 50 GHz. These power levels represent typical values on towers far away from people and street levels. As an example, at 10 m from the antenna, the received power is no more than 0.5 mW on the lowest presented, best propagating frequency of 500 MHz, while it is a fraction of a microwatt on the highest presented frequency of 50 GHz.

These values are inferior compared to the transmitted power of 5G handsets, which are typically up to 23 dBm or 0.2 W. In practice, the high-band base station power level is much lower, comparable to a wireless Wi-Fi home router, and its antenna can be installed, e.g., on a light pole within a city center. This is also the understanding of the industry, based on the available field tests such as the one Telstra Exchange presents [42]. As a conclusion of this case, we can see that the power levels of both 5G base station and mobile device are very low in any location users may find themselves.

8.6.4 Snapshot of Studies

In a variety of studies, such as the previously mentioned project of the FDA, typical statements reflect the need for further investigations. Meanwhile, evidence of the harmful effects of RF radiation is not strong enough to be considered causal.

To be on the safe side, some operators state that individuals who are concerned about RF radiation exposure can limit their exposure, including using an earpiece and limiting cellphone use, particularly among children. In general, though, the results do not seem to provide a credible foundation for making science-based recommendations for limiting human exposure to low-intensity RF fields as the Canadian regulator has stated:

> Despite the advent of numerous additional research studies on RF fields and health, the only established adverse health effects associated with RF field exposures in the frequency range from 3 kHz to 300 GHz relate to the occurrence of tissue heating and nerve stimulation (NS) from short-term (acute) exposures. At present, there is no scientific basis for the occurrence of acute, chronic and/or cumulative adverse health risks from RF field exposure at levels below the limits outlined in Safety Code 6. The hypotheses of other proposed adverse health effects occurring at levels below the exposure limits outlined in Safety Code 6 suffer from a lack of evidence of causality, biological plausibility and reproducibility and do not provide a credible foundation for making science-based recommendations for limiting human exposures to low-intensity RF fields.
>
> —The Minister of Public Works and Government Services, "Limits of Human Exposure to Radiofrequency Electromagnetic Fields in the Frequency Range from 3 kHz to 399 GHz; Safety Code 6"

As the GSMA indicates further, expert groups and public health agencies such as the World Health Organization (WHO) broadly agree that no health risks have been established from exposure to the low-level radio signals used for mobile communications. Based on experience with 3G and 4G networks and the results from 5G trials, the overall levels in the community will remain well below the international safety guidelines. In addition, compliance assessment of 5G network antennas and devices are dictated by international standards, which include new approaches for smart antennas and the use of new frequency ranges.

Regardless of the consensus of the scientific community, the topic still causes doubts, and there are many individual voices either against or in favor of the topic. There have also been either unintentional or possibly deliberate efforts to spread clear disinformation on the Internet based on rather non-scientific opinions, which can result in confusion by those not too familiar with the realities of non-ionizing electromagnetic radiation.

For those wanting to learn more on the topic, it is recommended to consult internationally recognized, high-quality, and non-biased studies and organizations that rely on scientifically justifiable reasoning and consensus. Table 8.10 summarizes some sources of information for further studies.

8.7 5G Measurements and Analytics

8.7.1 Key Measurement Types

In addition to measurements during the setup of the 5G network, one of the essential tasks of MNOs is to measure periodically the network performance in the operational phase. The results form an important base for constant optimization of the network.

Table 8.10 Resources for additional information on health aspects.

Source	Description
Joint Venture Silicon Valley	Bridging the gap; 21st Century Wireless Telecommunications Handbook [43]
World Health Organization	Establishing a dialogue on risks from electromagnetic fields [44]
US National Cancer Institute	Cellular phones and cancer risk [45]
Health Canada	Safety Code 6, 2015 [46]
Swisscom	5G Mobile Technology Fact Check [47]
GSMA	Safety of 5G Mobile Networks [48]
IEEE C95.1-2019	IEEE Standard for Safety Levels with Respect to Human Exposure to Electric, Magnetic, and Electromagnetic Fields, 0 Hz to 300 GHz (IEEE C95.1 and FCC OET-65, Electro-Magnetic Fields, frequencies in use, and the governing safety standards for those frequencies) [49]
FCC	FCC safety facts [50]

The measurement equipment for previous generation networks serves largely for 5G, too. Nevertheless, as 5G includes many new functionalities, new hardware is needed to support them. It might be possible to upgrade old equipment if the functionality is software based, which is the case for, e.g., a protocol analyzer with upgraded 5G protocol stacks. In other cases, if more processing power or extended frequency ranges are required, hardware upgrade is necessary to support wider bandwidths and increased data speeds.

Operators typically perform regular coverage area measurements to ensure that planned radio performance is achieved, and to monitor any potential issues. 5G is based on the already existing frequency bands as well as completely new, remarkably higher frequencies in the mmWave band.

The principles of radio measurements remain the same, and many of the old measurement types are still relevant in 5G such as quality (bit rates, bit error rates), radio coverage areas (received power level), and capacity. Nevertheless, 5G increases significantly radio bandwidths that new measurement equipment needs to support. The 5G bands will be supporting values typically from 500 MHz up to 2 GHz, and the novelty 5G bands are between 6 and 60 GHz.

For core and laboratory tests, wideband *signal generators* and *signal analyzers* are useful tools. Other measurement types relate to modulation, energy efficiency indicated by Peak-to-Average Power Ratio (PAPR), spectral efficiency indicated by the Out of Band leakage (OOB), Bit Error Rate (BER), MIMO performance, transmission latency, and quality of the synchronization, to mention a few. Table 8.11 summarizes some of the key measurement types in 5G.

The 5G era is approaching fast as NSA 5G deployments are currently taking place in many countries. The next phase is the SA option, which will provide a native, full 5G user experience.

Table 8.11 Examples of 5G measurement equipment

Type	Examples and suitability
Signal analyzer	Signal and spectral analyzers are designed to support frequency ranges of several tens of GHz, and in many cases, the range can be further extended by down-conversion. The demodulation of bandwidth may be several hundreds of GHz in basic models, while the range can be extended up to the maximum of 2 GHz for signal analysis by native hardware/software or by relying on, e.g., an external oscilloscope. The equipment may analyze digitally modulated single subcarriers at bit level or be limited to generic OFDM signal analysis. The supported frequency bands may vary from basic models' sub-6 GHz up to the highest foreseen bands in the 100 GHz range.
Signal generator	Vector signal generators serve in the simulation of wideband signals and may have native 2 GHz modulation bandwidth. The equipment may also be able to generate a variety of waveforms, apart from the 5G OFDM, and generation of inter-effects and multiple component carriers for carrier-aggregated simulations. Modulation schemes can also include 256QAM. Channel power level, co-channel power, and error vector magnitude may be included, too.
OTA	An Over-the-Air test chamber can be fixed or mobile for passive and active 5G antenna measurements, including RF patterns of complex antennas in the three-dimensional domain. The system may also support radio transceiver performance measurements. Typically, these types of measurements support all the defined RFs up to the maximum, around 87 GHz. The measurements indicate the impact of 5G RF OTA parameter values.
Network emulator	The network emulator can be useful for performance measurements of multiple subchannels of the OFDM, combined with a vector signal generator, as well as signal and spectrum analyzer.

Nevertheless, the schedule for the fully and globally interoperable IMT-2020 system, and its candidate technologies as evaluated by ITU-R, is still the original 2020 as the term indicates. IMT-2020 is a set of highly demanding requirements for next generation systems that will be forming the base for a connected society, and the 5G specifications as defined by 3GPP will definitely form part of this environment. It is worth remembering that the new frequency bands for 5G were discussed and decided at the ITU-R World Radio Conference in 2019 so we'll need to wait some time until the complete 5G system standards are ready.

Meanwhile, there are practical proofs of concept for 5G Release 16, as well as respective measurement equipment to make all the trials and pilots possible. The following presents a snapshot of recent activities in the field [51].

Anritsu has developed their Universal Wireless test platform to support new sub-6 GHz bands aimed at research of the NR of 5G.

Keysight has extended the PathWave software platform for 5G simulations, design, and test workflows. It works as a base for new innovation and product development for customers. In addition, Keysight has developed a low-frequency noise analyzer especially suitable for IoT sensor measurements.

National Instruments has productized the PXle-5840 radio tester that is suitable for NSA NR measurements up to 6 GHz frequency bands. The company has also cooperated with Samsung on 28 GHz 5G equipment and performance tests.

Rohde&Schwartz (R&S) has developed a movable OTA test chamber for 5G systems, covering test cases for antenna and transceiver measurements. R&S has also developed NR measurement equipment supporting wide bandwidths, network emulators, vector signal generators, and other relevant test devices for 5G.

Siemens has extended knowledge of 5G-related simulations, offering services for base station and silicon vendors for testing 5G functionality and performance. This has been possible by acquiring specialized companies such as Sarokal.

Testing of Release 16 performance is essential to expedite deployment of new 5G equipment as the second phase standards are approaching the final stage. There are still many challenges such as accuracy of the current measurements based on partially simulated environments but these can be managed by adjusting the final products accordingly when the standards are ready.

For more information on UE and Mobile Station OTA antenna performance and conformance testing, please refer to 3GPP TS 34.114.

8.7.2 In-Built Network Analytics

The 5G system includes an in-built capability for network analytics. 3GPP TS 23.288 defines the Network Data Analytics Function (NWDAF) as a part of the network data analytics architecture of 5G [52]. As stated in TS 23.288, the NWDAF provides analytics to 5GC network functions and Operations Administration and Maintenance (OAM). The results include statistical information of past events as well as predictive information. The type of analytics can be configured via different NWDAF instances. For the discovery procedure, NWDAF instances provide the list of Analytics IDs upon registering to the Network Repository Function (NRF) as part of the registration of the elements of the network function profile. The actual interactions between 5GC network functions and the NWDAF happen within a Public Land Mobile Network (PLMN).

The NWDAF is able to obtain data, via the Data Collection feature, from network functions that are AF, AMF, PCF, SMF, Unified Data Management (UDM), NRF, as well as OAM. The NWDAF has visibility to both radio and core networks, and it can post-process these data.

The input data for the NWDAF can include performance indicators of the UE via the respective network functions, as well as other metrics such as the number of UE at a given time within a certain location, and historic data during certain time periods. The post-processed data of the NWDAF works as a basis for the operator to understand the performance and functional indicators of the 5G network.

The NWDAF collects data from network functions via subscription. The NWDAF can thus subscribe and unsubscribe at the 5G core network function to establish the relation for notifications between the NWDAF and the subscribed function. In practice, the network functions rely on Event Exposure Service and other related services that 3GPP TS 23.502, Sections 4.15, 4.17, and 5.2 define.

In addition to the above-mentioned data collection from the network functions, the NWDAF is also capable of collecting management data from the OAM services of the operator. These data include the NG-RAN and 5G core network performance measurements based on 3GPP TS 28.552, and the end-to-end Key Performance Indicators (KPIs) of 5G based on 3GPP TS 28.554.

The OAM relation with the NWDAF is based on generic performance assurance and fault supervision management services as defined in TS 28.532, Performance Management (PM) services as defined in TS 28.550, and Fault Supervision (FS) services as defined in TS 28.545.

The respective analytics are able to provide network function analytics related to, e.g., virtual resource utilization and resource configuration information. The OAM services can also provide the NWDAF with UE-related insights such as measurement collection for Minimization of Drive Tests (MDT) as defined in 3GPP TS 37.320.

Among other information, the NWDAF can also provide information on the slice load to a network function on an NSI level. In practice, the NWDAF can notify and provide to the network functions (which are subscribed to the NWDAF) the slice-specific network status and analytics by relying on Analytics ID, Single Network Slice Selection Assistance Information (S-NSSAI), and NSI ID.

8.7.3 Minimization of Drive Tests

3GPP TS 28.552 presents the management and orchestration of 5G performance measurements. As summarized in Ref. [53] and detailed in 3GPP 37.320, MDT refers to combined reporting of both UL and DL radio quality together with a set of call events such as failed call attempts, UP QoS, and the geographical position of the UE. The reporting procedure contains steps for the UE to send an MDT ticket accompanied by DL radio quality and call events, with optional location information, and for the LTE eNB or 5G gNB to add cell-based UL interference and UP QoS per QoS Class Identifier (QCI) per call. Upon completing the addition, the base station sends the combined UL/DL measurements and events to a centralized server as depicted in Figure 8.18.

Typical use cases of MDT are to detect UL and DL coverage outage areas and weak signals, to detect UL/DL interference, and to detect excess coverage scenarios (as an example, to detect issues for UE engaging in a random access procedure far away from the base station). The MDT also works for exploring user radio QoS experience via IP throughput and latency, and helps operators to plan small cell deployment via detecting traffic hotspots.

In practice, it can be expected that around 10% of all UE may support and use the MDT feature in the network. The limiting factor is the amount of legacy UE base that does not support MDT, as well as potential privacy concerns in UE reporting location (which does not allow the activation of the feature in some regions).

Immediate MDT reporting is the simple form of MDT reporting. It means that each RRC measurement report is translated into MDT. The 3GPP defines the respective procedure,

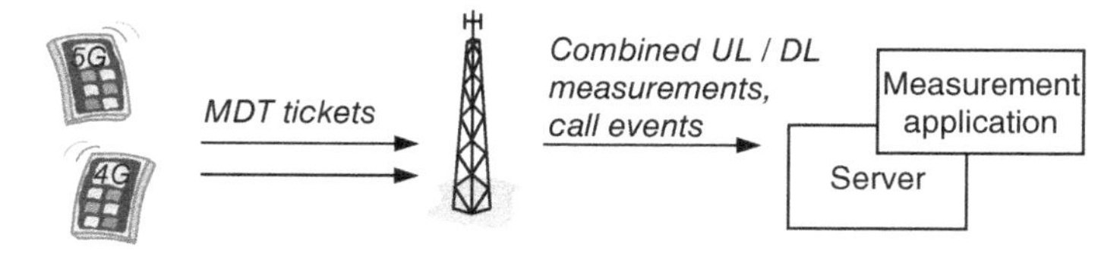

Figure 8.18 The principle of MDT reporting.

but content coding is vendor specific. As an example of direct reporting, a handover scenario triggers a series of UP reports of data volume and packet delay and loss that the UE sends to the base station, which is further delivered to the server.

The other variant is logged measurement reporting. This functions in such a way that, after successful RRC establishment, the base station sends a configuration log to the UE with accompanying key parameters. Posteriorly, the UE informs the base station only in certain situations such as during the RRC re-establishment procedure about the available logged measurements at that moment.

3GPP Release 16 includes the possibility of including wireless LAN and Bluetooth Received Signal Strength Indicator (RSSI) measurement reports to the MDT. Furthermore, Release 16 adds a new NR beam-level measurement to the UE, as well as WLAN Roundtrip Time (RTT) measurement.

As some of the functional Release 16 additions, beam recovery failure triggers a Radio Link Failure (RLF) report, Absolute Radio-Frequency Channel Number New Radio (ARFCN-NR) and subcarrier spacing are reported, and a Random Access (RA) report is delivered per RA attempt. For more details on the RRC procedures, please see 3GPP TS 38.331.

References

1 Tabbane, S., "5G Networks and 3GPP Release 15: ITU PITA Workshop on Mobile Network Planning and Security," ITU, Nadi, Fiji Islands, 23–25 October 2019.

2 3GPP, "TS 23.501 V16.5.1, System Architecture for the 5G System (5GS), Stage 2, Release 16," 3GPP, August 2020.

3 3GPP, "Ts 38.401, V16.2.0," 3GPP, June 2020.

4 Ericsson, "Simplifying the 5G Ecosystem by Reducing Architecture Options," 30 November 2018. [Online]. Available: https://www.ericsson.com/en/reports-and-papers/ericsson-technology-review/articles/simplifying-the-5g-ecosystem-by-reducing-architecture-options. [Accessed 9 September 2020].

5 Ericsson, "5G Deployment Considerations," Ericsson. [Online]. Available: https://www.ericsson.com/en/networks/trending/insights-and-reports/5g-deployment-considerations. [Accessed 5 July 2019].

6 Sirotkin, S., "5G Standards: 3GPP Release 15, 16, and Beyond," 3GPP, 2019. [Online]. Available: https://www.3gpp.org/ftp/Information/presentations/presentations_2019/2019_05_Wireless-Russia-Sasha-Sirotkin.pdf. [Accessed 1 September 2020].

7 GSMA, "Road to 5G: Introduction and Migration," 3GPP, April 2018. [Online]. Available: https://www.gsma.com/futurenetworks/wp-content/uploads/2018/04/Road-to-5G-Introduction-and-Migration_FINAL.pdf. [Accessed 9 September 2020].

8 ETSI, "ETSI TR 138 913 V14.2.0 (2017-05) Study on Scenarios and Requirements for Next Generation Access Technologies," ETSI, May 2017.

9 Kyösti, P., "Radio Channel Modelling for 5G Telecommunication System Evaluation and over the Air Testing," University of Oulu, Oulu, 2018.

10 Larsson, E.G., Edfors, O., Tufvesson, F. et al., "Massive MIMO for Next Generation Wireless Systems," *IEEE Communications Magazine* 52(2): 186–195, 2014.
11 Verizon, "Physical Layer Procedures, Technical Report TS V5G.213 V1.0," Verizon 5G TF, 2016. [Online]. Available: http://5gtf.net/V5G_213_v1p0.pdf. [Accessed 30 July 2018].
12 Techplayon, "5G Network RF Planning – Link Budget Basics," Techplayon, 19 November 2019. [Online]. Available: http://www.techplayon.com/5g-network-rf-planning-link-budget-basics. [Accessed 30 September 2020].
13 ITU, "Report ITU-R M.2376-0, Technical Feasibility of IMT in Bands above 6 GHz," ITU, July 2015.
14 ITU, "Handbook: ITU-R Propagation Prediction Methods for Interference and Sharing Studies," ITU, 2012. [Online]. Available: https://www.itu.int/dms_pub/itu-r/opb/hdb/R-HDB-58-2012-OAS-PDF-E.pdf. [Accessed 25 September 2020].
15 Wang, H., Zhang, P., Li, J. et al., "Radio Propagation and Wireless Coverage of LSAA-Based 5G Millimeter-Wave Mobile Communication Systems," China Communications, 2019.
16 ITU, "Recommendation ITU-R P.1238-9 (06/2017), Propagation Data and Prediction Methods for the Planning of Indoor Radiocommunication Systems and Radio Local Area Networks in the Frequency Range 300 MHz to 100 GHz," ITU, June 2017. [Online]. Available: https://www.itu.int/dms_pubrec/itu-r/rec/p/R-REC-P.1238-9-201706-I!!PDF-E.pdf. [Accessed 25 September 2020].
17 ITU, "HANDBOOK: Terrestrial Land Mobile Radiowave Propagation in the VHF/UHF Bands," ITU, 2002.
18 3GPP, "TS 36.300 V16.1.0 (2020-03), E-UTRA and E-UTRAN, Overall Description, Stage 2, (Release 16)," 3GPP, March 2020.
19 O-RAN Alliance, "O-RAN Use Cases and Deployment Scenarios towards Open and Smart RAN (White Paper)," O-RAN Alliance, February 2020.
20 CPRI, "CPRI, Common Public Radio Interface," CPRI. [Online]. Available: http://www.cpri.info. [Accessed 18 September 2020].
21 CPRI, "CPRI Specification V7.0 (2015-10-09), CPRI Interface," CPRI, October 2015.
22 Knopp, R., Nikaein, N., Bonnet, C. et al., "Prototyping of Next Generation Fronthaul Interfaces (NGFI) Using OpenAirInterface," Open Air Interface. [Online]. Available: https://www.openairinterface.org/?page_id=1695. [Accessed 18 September 2020].
23 CPRI, "eCPRI Specification V2.0 (2019-05-10), eCPRI Interface," CPRI, May 2019.
24 IEEE, "1914.1-2019 – IEEE Standard for Packet-based Fronthaul Transport Networks." [Online]. Available: https://standards.ieee.org/standard/1914_1-2019.html. [Accessed 18 September 2020].
25 Parallel Wireless, "5G NR Logical Architecture and Its Functional Splits," Parallel Wireless. [Online]. Available: https://www.parallelwireless.com/wp-content/uploads/5GFunctionalSplits.pdf. [Accessed 15 September 2020].
26 3GPP, "TR 38.801 V14.0.0 (2017-03), Study on New Radio Access Technology, Radio Access Architecture and Interfaces, Release 14," 3GPP, March 2017.
27 Głąbowski, M., Gacanin, H., Moschiolos, I. et al., "Editorial Design, Dimensioning, and Optimization of 4G/5G Wireless Communication Networks," 2017.

28 Basta, A., Blenk A., Hoffmann, K. et al., "Towards a Cost Optimal Design for a 5G Mobile Core Network Based on SDN and NFV," *IEEE Transactions on Network and Service Management* 14(4), 2017.
29 Gebert, S., Zinner, T., Tran-Gia, P. et al. "Demonstrating the Optimal Placement of Virtualized Cellular Network Functions in Case of Large Crowd Events," *ACM SIGCOMM Computer Communication Review* 44(4): 359–360, 2014.
30 Suksomboon, K., Fukushima, M., Hayashi, M. et al.,, "LawNFO: A Decision Framework for Optimal Locationaware Network Function Outsourcing," *Proc. 1st IEEE Conf. Netw. Softwarization (NetSoft)*, London, UK, 2015.
31 Shi, R., Zhang, J., Chu, W. et al., "MDP and Machine Learning-based Cost-optimization of Dynamic Resource Allocation for Network Function Virtualization," Proc. IEEE Int. Conf. Services Comput. (SCC), New York, NY, USA, 2015.
32 ETSI, "MEC Deployments in 4G and Evolution towards 5G," ETSI, February 2018. [Online]. Available: https://www.etsi.org/images/files/ETSIWhitePapers/etsi_wp24_MEC_deployment_in_4G_5G_FINAL.pdf. [Accessed 24 September 2020].
33 ITU-T, "ITU-T Technical Report GSTR-TN5G, Transport Network Support of IMT-2020/5G," ITU-T, 9 February 2018. [Online]. Available: https://www.itu.int/dms_pub/itu-t/opb/tut/T-TUT-HOME-2018-PDF-E.pdf. [Accessed 13 September 2020].
34 Jordan, E., "The Ultimate Guide to Open RAN: Open RAN Components and RAN Functional Splits," The Fast Mode, 2020. [Online]. Available: https://www.thefastmode.com/expert-opinion/17921-the-ultimate-guide-to-open-ran-open-ran-components-and-ran-functional-splits. [Accessed 13 September 2020].
35 Trivisonno, R., Condoluci, M., An, X. et al., "mIoT Slice for 5G Systems," Design and Performance Evaluation.
36 GSMA, "NG.116 Generic Network Slice Template V3.0," GSMA, 22 October 2019. [Online]. Available: https://www.gsma.com/newsroom/resources/ng-116-generic-network-slice-template-v3-0. [Accessed 10 September 2020].
37 GSMA, "Network Slicing Use Case Requirements," GSMA, April 2018.
38 GSMA, "NG.116: Neywork Slice Template, V3.0," GSMA, 2020.
39 3GPP, "TS 28.530 V16.2.0, Concepts, Use Cases and Requirements of Management and Orchestration, Release 16," 3GPP, July 2020.
40 NASA, "What Are the Spectrum Band Designators and Bandwidths?" NASA, 2 September 2018. [Online]. Available: https://www.nasa.gov/directorates/heo/scan/communications/outreach/funfacts/txt_band_designators.html. [Accessed 20 September 2020].
41 National Toxicology Program, "Cellphone Radio Frequency Radiation Studies," National Institute of Environmental Health Sciences NIH-HHS, August 2020. [Online]. Available: https://www.niehs.nih.gov/health/materials/cell_phone_radiofrequency_radiation_studies_508.pdf. [Accessed 20 September 2020].
42 Wood, M., "5 Surveys of 5G Show EME Levels Well below Safety Limits," Telstra Exchange, 8 July 2019. [Online]. Available: https://exchange.telstra.com.au/5-surveys-of-5g-show-eme-levels-well-below-safety-limits. [Accessed 20 September 2020].
43 Joint Venture Silion Valley, "Bridging the Gap," 2020. [Online]. Available: https://jointventure.org/images/stories/pdf/JVSV_Wireless-Telecommunications-Handbook_2ndEd_DEC2019.pdf.
44 WHO, "Electromagnetic Fields (EMF)," 31 July 2019. [Online]. Available: www.who.int/peh-emf/en.

45 National Cancer Institute, "Cellular Phones and Cancer Risk," National Cancer Institute, 9 January 2019. [Online]. Available: https://www.cancer.gov/about-cancer/causes-prevention/risk/radiation/cell-phones-fact-sheet. [Accessed 4 July 2019].

46 Health Canada, "Limits of Human Exposure to Radiofrequency Electromagnetic Energy in the Frequency Range from 3 kHz to 300 GHz," Health Canada, 18 April 2018. [Online]. Available: https://www.canada.ca/en/health-canada/services/environmental-workplace-health/consultations/limits-human-exposure-radiofrequency-electromagnetic-energy-frequency-range-3-300.html. [Accessed 4 July 2019].

47 Swisscom, "5G Mobile Technology Fact Check," Swisscom, 27 March 2019. [Online]. Available: https://www.swisscom.ch/en/about/news/2019/03/27-5g-mobile-technology-fact-check.html. [Accessed 4 July 2019].

48 GSMA, "Safety of 5G Mobile Networks," July 2019. [Online]. Available: https://www.gsma.com/publicpolicy/wp-content/uploads/2019/06/GSMA_Safety-of-5G-Mobile-Networks_July-2019.pdf.

49 IEEE, "C95.1-2019 – IEEE Standard for Safety Levels with Respect to Human Exposure to Electric, Magnetic, and Electromagnetic Fields, 0 Hz to 300 GHz," IEEE, 4 October 2019. [Online]. Available: https://ieeexplore.ieee.org/document/8859679. [Accessed 20 September 2020].

50 FCC, "RF Safety FAQ," 25 November 2015. [Online]. Available: https://www.fcc.gov/engineering-technology/electromagnetic-compatibility-division/radio-frequency-safety/faq/rf-safety.

51 Penttinen, J., "5G Testereihin Monta Kisaajaa (Finnish Article on 5G Testers with English Summaries)," 2 September 2017. [Online]. Available: https://issuu.com/uusiteknologia.fi/docs/2_2017/42.

52 3GPP, "TS 23.288 V16.4.0 (2020-07), Architecture Enhancements for 5G System to Support Network Data Analytics Services, Release 16," 3GPP, July 2020.

53 Kreher, R., "What Is 3GPP Minimization of Drive Tests?" The 3G4G Blog, 28 August 2020. [Online]. Available: https://blog.3g4g.co.uk/2020/08/3gpp-mdt-how-it-works-and-what-is-new.html. [Accessed 27 September 2020].

Appendix

Table 1 summarizes the main thematic categories of the 3GPP specifications and reports relevant to 5G. As can be noted, the foundational 5G radio aspects are found in the 38-series. The core aspects are described in a variety of specifications such as 21 (requirements, defining Stage 1), 22 (service aspects, defining Stage 2), 23 (technical realization, defining Stage 3), 24 (signaling protocols), and 33 (security aspects) among others.

On the functional and architectural descriptions of 5G and other generations, the foundational specifications are 22, 23, and 24.

5G Second Phase Explained: The 3GPP Release 16 Enhancements, First Edition. Jyrki T.J. Penttinen.

Table 1 Snapshot of the 3GPP Technical Specifications (TS) and Technical Reports (TR) relevant to 5G. Series categorization and some of the most relevant documents related to this book defining Release 16 of 5G.

Technical area	Series	Key documents
Requirements	21	TR 21.914, Release 14 description
		TR 21.915, Release 15 description
		TR 21.916, Release 16 description
Service aspects, Stage 1	22	TR 22.888, Study on enhancements for Machine-Type Communications (MTC)
		TR 22.889, Study on Future Railway Mobile Communication System (FRMCS)
		TR 22.924, Charging and accounting mechanisms
		TR 22.925, Quality of Service (QoS) and network performance
		TR 22.926, Guidelines for extra-territorial 5G systems
		TR 22.934, Feasibility study on 3GPP system to Wireless Local Area Network (WLAN) interworking
		TR 22.940, IP Multimedia Subsystem (IMS) messaging
		TR 22.951, Service aspects and requirements for network sharing
		TR 22.973, IMS Multimedia Telephony service
Technical realization, Stage 2	23	TS 23.273, 5G System (5GS) Location Services (LCS); Stage 2
		TS 23.501, System architecture for the 5G System (5GS)
		TS 23.502, Procedures for the 5G System (5GS)
		TS 23.503, Policy and charging control framework for the 5G System (5GS); Stage 2
		TR 23.700-20, Study on enhanced support of Industrial Internet of Things (IIoT) in the 5G System (5GS)
		TR 23.700-24, Study on support of the 5GMSG (Message Service for MIoT over 5G System) Service
		TR 23.716, Study on the wireless and wireline convergence for the 5G System (5GS) architecture
		TR 23.724, Study on Cellular Internet of Things (CIoT) support and evolution for the 5G System (5GS)
		TR 23.725, Study on enhancement of Ultra-Reliable Low-Latency Communication (URLLC) support in the 5G Core network (5GC)
		TR 23.726, Study on enhancing topology of the Service Management Function (SMF) and the User Plane Function (UPF) in 5G networks
		TR 23.727, Study on application awareness interworking between LTE and NR

		TR 23.730, Study on extended architecture support for Cellular Internet of Things (CIoT)
		TR 23.731, Study on enhancement to the 5GC LoCation Services (LCS)
		TR 23.734, Study on enhancement of 5G System (5GS) for vertical and Local Area Network (LAN) services
		TR 23.737, Study on architecture aspects for using satellite access in 5G
		TR 23.745, Study on application support layer for Factories of the Future (FotF) in the 5G network
		TR 23.748, Study on enhancement of support for Edge Computing in 5G Core network (5GC)
		TR 23.752, Study on system enhancement for Proximity based Services (ProSe) in the 5G System (5GS)
		TR 23.756, Study for single radio voice continuity from 5G to 3G
		TR 23.774, Study on mission critical services over 5G multicast broadcast system
		TR 23.783, Study on Mission Critical (MC) services support over the 5G System (5GS)
		TR 23.791, Study of enablers for Network Automation for 5G
		TR 23.793, Study on access traffic steering, switch and splitting support in the 5G System (5GS) architecture
		TR 23.794, Study on enhanced IP Multimedia Subsystem (IMS) to 5GC integration
		TR 23.892, IP Multimedia Subsystem (IMS) centralized services
		TR 23.948, Deployment guidelines for typical edge computing use cases in 5G Core network (5GC)
Signaling protocols, stage 3, user equipment to network	24	TS 24.193, 5G System; Access Traffic Steering, Switching and Splitting (ATSSS); Stage 3
		TS 24.501, Non-Access-Stratum (NAS) protocol for 5G System (5GS); Stage 3
		TS 24.502, Access to the 3GPP 5G Core Network (5GCN) via non-3GPP access networks
		TS 24.519, 5G System (5GS); Time-Sensitive Networking (TSN) Application Function (AF) to Device-Side TSN Translator (DS-TT) and Network-Side TSN Translator (NW-TT) protocol aspects; Stage 3
		TS 24.526, User Equipment (UE) policies for 5G System (5GS); Stage 3
		TS 24.571, 5G System (5GS); Control plane Location Services (LCS) procedures; Stage 3
		TS 24.587, Vehicle-to-Everything (V2X) services in 5G System (5GS); Stage 3
		TS 24.588, Vehicle-to-Everything (V2X) services in 5G System (5GS); User Equipment (UE) policies; Stage 3
		TR 24.890, CT WG1 aspects of 5G System Phase 1

(Continued)

Technical area	Series	Key documents
Radio aspects	25	The 5G radio aspects are presented in TS series 38. For the interoperability aspects (inter-RAT handovers etc.), the following specifications form the key base. TS 25.101, User Equipment (UE) radio transmission and reception (FDD) TS 25.102, User Equipment (UE) radio transmission and reception (TDD) TS 25.104, Base Station (BS) radio transmission and reception (FDD) TS 25.105, Base Station (BS) radio transmission and reception (TDD) TS 25.113, Base Station (BS) and repeater electromagnetic compatibility (EMC)
CODECs	26	TS 26.117, 5G Media Streaming (5GMS); speech and audio profiles TS 26.501, 5G Media Streaming (5GMS); general description and architecture TS 26.511, 5G Media Streaming (5GMS); profiles, codecs and formats TS 26.512, 5G Media Streaming (5GMS); protocols TR 26.802, 5G Multimedia Streaming (5GMS); multicast architecture TR 26.803, 5G Media Streaming (5GMS); architecture extensions TR 26.891, 5G enhanced mobile broadband; media distribution
Data	27	TS 27.007, AT command set for User Equipment (UE)
Signaling protocols, Stage 3; RSS-CN, OAM&P and Charging aspects	28	This series presents some items transferred from the 32 series, too. TS 28.554, Management and orchestration; 5G end to end Key Performance Indicators (KPI) TS 28.555, Management and orchestration; network policy management for 5G mobile networks TS 28.556, Management and orchestration; network policy management for 5G mobile networks; Stage 2 and stage 3 TS 28.557, Management and orchestration; management of non-public networks; Stage 1 and stage 2 TR 28.800, Study on management and orchestration architecture of next generation networks and services TR 28.801, Telecommunication management; study on management and orchestration of network slicing for next generation network TR 28.802, Telecommunication management; study on management aspects of next generation network architecture and features TR 28.803, Telecommunication management; study on management aspects of edge computing

		TR 28.804, Telecommunication management; study on tenancy concept in 5G networks and network slicing management TR 28.808, Study on management and orchestration aspects of integrated satellite components in a 5G network TR 28.811, Network slice management enhancement TR 28.813, Study on new aspects of Energy Efficiency (EE) for 5G TR 28.861, Telecommunication management; study on the Self-Organizing Networks (SON) for 5G networks TR 28.890, Management and orchestration; study on integration of Open Network Automation Platform (ONAP) and 3GPP management for 5G networks
Signaling protocols, Stage 3. Intra-fixed-network aspects	29	TS 29.500, 5G System; Technical Realization of Service Based Architecture; Stage 3 TS 29.501, 5G System; Principles and Guidelines for Services Definition; Stage 3 TS 29.502, 5G System; Session Management Services; Stage 3 TS 29.503, 5G System; Unified Data Management Services; Stage 3 TS 29.504, 5G System; Unified Data Repository Services; Stage 3 TS 29.505, 5G System; Usage of the Unified Data Repository services for Subscription Data; Stage 3 TS 29.507, 5G System; Access and Mobility Policy Control Service; Stage 3 TS 29.508, 5G System; Session Management Event Exposure Service; Stage 3 TS 29.509, 5G System; Authentication Server Services; Stage 3 TS 29.510, 5G System; Network function repository services; Stage 3 TS 29.511, 5G System; Equipment Identity Register Services; Stage 3 TS 29.512, 5G System; Session Management Policy Control Service; Stage 3 TS 29.513, 5G System; Policy and Charging Control signaling flows and QoS parameter mapping; Stage 3 TS 29.514, 5G System; Policy Authorization Service; Stage 3 TS 29.515, 5G System; Gateway Mobile Location Services; Stage 3 TS 29.517, 5G System; Application Function (AF) event exposure service; Stage 3 TS 29.518, 5G System; Access and Mobility Management Services; Stage 3 TS 29.519, 5G System; Usage of the Unified Data Repository Service for Policy Data, Application Data and Structured Data for Exposure; Stage 3 TS 29.520, 5G System; Network Data Analytics Services; Stage 3 TS 29.521, 5G System; Binding Support Management Service; Stage 3

(*Continued*)

Technical area	Series	Key documents
		TS 29.522, 5G System; Network Exposure Function Northbound APIs; Stage 3
		TS 29.523, 5G System; Policy Control Event Exposure Service; Stage 3
		TS 29.524, 5G System; cause codes mapping between 5GC interfaces; Stage 3
		TS 29.525, 5G System; UE Policy Control Service; Stage 3
		TS 29.526, 5G System; Network Slice-Specific Authentication and Authorization (NSSAA) services; Stage 3
		TS 29.531, 5G System; Network Slice Selection Services; Stage 3
		TS 29.540, 5G System; SMS Services; Stage 3
		TS 29.541, 5G System; Network Exposure (NE) function services for Non-IP Data Delivery (NIDD); Stage 3
		TS 29.542, 5G System; session management services for Non-IP Data Delivery (NIDD); Stage 3
		TS 29.544, 5G System; Secured Packet Application Function (SP-AF) services; Stage 3
		TS 29.550, 5G System; steering of roaming application function services; Stage 3
		TS 29.551, 5G System; Packet Flow Description Management Service; Stage 3
		TS 29.554, 5G System; Background Data Transfer Policy Control Service; Stage 3
		TS 29.561, 5G System; Interworking between 5G Network and external Data Networks; Stage 3
		TS 29.562, 5G System; Home Subscriber Server (HSS) services for interworking with the IP Multimedia Subsystem (IMS); Stage 3
		TS 29.563, 5G System; Home Subscriber Server (HSS) services for interworking with Unified Data Management (UDM); Stage 3
		TS 29.571, 5G System; Common Data Types for Service Based Interfaces; Stage 3
		TS 29.572, 5G System; Location Management Services; Stage 3
		TS 29.573, 5G System; Public Land Mobile Network (PLMN) Interconnection; Stage 3
		TS 29.591, 5G System (5GS); Network exposure function southbound services; Stage 3
		TS 29.594, 5G System; Spending Limit Control Service; Stage 3
		TS 29.673, 5G System; UE radio capability management services; Stage 3
		TR 29.843, Study on Load and Overload Control of 5GC Service Based Interface
		TR 29.890, CT WG3 aspects of 5G System Phase 1
		TR 29.891, 5G System – Phase 1 CT WG4 Aspects
		TR 29.892, Study on User Plane Protocol in 5GC
		TR 29.893, Study on IETF QUIC Transport for 5GC Service Based Interfaces

Program management	30	N/A
Subscriber Identity Module (SIM), Universal SIM (USIM), IC Cards	31	Test specifications. TS 31.101, UICC-terminal interface; physical and logical characteristics TS 31.102, Characteristics of the Universal Subscriber Identity Module (USIM) application TS 31.103, Characteristics of the IP Multimedia Services Identity Module (ISIM) application TS 31.104, Characteristics of the Hosting Party Subscription Identity Module (HPSIM) application TS 31.110, Numbering system for telecommunication IC card applications TS 31.111, Universal Subscriber Identity Module (USIM) Application Toolkit (USAT)
OAM&P and Charging	32	TS 32.255, Telecommunication management; charging management; 5G data connectivity domain charging; Stage 2 TS 32.256, Charging management; 5G connection and mobility domain charging; Stage 2 TS 32.290, Telecommunication management; charging management; 5G system; Services, operations and procedures of charging using Service Based Interface (SBI) TS 32.291, Telecommunication management; charging management; 5G system, charging service; Stage 3 TR 32.871, Study on policy management for mobile networks based on Network Function Virtualization (NFV) scenarios TR 32.899, Telecommunication management; charging management; study on charging aspects of 5G system architecture phase 1 TR 32.972, Telecommunication management; study on system and functional aspects of energy efficiency in 5G networks
Access requirements and test specifications Security aspects	33	TS 33.310, Network Domain Security (NDS); Authentication Framework (AF) TS 33.501, Security architecture and procedures for 5G System (as per ETSI TS 133 501) TS 33.511, Security Assurance Specification (SCAS) for the next generation Node B (gNodeB) network product class TS 33.512, 5G Security Assurance Specification (SCAS); Access and Mobility management Function (AMF) TS 33.513, 5G Security Assurance Specification (SCAS); User Plane Function (UPF) TS 33.514, 5G Security Assurance Specification (SCAS) for the Unified Data Management (UDM) network product class TS 33.515, 5G Security Assurance Specification (SCAS) for the Session Management Function (SMF) network product class TS 33.516, 5G Security Assurance Specification (SCAS) for the Authentication Server Function (AUSF) network product class TS 33.517, 5G Security Assurance Specification (SCAS) for the Security Edge Protection Proxy (SEPP) network product class

(Continued)

Technical area	Series	Key documents
		TS 33.518, 5G Security Assurance Specification (SCAS) for the Network Repository Function (NRF) network product class TS 33.519, 5G Security Assurance Specification (SCAS) for the Network Exposure Function (NEF) network product class TS 33.520, 5G Security Assurance Specification (SCAS); Non-3GPP InterWorking Function (N3IWF) TS 33.521, 5G Security Assurance Specification (SCAS); Network Data Analytics Function (NWDAF) TS 33.522, 5G Security Assurance Specification (SCAS); Service Communication Proxy (SECOP) TS 33.535, Authentication and key management for applications based on 3GPP credentials in the 5G System (5GS) TR 33.809, Study on 5G security enhancements against false base stations TR 33.811, Study on security aspects of 5G network slicing management TR 33.814, Study on the security of the enhancement to the 5G Core (5GC) location services TR 33.819, Study on security enhancements of 5G System (5GS) for vertical and Local Area Network (LAN) services TR 33.841, Study on the support of 256-bit algorithms for 5G TR 33.842, Study on Lawful Interception (LI) service in 5G TR 33.845, Study on storage and transport of 5GC security parameters for ARPF authentication TR 33.846, Study on authentication enhancements in the 5G System (5GS) TR 33.848, Study on security impacts of virtualization TR 33.855, Study on security aspects of the 5G Service Based Architecture (SBA) TR 33.856, Study on security aspects of single radio voice continuity from 5G to UTRAN TR 33.861, Study on evolution of cellular IoT security for the 5G System
UE and (U)SIM test specifications	34	TS 34.124, Electromagnetic compatibility (EMC) requirements for mobile terminals and ancillary equipment (pre-5G) TS 34.229-5, Internet Protocol (IP) multimedia call control protocol based on Session Initiation Protocol (SIP) and Session Description Protocol (SDP); User Equipment (UE) conformance specification; Part 5: Protocol conformance specification using 5G System (5GS) TR 34.907, Report on electrical safety requirements and regulations (pre-5G) TR 34.925, Specific Absorption Rate (SAR) requirements and regulations in different regions (pre-5G) TR 34.926, Electromagnetic compatibility (EMC); Table of international requirements for mobile terminals and ancillary equipment (pre-5G)

Security algorithms	35	The TS/TR 35 series presents algorithms for authentication, confidentiality, integrity, and key generation functions in pre-5G environment. For the 5G security, the TS 33.501, security architecture and procedures for 5G System, serves as a starting point. Please note these algorithms may be subject to export licensing conditions.
LTE (Evolved UTRA), LTE-Advanced, LTE-Advanced Pro radio technology	36	TS 36.300, Evolved Universal Terrestrial Radio Access (E-UTRA) and Evolved Universal Terrestrial Radio Access Network (E-UTRAN); overall description; Stage 2 TR 36.776, Evolved Universal Terrestrial Radio Access (E-UTRA); study on LTE-based 5G terrestrial broadcast TR 36.976, Overall description of LTE-based 5G broadcast
Multiple radio access technology aspects	37	TS 37.104, NR, E-UTRA, UTRA and GSM/EDGE; Multi-Standard Radio (MSR) Base Station (BS) radio transmission and reception TS 37.113, NR, E-UTRA, UTRA and GSM/EDGE; Multi-Standard Radio (MSR) Base Station (BS) Electromagnetic Compatibility (EMC) TS 37.141, NR, E-UTRA, UTRA and GSM/EDGE; Multi-Standard Radio (MSR) Base Station (BS) conformance testing TS 37.324, Evolved Universal Terrestrial Radio Access (E-UTRA) and NR; Service Data Adaptation Protocol (SDAP) specification TS 37.340, NR; Multi-connectivity; overall description; Stage-2 TR 37.716-00-00, Band combinations for Stand-Alone (SA) NR Supplementary UpLink (SUL), Non-Stand-Alone (NSA) NR SUL, NSA NR SUL with UL sharing from the UE perspective (ULSUP) TR 37.716-11-11, E-UTRA (Evolved Universal Terrestrial Radio Access) – NR Dual Connectivity (EN-DC) of 1 band LTE (1 Down Link (DL)/1 Up Link (UL)) and 1 NR band (1 DL/1 UL) TR 37.716-21-11, E-UTRA (Evolved Universal Terrestrial Radio Access) – NR Dual Connectivity (EN-DC) of 2 bands LTE inter-band Carrier aggregation (CA) (2 Down Link (DL)/1 Up Link (UL)) and 1 NR band (1 DL/1 UL) TR 37.716-21-21, E-UTRA (Evolved Universal Terrestrial Radio Access) – NR Dual Connectivity (EN-DC) of x bands (x=1, 2, 3, 4) LTE inter-band Carrier aggregation (CA) (x Down Link (DL)/1 Up Link (UL)) and 2 NR bands NR inter-band CA (2 DL/1 UL) TR 37.716-31-11, E-UTRA (Evolved Universal Terrestrial Radio Access) – NR Dual Connectivity (EN-DC) of 3 bands LTE inter-band Carrier aggregation (CA) (3 Down Link (DL)/1 Up Link (UL)) and 1 NR band (1 DL/1 UL) TR 37.716-33, E-UTRA (Evolved Universal Terrestrial Radio Access) – NR Dual Connectivity (EN-DC) of 3 bands Down Link (DL) and 3 band Up Link (UL)

(Continued)

Technical area	Series	Key documents
		TR 37.716-41-11, E-UTRA (Evolved Universal Terrestrial Radio Access) – NR Dual Connectivity (EN-DC) of 4 bands LTE inter-band Carrier aggregation (CA) (4 Down Link (DL)/1 Up Link (UL)) and 1 NR band (1 DL/1 UL)
		TR 37.716-41-22, Dual Connectivity (EN-DC) of LTE inter-band CA xDL/1UL bands (x = 2,3,4) and NR FR1 1DL/1UL band and NR FR2 1DL/1UL band
		TR 37.716-51-11, E-UTRA (Evolved Universal Terrestrial Radio Access) - NR Dual Connectivity (EN-DC) of 5 bands LTE inter-band Carrier aggregation (CA) (5 Down Link (DL)/1 Up Link (UL)) and 1 NR band (1 DL/1 UL)
		TR 37.815, Study on high power User Equipment (UE) (power class 2) for E-UTRA (Evolved Universal Terrestrial Radio Access) – NR Dual Connectivity (EN-DC) (1 LTE FDD band + 1 NR TDD band)
		TR 37.816, Study on RAN-centric data collection and utilization for LTE and NR
		TR 37.822, Study on next generation Self-Optimizing Network (SON) for UTRAN and E-UTRAN
		TR 37.823, Coexistence between LTE-MTC and NR
		TR 37.824, Coexistence between NB-IoT and NR
		TR 37.825, High power User Equipment (UE) (power class 2) for E-UTRA (Evolved Universal Terrestrial Radio Access) – NR Dual Connectivity (EN-DC) (1 LTE TDD band + 1 NR TDD band)
		TR 37.863-01-01, E-UTRA (Evolved Universal Terrestrial Radio Access) – NR Dual Connectivity (EN-DC) of LTE 1 Down Link (DL)/1 Up Link (UL) and 1 NR band
		TR 37.863-02-01, E-UTRA (Evolved Universal Terrestrial Radio Access) – NR Dual Connectivity (EN-DC) of LTE 2 Down Link (DL)/1 Up Link (UL) and 1 NR band
		TR 37.863-03-01, E-UTRA (Evolved Universal Terrestrial Radio Access) – NR Dual Connectivity (EN-DC) of LTE 3 Down Link (DL)/1 Up Link (UL) and 1 NR band
		TR 37.863-04-01, E-UTRA (Evolved Universal Terrestrial Radio Access) – NR Dual Connectivity (EN-DC) of LTE 4 Down Link (DL)/1 Up Link (UL) and 1 NR band
		TR 37.863-05-01, E-UTRA (Evolved Universal Terrestrial Radio Access) – NR Dual Connectivity (EN-DC) of LTE 5 Down Link (DL)/1 Up Link (UL) and 1 NR band
		TR 37.864-11-22, E-UTRA (Evolved Universal Terrestrial Radio Access) – NR Dual Connectivity (EN-DC) of LTE 1 Down Link (DL)/1 Up Link (UL) and inter-/intra-band NR 2 Down Link (DL)/2 Up Link (UL) bands (Frequency Range 1 (FR1) + Frequency Range 2 (FR2))
		TR 37.864-41-21, E-UTRA (Evolved Universal Terrestrial Radio Access) - NR Dual Connectivity (EN-DC) of LTE x Down

		Link (DL)/1 Up Link (UL) (x = 1, 2, 3, 4) and Inter-band NR 2 Down Link (DL)/1 Up Link (UL) TR 37.865-01-01, NR Carrier Aggregation for intra-band (m Down Link (DL)/1 Up Link (UL) bands) and inter-band (n Down Link (DL)/1 Up Link (UL) bands) TR 37.866-00-02, NR Carrier Aggregation for inter-band n Down Link (DL)/2 Up Link (UL) bands TR 37.872, Supplementary uplink (SUL) and LTE-NR co-existence TR 37.873, Study on optimizations of UE radio capability signaling; NR/Evolved Universal Terrestrial Radio Access Network (E-UTRAN) aspects TR 37.876, Study on eNB(s) Architecture Evolution for E-UTRAN and NG-RAN TR 37.885, Study on evaluation methodology of new Vehicle-to-Everything (V2X) use cases for LTE and NR TR 37.890, Feasibility Study on 6 GHz for LTE and NR in Licensed and Unlicensed Operations TR 37.901-5, Study on 5G NR User Equipment (UE) application layer data throughput performance TR 37.985, Overall description of Radio Access Network (RAN) aspects for Vehicle-to-everything (V2X) based on LTE and NR
Radio technology beyond LTE	38	TS 38.101-1, NR; User Equipment (UE) radio transmission and reception; Part 1: Range 1 Standalone TS 38.101-2, NR; User Equipment (UE) radio transmission and reception; Part 2: Range 2 Standalone TS 38.101-3, NR; User Equipment (UE) radio transmission and reception; Part 3: Range 1 and Range 2 Interworking operation with other radios TS 38.101-4, NR; User Equipment (UE) radio transmission and reception; Part 4: Performance requirements TS 38.104, NR; Base Station (BS) radio transmission and reception TS 38.113, NR; Base Station (BS) ElectroMagnetic Compatibility (EMC) TS 38.124, NR; electromagnetic compatibility (EMC) requirements for mobile terminals and ancillary equipment; TS 38.133, NR; requirements for support of radio resource management TS 38.141-1, NR; Base Station (BS) conformance testing Part 1: Conducted conformance testing TS 38.141-2, NR; Base Station (BS) conformance testing Part 2: Radiated conformance testing TS 38.171, NR; requirements for support of Assisted Global Navigation Satellite System (A-GNSS) TS 38.174, NR; Integrated Access and Backhaul (IAB) radio transmission and reception TS 38.201, NR; physical layer; general description TS 38.202, NR; services provided by the physical layer

(*Continued*)

Technical area	Series	Key documents
		TS 38.211, NR; physical channels and modulation
		TS 38.212, NR; multiplexing and channel coding
		TS 38.213, NR; physical layer procedures for control
		TS 38.214, NR; physical layer procedures for data
		TS 38.215, NR; physical layer measurements
		TS 38.300, NR; overall description; Stage-2
		TS 38.304; NR; User Equipment (UE) procedures in idle mode and in RRC Inactive state
		TS 38.305, NG Radio Access Network (NG-RAN); Stage 2 functional specification of User Equipment (UE) positioning in NG-RAN
		TS 38.306, NR; User Equipment (UE) radio access capabilities
		TS 38.307, NR; requirements on User Equipments (UEs) supporting a release-independent frequency band
		TS 38.314, NR; layer 2 measurements
		TS 38.321, NR; Medium Access Control (MAC) protocol specification
		TS 38.322, NR; Radio Link Control (RLC) protocol specification
		TS 38.323, NR; Packet Data Convergence Protocol (PDCP) specification
		TS 38.331, NR; Radio Resource Control (RRC); protocol specification
		TS 38.340, NR; Backhaul Adaptation Protocol (BAP) specification
		TS 38.401, NG-RAN; architecture description
		TS 38.410, NG-RAN; NG general aspects and principles
		TS 38.411, NG-RAN; NG layer 1
		TS 38.412, NG-RAN; NG signaling transport
		TS 38.413, NG-RAN; NG Application Protocol (NGAP)
		TS 38.414, NG-RAN; NG data transport
		TS 38.415, NG-RAN; PDU session user plane protocol
		TS 38.420, NG-RAN; Xn general aspects and principles
		TS 38.421, NG-RAN; Xn layer 1
		TS 38.422, NG-RAN; Xn signaling transport
		TS 38.423, NG-RAN; Xn Application Protocol (XnAP)

TS 38.424, NG-RAN; Xn data transport

TS 38.425, NG-RAN; NR user plane protocol

TS 38.455, NG-RAN; NR Positioning Protocol A (NRPPa)

TS 38.460, NG-RAN; E1 general aspects and principles

TS 38.461, NG-RAN; E1 layer 1

TS 38.462, NG-RAN; E1 signaling transport

TS 38.463, NG-RAN; E1 Application Protocol (E1AP)

TS 38.470, NG-RAN; F1 general aspects and principles

TS 38.471, NG-RAN; F1 layer 1

TS 38.472, NG-RAN; F1 signaling transport

TS 38.473, NG-RAN; F1 Application Protocol (F1AP)

TS 38.474, NG-RAN; F1 data transport

TS 38.508-1, 5GS; User Equipment (UE) conformance specification; Part 1: Common test environment

TS 38.508-2, 5GS; User Equipment (UE) conformance specification; Part 2: Common Implementation Conformance Statement (ICS) performance

TS 38.509, 5GS; Special conformance testing functions for User Equipment (UE)

TS 38.521-1, NR; User Equipment (UE) conformance specification; radio transmission and reception; Part 1: Range 1 standalone

TS 38.521-2, NR; User Equipment (UE) conformance specification; radio transmission and reception; Part 2: Range 2 standalone

TS 38.521-3, NR; User Equipment (UE) conformance specification; radio transmission and reception; Part 3: Range 1 and Range 2 Interworking operation with other radios

TS 38.521-4, NR; User Equipment (UE) conformance specification; radio transmission and reception; Part 4: Performance

TS 38.522, NR; User Equipment (UE) conformance specification; applicability of radio transmission, radio reception and radio resource management test cases

TS 38.523-1, 5GS; User Equipment (UE) conformance specification; Part 1: Protocol

TS 38.523-2, 5GS; User Equipment (UE) conformance specification; Part 2: Applicability of protocol test cases

TS 38.523-3, 5GS; User Equipment (UE) conformance specification; Part 3: Protocol Test Suites

TS 38.533, NR; User Equipment (UE) conformance specification; Radio Resource Management (RRM)

(Continued)

Technical area	Series	Key documents
		TR 38.716-01-01, NR intra-band Carrier Aggregation (CA) Rel-16 for xCC Down Link (DL)/yCC Up Link (UL) including contiguous and non-contiguous spectrum ($x \geq y$)
		TR 38.716-02-00, NR inter-band Carrier Aggregation (CA)/Dual Connectivity (DC) Rel-16 for 2 bands Down Link (DL)/x bands Up Link (UL)
		TR 38.716-03-01, NR inter-band Carrier Aggregation (CA)/Dual Connectivity (DC) Rel-16 for 3 bands Down Link (DL)/1 bands Up Link (UL)
		TR 38.716-03-02, NR inter-band Carrier Aggregation (CA)/Dual Connectivity (DC) Rel-16 for 3 bands Down Link (DL)/2 bands Up Link (UL)
		TR 38.716-04-01, NR inter-band Carrier Aggregation (CA) Rel-16 for 4 bands Down Link (DL)/1 bands Up Link (UL)
		TR 38.801, Study on new radio access technology: Radio access architecture and interfaces
		TR 38.802, Study on new radio access technology: Physical layer aspects
		TR 38.803, Study on new radio access technology: Radio Frequency (RF) and co-existence aspects
		TR 38.804, Study on new radio access technology: Radio interface protocol aspects
		TR 38.805, Study on new radio access technology; 60 GHz unlicensed spectrum
		TR 38.806, Study of separation of NR Control Plane (CP) and User Plane (UP) for split option 2
		TR 38.807, Study on requirements for NR beyond 52.6 GHz
		TR 38.808, Study on supporting NR from 52.6 GHz to 71 GHz
		TR 38.809, NR; background for integrated access and backhaul radio transmission and reception
		TR 38.810, NR; study on test methods
		TR 38.811, Study on New Radio (NR) to support non-terrestrial networks
		TR 38.812, Study on Non-Orthogonal Multiple Access (NOMA) for NR
		TR 38.813, New frequency range for NR (3.3-4.2 GHz)
		TR 38.814, New frequency range for NR (4.4-5.0 GHz)
		TR 38.815, New frequency range for NR (24.25-29.5 GHz)
		TR 38.816, Study on Central Unit (CU) – Distributed Unit (DU) lower layer split for NR
		TR 38.817-01, General aspects for User Equipment (UE) Radio Frequency (RF) for NR
		TR 38.817-02, General aspects for Base Station (BS) Radio Frequency (RF) for NR
		TR 38.818, General aspects for Radio Resource Management (RRM) and demodulation for NR

TR 38.819, LTE Band 65 for NR (n65)
TR 38.820, Study on the 7 to 24 GHz frequency range for NR
TR 38.821, Solutions for NR to support Non-Terrestrial Networks (NTN)
TR 38.822, NR; User Equipment (UE) feature list
TR 38.823, Study of further enhancement for disaggregated gNB
TS 38.824, Study on physical layer enhancements for NR ultra-reliable and low latency case (URLLC)
TR 38.825, Study on NR industrial Internet of Things (IoT)
TR 38.826, Study on evaluation for 2 receiver exception in Rel-15 vehicle mounted User Equipment (UE) for NR
TR 38.827, Study on radiated metrics and test methodology for the verification of multi-antenna reception performance of NR User Equipment (UE)
TR 38.828, Cross Link Interference (CLI) handling and Remote Interference Management (RIM) for NR
TR 38.830, Study on NR coverage enhancements
TR 38.831, User Equipment (UE) Radio Frequency (RF) requirements for Frequency Range 2 (FR2)
TR 38.832, Study on enhancement of Radio Access Network (RAN) slicing for NR
TR 38.836, Study on NR sidelink relay
TR 38.838, Study on XR (Extended Reality) evaluations for NR
TR 38.840, Study on User Equipment (UE) power saving in NR
TR 38.855, Study on NR positioning support
TR 38.856, Study on local NR positioning in NG-RAN
TR 38.857, Study on NR positioning enhancements
TR 38.866, Study on remote interference management for NR
TR 38.873, Time Division Duplex (TDD) operating band in Band n48
TR 38.874, NR; Study on integrated access and backhaul
TR 38.875, Study on support of reduced capability NR devices
TR 38.883, Study on support of NR downlink 256 Quadrature Amplitude Modulation (QAM) for frequency range 2 (FR2)
TR 38.884, Study on enhanced test methods for Frequency Range 2 (FR2) NR User Equipment (UE)
TR 38.885, Study on NR Vehicle-to-Everything (V2X)
TR 38.886, V2X Services based on NR; User Equipment (UE) radio transmission and reception

(*Continued*)

Technical area	Series	Key documents
		TR 38.887, TDD operating band in Band n259
		TR 38.888, Adding wider channel bandwidth in NR band n28
		TR 38.889, Study on NR-based access to unlicensed spectrum
		TR 38.900, Study on channel model for frequency spectrum above 6 GHz
		TR 38.901, Study on channel model for frequencies from 0.5 to 100 GHz
		TR 38.903, NR; Derivation of test tolerances and measurement uncertainty for User Equipment (UE) conformance test cases
		TR 38.905, NR; Derivation of test points for radio transmission and reception User Equipment (UE) conformance test cases
		TR 38.912, Study on New Radio (NR) access technology
		TR 38.913, Study on scenarios and requirements for next generation access technologies

Index

5G Second Phase Explained: The 3GPP Release 16 Enhancements, First Edition. Jyrki T.J. Penttinen.